Solving Problems in
Surveying

Other titles in the Series

Solving Problems in Surveying

A. Bannister MC BSc MSc CEng FICE
Formerly Reader in Civil Engineering,
University of Salford

R. Baker BSc MSc CEng MICE MIWEM
Lecturer in the Department of Civil Engineering,
University of Salford

Copublished in the United States with
John Wiley & Sons, Inc., New York

Longman Scientific & Technical
Longman Group UK Limited
Longman House, Burnt Mill, Harlow
Essex CM20 2JE, England
and Associated Companies throughout the world

Copublished in the United States with
John Wiley & Sons, Inc., 605 Third Avenue, New York,
NY 10158

First published 1989
Reprinted 1991

British Library Cataloguing in Publication Data
Bannister, A. (Arthur)
  Solving problems in surveying
  1. Surveying
  I. Title   II. Baker, R.
  526.9

ISBN 0-582-01622-3

Library of Congress Cataloging-in-Publication Data
Bannister, A. (Arthur)
  Solving problems in surveying / A. Bannister, R. Baker.
    p.   cm.
  Includes index.
  ISBN 0-470-21426-0 (Wiley)
  1. Surveying—Problems, exercises, etc.   I. Baker, R.,
  1956–   II. Title.
  TA537.B36   1989   89-1255
  526.9—dc20   CIP

Printed in Malaysia
by Percetakan Jiwabaru Sdn. Bhd.,
Bangi, Selangor Darul Ehsan

Contents

Contents

Preface

This book has been written to assist students who are preparing for examinations in surveying. Although the book is devoted primarily to the solution of problems typically encountered in such examinations, each chapter commences with some basic theory. Frequently, the solutions themselves are introduced with mention of the salient features involved, before being developed in accordance with the theoretical principles. Some simple computer programs in BASIC are included, covering topics which are frequently encountered.

An appreciable number of the worked examples and problems have been taken from past examination papers, and their sources have been duly indicated. It must be emphasized that although permission has been given for the reproduction of questions as set, the solutions are ours and responsibility for their validity rests entirely on our shoulders. Thanks are due to the Senates of the University of London, University of Bradford and University of Salford, to Professor Cusens of the University of Leeds, the Council of Engineering Institutions and the Engineering Council and Councils of the Institution of Civil Engineers, Structural Engineers and the Royal Institution of Chartered Surveyors for their kind permission to reprint questions set in their examination papers.

A. Bannister
R. Baker

July 1989

1

Levelling

Datum

Levelling is a means by which a comparison of the heights of points on the earth's surface may be made. A datum is required and in the United Kingdom the normal adopted is mean sea level, as measured at Newlyn between 1915 and 1921. The Ordnance Survey have established a network of marks (bench marks) on permanent features at close intervals over the country. The height (reduced level) of each mark above Ordnance Datum has been determined and is available to the engineer on certain Ordnance sheets. It is of course quite feasible to set up one's own temporary bench mark on a site, which can be levelled in to a nearby Ordnance Survey bench mark or which can have a 'site datum' assigned to it.

Equipment

The basic items of equipment are the optical level and a graduated staff, although when long sight distances are involved a theodolite and target can be used. This latter method is termed trigonometrical levelling.

Surveyor's telescope

A telescope is an essential feature of both the level and theodolite since its overall magnification facilitates the pointing on to a distant target under satisfactory atmospheric conditions. The optics include a diaphragm or reticule, often referred to as the crosshairs, which gives an internally fixed aiming mark. The line of sight of the telescope is given by the line containing the intersection of the vertical and horizontal crosshairs and the optical centre of the object lens of the telescope. The image of the target has to lie in the plane of the crosshairs, and in modern instruments this is effected by means of a movable lens (Fig. 1.1).

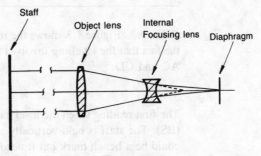

Staff Object lens Internal Focusing lens Diaphragm

Figure 1.1

Level line

A level in proper adjustment, and correctly set up, will produce a horizontal sight line which is at right angles to the direction of gravity. The sight line is tangential to the level line at instrument height; this line follows a constant height above mean sea level and hence is a curved line, as shown in Fig. 1.2. Over short distances, such as those met in civil engineering work, the two lines can be taken to coincide.

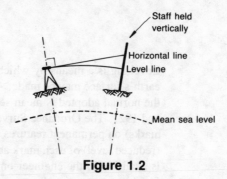

Figure 1.2

Long sight lines

Over long distances, a correction is required to reduce staff readings given by the horizontal sight line to the level line equivalent. Refraction of the sight line has also to be taken into account. The technique of reciprocal levelling can be adopted to eliminate the need for the corrections when long sights are being used.

1.1 Booking and reducing levels

> A properly adjusted tilting level was set up at a point P and the following consecutive readings were taken on a staff positioned at points A, B and C, respectively: 0.663, −0.841, 0.939.
>
> The level was then moved to point Q and further readings at C and D, respectively, were taken as follows: 1.198, 1.100.
>
> Use this example to explain what is meant by the terms backsight, foresight, intersight and change point. Book, reduce and check the levels using both standard methods, given that the reduced level of A was 94.115 m AOD. (Note that the reading at B was taken on an inverted staff and has therefore been recorded with a negative sign.)
>
> What are the advantages and disadvantages that you would associate with each method of booking? [Salford]

Solution. Figure 1.3 shows the readings listed in the question and illustrates the fact that the levelling involves the interconnection of separate groups, i.e. AC and CD.

Backsight

The first reading when the level has been set up (at P) is known as a backsight (BS). The staff is held vertically at a point (A) of known reduced level; this could be a bench mark but it need not be. When the instrument position has to be changed, the first sight taken in the next section is also a backsight.

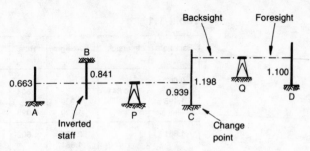

Figure 1.3

Foresight

The last staff reading from a level station on to a staff held vertically at a point is termed a foresight (FS). It is thus the last reading taken within a section of levels before moving the instrument, and also is the last sighting made over the whole series of levels. It is good practice to finish at a point of known reduced level or to return to the start point so that an appraisal of the overall accuracy of the work can be assessed from any misclosure.

Intersight

The term 'intersight' covers all sightings and consequent staff readings made between the backsight and foresight within each section. In Fig. 1.3 the inverted staff reading is an intersight.

Change point

To allow the levelling to cover an appreciable distance or to cater for difficult terrain, the level position may have to be changed frequently. In order to relate the various sections it is essential that a foresight and backsight be made on a staff held at a change point. Point C in Fig. 1.3 fulfils this purpose. Change points need to be firm features so that there is no relative displacement of the staff when its face is turned towards the new position of the instrument. When selecting such points it is good practice to try to arrange the lengths of the backsight and foresight to be equal in order to eliminate collimation error and any error due to curvature and refraction. Although modern levels allow estimation of the staff reading to 1 mm over distances upwards from 100 m, it is commonplace to have maximum sight distances of the order of 50 m.

Booking and reducing

The level book will have columns headed 'Backsight', 'Intersight' and 'Foresight', together with others covering 'Reduced Level', 'Distance' and 'Remarks'. It will also have provision for one of the two possible systems by which the levels of specific points are calculated, namely the *rise and fall method* and the *height of collimation method*.

Rise and fall method

Note that except for the change point, each staff reading is written on a separate line so that each staff position has its unique reduced level. This remains true at the change point since the staff does not move and the backsight from Q is thereby related to the reduced level derived from the foresight from P. Hence

Table 1.1

Backsight	Intersight	Foresight	Rise	Fall	Reduced level	Distance	Remarks
0.663					94.115		Station A
	−0.841		1.504		95.619		B
1.198		0.939		1.780	93.839		Change Point C
		1.100	0.098		93.937		D
1.861		2.03~~9~~	1.602	1.78~~0~~	94.11~~5~~		
		1.8~~6~~1		1.6~~0~~2	93.9~~3~~7		
		~~0~~.178		~~0~~.178	~~0~~.178		

the two readings are entered on the same line. Successive staff readings are used in calculating rise or fall between the points. From the booking we have:

$$A \text{ to } B \qquad 0.663 - (-0.841) = +1.504, \qquad \text{i.e. a rise,}$$
$$B \text{ to } C \qquad -0.841 - 0.939 = -1.780, \qquad \text{i.e. a fall,}$$
$$C \text{ to } D \qquad 1.198 - 1.100 = +0.098, \qquad \text{i.e. a rise.}$$

Notice that a decrease in staff reading implies a rise from the first point to the second, i.e. C to D.

The arithmetic involved in the reduction of the levels can be checked using the following rule:

$$(\Sigma FS - \Sigma BS) = (\Sigma fall - \Sigma rise) = (\text{final level} - \text{first level}).$$

Height of collimation method

In so far as the booking of backsights, etc., is concerned the two systems are identical but the method of the reduction of levels is different. First, the height of the collimation of the instrument for each of its positions is determined, i.e. for the setting at P:

Table 1.2

Backsight	Intersight	Foresight	Height of collimation	Reduced level	Distance	Remarks
0.663			94.778	94.115		Station A
	−0.841			95.619		B
1.198		0.939	95.037	93.839		Change point C
		1.100		93.937		D
1.861		2.03~~9~~		94.11~~5~~		
		1.8~~6~~1		93.937		
		~~0~~.178		~~0~~.178		

The height of collimation = reduced level of A + staff reading
(backsight) on A
= 94.115 + 0.663
= 94.778 m above datum.

All reduced levels for this set up are obtained from the expression

Height of collimation − Staff reading,

i.e. for B	94.778 − (−0.841)	= 95.619 m
for C	94.778 − 0.939	= 93.839 m.

When the instrument is transferred to Q,

the height of collimation = reduced level of C + staff reading
(backsight) on C
= 93.839 + 1.198 = 95.037 m.

Therefore the reduced level of D = 95.037 − 1.100 = 93.937 m.

Comparison of methods

Less arithmetic is required in the reduction of levels with the collimation method than with the rise and fall method, in particular when large numbers of intermediate sights are involved. Moreover, the rule quoted for the rise and fall method gives an arithmetical check on all the levels reduced, whereas only change points are checked under the collimation method. However, a rule can be established to give an arithmetic check for all reduced levels obtained by the collimation method. Consider the levelling between stations A and B shown in Table 1.3.

Table 1.3

Backsight	Intersight	Foresight	Height of collimation	Reduced level	Remarks
b			$b+x$	x	Station A
	i_1			$b+x-i_1$	
	i_2			$b+x-i_2$	
	.			.	
	.			.	
	.			.	
	i_n			$b+x-i_n$	
		f		$b+x-f$	Station B

The sum of all reduced levels except the known level at station A

$$= (b + x - f) + \Sigma(b + x - i)$$
$$= (n + 1)(b + x) - \Sigma i - f$$

in which n is the number of intermediate sights between A and B, and $(b+x)$ is the height of collimation of the level.

This analysis can be readily extended to cover a number of consecutive sections to produce the general rule:

Σ All reduced levels except the first = Σ (Each height of collimation times the number of intermediate and foresights related thereto) $- \Sigma IS - \Sigma FS$.

In this example, in which there is one change point, we have

Σ Each height of collimation × Relevant number of intersights and foresights = $(94.778 \times 2) + (95.037 \times 1)$

$$= 284.593 \text{ m.}$$
$$\Sigma IS = -0.841 \text{ m.}$$
$$\Sigma FS = 2.039 \text{ m.}$$

Whence $284.593 - (-0.841) - 2.039 = 283.395$

$= \Sigma$ All reduced levels except the first.

1.2 Apportioning closing errors and calculating gradients

Reduce the levels below by the collimation method, apply appropriate checks. I is an OBM of specified level 56.174 m AOD. Adjust the calculated levels so that the results at I agree with the specified level. If the distance from A to H is 220 m calculate the mean gradient between those points. Outline the practical precautions which must be taken for accurate levelling.

Table 1.4

BS	IS	FS	Remarks
0.599			OBM 58.031 m AOD
2.587		3.132	A
	1.565		B
	1.911		C
	0.376		D
2.244		1.522	E
	3.771		F
1.334		1.985	G
	0.601		H
		2.002	I

[Bradford]

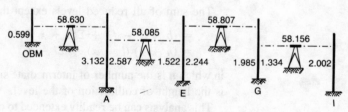

Figure 1.4

Solution Evaluate the reduced levels of the stations.

Table 1.5

Backsight	Intersight	Foresight	Height of collimation	Reduced level	Remarks
0.599			58.630	58.031	OBM
2.587		3.132	58.085	55.498	A cp.
	1.565			56.520	B
	1.911			56.174	C
	0.376			57.709	D
2.244		1.522	58.807	56.563	E cp.
	3.771			55.036	F
1.334		1.985	58.156	56.822	G cp.
	0.601			57.555	H
		2.002		56.154	I
6.764		8.641		58.031	
		6.764		56.154	
		1.877		1.877	

The next task is to determine the corrections to be applied. It will be noted that in the table there is agreement between (ΣBS $-$ ΣFS) and (final reduced level $-$ first reduced level). In addition, ΣIS = 8.224 m, and so as a further check, we have

$$(58.630 \times 1) + (58.085 \times 4) + (58.807 \times 2) + (58.156 \times 2) = 524.896 \text{ m}$$
Therefore $524.896 - 8.224 - 8.641 = 508.031$
$$= \Sigma \text{ all reduced levels except the first.}$$

However, the observations have been such that an error of -0.020 m has arisen during the levelling since the level of I should be 56.174 m. Thus an overall correction of $+0.020$ m has to be applied. In this example there have been four instrument stations and it is reasonable to assume that similar errors have arisen in each section, so we can apply a correction of $+0.005$ m for each instrument station progressively from A to I as in Table 1.6.

As an alternative, $+10$ mm could be applied to the backsights and -10 mm to the foresights, being shared out to individual readings as either 3 mm or 2 mm, to give the total of 20 mm.

We now find the gradient of line AH. The corrected, reduced levels of stations A and H are 55.503 m and 57.575 m, respectively, giving a rise of 2.072 m over a distance of 220 m. Hence the gradient is **1 in 220/2.072** or **1 in 106.2.**

Table 1.6

Station	Observed reduced level	Correction	Corrected reduced level
OBM	58.031	—	58.031
A	55.498	+ 0.005	55.503
B	56.520	+ 0.010	56.530
C	56.174	+ 0.010	56.184
D	57.709	+ 0.010	57.719
E	56.563	+ 0.010	56.573
F	55.036	+ 0.015	55.051
G	56.822	+ 0.015	56.837
H	57.555	+ 0.020	57.575
I	56.154	+ 0.020	56.174

1.3 Two-peg test

(a) Explain how you would check a level for adjustment using the 'classical' two-peg test.

(b) Four stations P, Q, R and S were set out in a straight line such that PQ = QR = RS = 30 m. A tilting level was set up at P and readings of 2.148 m and 1.836 m were observed on a staff held vertically at Q and R, respectively. The level was then set up at S and readings of 2.013 m and 1.755 m were observed on the staff held vertically in turn at stations Q and R. The bubble was adjusted to be at the centre of its run for each reading.

Check the overall adjustment of the level.

Solution. (a) Figure 1.5 shows the method of conducting the two-peg test for checking the adjustment of a level. Two rigid points A and B are marked on the ground (the pegs) and the instrument is set up exactly between them at point C. Readings are taken on to the staff held at A and B, and the difference between them gives the difference in level of the pegs. The equality in length of the backsights and foresights ensures that any instrumental error is equal on both readings. The instrument is then moved so that it is outside the line of the pegs and close to one peg. Readings are again taken on to the staff held at points A and B, any discrepancy between the level difference given by the first readings compared to that from the second readings is due to the instrument being out of adjustment.

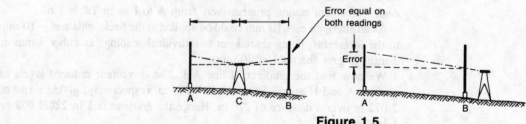

Figure 1.5

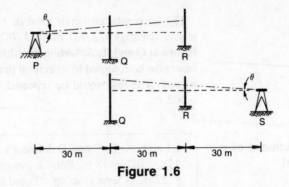

|← 30 m →|← 30 m →|← 30 m →|

Figure 1.6

(*b*) This example shows an alternative method of carrying out the test. First determine the apparent differences in level between Q and R. From Fig. 1.6,

(*a*) 2.148 − 1.836 = 0.312 m
(*b*) 2.013 − 1.755 = 0.258 m.

Since the two differences do not agree the horizontal axis of the telescope, called its line of collimation, is not parallel to the bubble tube axis. To determine the inclination of the line of collimation it is assumed that the line of collimation is directed upwards, by θ, from the horizontal so for a horizontal sight line the corrected staff readings are given in Table 1.7.

The two corrected differences in level are therefore

$$(2.148 - 30\theta) - (1.836 - 60\theta)$$
$$(2.013 - 60\theta) - (1.755 - 30\theta).$$

These must be equal

$$\therefore \ 0.312 + 30\theta = 0.258 - 30\theta$$
$$\theta = -0.0009 \ \text{radian}.$$

Thus the line of collimation is pointed downwards rather than upwards, as assumed in Fig. 1.6.

We now adjust the level. With the level at S the corrected staff readings are:

Staff at Q   2.013 + 0.054 = 2.067 m
Staff at R   1.755 + 0.027 = 1.782 m
True difference   = 0.285 m

which is half the sum of the apparent differences.

Table 1.7

Instrument Station	Staff position	
	Q	R
P	2.148−30θ	1.836−60θ
S	2.013−60θ	1.755−30θ

The horizontal crosshair must be positioned by means of the tilting screw to give readings of 2.067 m and 1.782 m, respectively, on the staff when held in turn at Q and R. Actuating the tilting screw will displace the bubble which must now be returned to its central position by means of the relevant adjusting screws. The test would be repeated to ensure that readings conform to say 0.002 m.

1.4 Adjustment of a tilting level

Points A, B, C and D define a rectangle with AB = DC = 20 m and AD = BC = 48 m. Point E lies on AD, such that AE = 15 m and ED = 33 m. A level is set up at E and staff readings of 1.735 m and 0.688 m are obtained at A and B respectively. The level is then moved to D and staff readings of 2.307 m, 1.248 m and 1.546 m are obtained at A, B and C respectively. Assuming the height of A to be 100 m, calculate the heights of B and C.   [London]

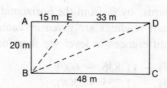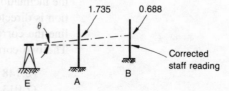

Figure 1.7

Solution. Determine error in line of sight of level.

With the instrument established at E the measured difference in level between A and B is 1.735 m − 0.688 m = 1.047 m (B is higher than A). When the instrument is moved to D the rise from A to B is found to be (2.307 − 1.248) = 1.059 m (B is higher than A). The two values should be equal and hence the line of sight does not appear to be horizontal when the levelling bubble has been brought to the centre of its run (assuming a tilting level).

Instrument at E, the corrected staff reading on

$$A = 1.735 \text{ m} - 15\theta$$
$$B = 0.688 \text{ m} - 25\theta$$

Instrument at D, the corrected staff reading on

$$A = 2.307 \text{ m} - 48\theta$$
$$B = 1.248 \text{ m} - 52\theta$$

Whence

$$(1.735 - 15\theta) - (0.688 - 25\theta) = (2.307 - 48\theta) - (1.248 - 52\theta)$$
$$1.047 + 10\theta = 1.059 + 4\theta$$
$$\theta = 0.002 \text{ radian.}$$

The corrected staff readings are given in Table 1.8.

Table 1.8

Staff station	From E	From D
A	1.735−0.030 = 1.705	2.307−0.096 = 2.211
B	0.688−0.050 = 0.638	1.248−0.104 = 1.144
Difference	1.067 m	1.067 m

Therefore corrected level difference = 1.067 m

Reduced level of B $\quad = 100.000 + 1.067$

$\quad\quad\quad = \mathbf{101.067\ m.}$

With respect to C the sighting distance was 20 m,

Therefore the corrected staff reading $= 1.546 - 20\theta$

$\quad\quad\quad\quad = 1.506\ m,$

and the difference in level between A and C is

$$2.211 - 1.506 = 0.705\ m,$$

giving

Reduced level of C = **100.705 m.**

1.5 Corrections for long sights

(a) What corrections should be applied to staff readings when levelling over long distances? Derive an expression giving the combined correction and determine the distance for which its value is 5 mm.

(b) In levelling across a wide river reciprocal observations gave the following results for staffs held vertically at points P and Q from level stations A and B respectively. A and P were near to each other on one bank, whilst B and Q were similarly situated on the other bank.

Reading of staff at P from A = 1.495 m.
Reading of staff at P from B = 1.730 m.
Reading of staff at Q from A = 1.180 m.
Reading of staff at Q from B = 1.405 m.

If the reduced level of P is 40.74 m above datum obtain that of Q.

[I. Struct. E.]

Solution. (a) In Figure 1.8 the level has been set up at MSL and sighted towards a staff held vertically at a distance IH away. IH is the line containing the instrument axis, but the line of sight has been refracted to give J as the staff reading. The two corrections which have to be combined to give the required correction, JK, are the correction for curvature, HK, and the correction for refraction of the line of sight, HJ.

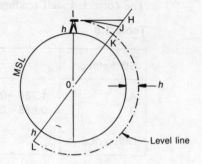

Figure 1.8

Using Pythagoras's theorem in triangle IOH

$$IH^2 = HO^2 - IO^2$$
$$= (HK + KO)^2 - IO^2$$
$$= HK^2 + 2HK \times IO \quad (\text{since } KO = IO)$$
$$= HK \, (HK + KL).$$

Now $KL = (2R + 2h)$, and R, the radius of the earth, is approximately 6.37 km, whilst h will be of the order of 1.4 m. When using a typical 4 m staff HK will not exceed 2.6 m, and so we can write $(HK + KL) = 2R$ for all practical purposes.

Therefore the correction for curvature $= HK = \dfrac{IH^2}{2R}$.

The line of sight is taken to be refracted downwards to meet the staff at J, and if we assume this line to be of radius R_1, using the same argument as previously we can write

$$HJ = \frac{IH^2}{2R_1}$$

Combined correction $JK = \dfrac{IH^2}{2R} - \dfrac{IH^2}{2R_1} = \dfrac{IH^2}{2R}\left(1 - \dfrac{R}{R_1}\right)$

Now $\dfrac{IH^2}{2R} = \dfrac{IH^2}{12\,740}$ (km) $= \dfrac{1000}{12\,740}\,IH^2$ (m)

$$= 0.078 \, IH^2 \text{ (m) with IH expressed in kilometres.}$$

If $\dfrac{R}{R_1}$ is taken to be of the order of $\frac{1}{7}$

$$JK = 0.078 \, IH^2 \times \tfrac{6}{7}$$
$$= 0.067 \, IH^2 \text{ (m)}$$
$$= 0.067 \, d^2 \text{ (m)}.$$

where d = IH (km).

If JK = 5 mm = 0.005 m
$$0.005 = 0.067d^2$$

Therefore
$$d = 0.273 \text{ km} = \textbf{273 m.}$$

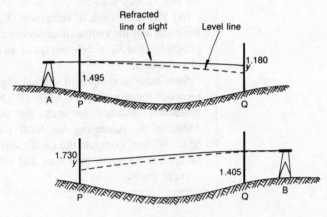

Figure 1.9

(b) Figure 1.9 shows the level stations and staff stations on each bank. The line of sight and the level line can be considered the same between A and P, B and Q, respectively, since the points are close together. Assuming no change in refraction conditions when observing from A and then from B with the same level, the level lines will give staff readings y below the actual staff readings at Q and P, respectively.

The apparent differences in level are $(1.495 - 1.180)$ m and $(1.730 - 1.405)$ m, respectively. The corrected differences (H) in level are: $[1.495 - (1.180 - y)]$ and $[1.730 - (1.405 + y)]$, respectively. These differences must be equal, and so

$$H = 1.495 - 1.180 + y = 1.730 - 1.405 - y.$$
Therefore $2H = (1.495 - 1.180) + (1.730 - 1.405)$

or

$$H = \frac{\text{Sum of apparent differences}}{2}$$

$$= 0.320 \text{ m.}$$

Therefore the reduced level of Q = 40.74 + 0.32
$$= \textbf{41.06 m.}$$

The reader should check that taking readings from level stations A and B on to staffs held at P and Q will also eliminate instrument error. It is good practice to read from A, then from B, and finally from A again to check the refraction conditions.

If two levels be used for just one pair of observations they must be in good adjustment.

1.6 Trigonometrical levelling

Define

(*i*) The coefficient of refraction (K_1) in terms of the angle of refraction and the angle subtended at the centre of the spheroid by the arc joining the two survey stations.

(*ii*) The coefficient of refraction (K_2) in terms of the local radius of the earth and the radius of curvature of the ray path. Prove from first principles that $K_2 = 2K_1$ and quote an average value for the coefficient of refraction.

Simultaneous reciprocal vertical angles have been observed, and the recorded mean values are $+06° 30' 59.7''$ and $-06° 31' 37.5''$, the horizontal distance between the two observation stations being 1389.396 m. Assuming the local mean radius of the earth to be 6383.393 km, compute the coefficient of refraction, the corrections to be applied for earth curvature and refraction and the corrected mean vertical angle.   [CEI]

Introduction.   The theory of trigonometrical levelling is as follows. Figure 1.10 shows two stations A and B whose height is to be established by reciprocal observations from A and from B on to signals at B and A, respectively. Vertical angles of α (elevation) and β (depression) are therefore measured. The line of sight will have been refracted and in Fig. 1.10 the tangent to the line of sight makes an angle of r (the angle of refraction) with the direct line AB. The vertical angle α is measured with respect to this tangent and to the horizontal at A. Similarly from B the angle of depression β is measured from the horizontal to the tangent to the line of sight.

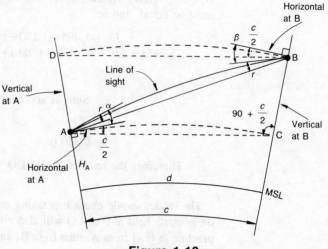

Figure 1.10

To compare the difference in height between A and B triangle ABC needs to be solved for BC. It will be noted that C lies on the arc through A, which is parallel to the mean sea level surface. d is the geodetic or spheroidal distance between A and B and could be deduced from National Grid coordinates. Angle BÂC between AB and chord AC is related to the angle c between the two verticals at A and B which meet at the earth's centre, since AC makes an angle of $c/2$ with the horizontal at A.

Thus in triangle ABC

$$BC = AC \frac{\sin (\alpha + c/2 - r)}{\sin [180 - (90 + c/2) - (\alpha + c/2 - r)]}$$

$$= AC \frac{\sin (\alpha + c/2 - r)}{\cos (\alpha + c - r)}.$$

As will be seen from Fig. 1.10 the correction for curvature and refraction at A and B is $(c/2 - r)$ and this refers angles α and β to chords AC and BD, respectively.

Thus elevation angle   BÂC $= \alpha + c/2 - r$
and depression angle   DB̂A $= \beta - c/2 + r$
but   BÂC $=$ DB̂A since AC and BD are parallel.

Therefore   BÂC $= \dfrac{\alpha + \beta}{2} =$ DB̂A.

This is the corrected vertical angle of elevation at A (or depression at B) and is the mean of the two measurements. In addition, since c and r are not usually very large we can write

$$BC = AC \tan (\alpha + c/2 - r)$$

$$= AC \tan \frac{(\alpha + \beta)}{2}.$$

Therefore height of B $= H_A + AC \tan \dfrac{(\alpha + \beta)}{2}.$

Note that it is assumed that α is elevation and β is depression in the term $(\alpha + \beta)/2$. In practice, only magnitudes need to be considered, not signs, providing one angle is elevation and the other is depression.

Solution   Calculate the refraction correction.

Put $K_1 = \dfrac{r}{c}$

and $K_2 = \dfrac{R}{R_1}$

in which r is the angle of refraction, R is the mean radius of the earth, R_1 is the mean radius of the sight line, and c is the angle contained between the two verticals.

In Example 1.5 the combined correction JK was evaluated as

$$JK = \frac{IH^2}{2R}\left(1 - \frac{R}{R_1}\right)$$

$$= \frac{d^2}{2R}(1 - K_2)$$

in which $IH = d$. Now from Fig. 1.8 in which $JK = IJ \times J\hat{I}K$, in which $J\hat{I}K$ is the angle between the level line and the refracted line of sight.

Then
$$J\hat{I}K = \frac{d}{2R}(1 - K_2)$$

From above $c/2 - r = c/2 - K_1 c$
$$= c/2\,(1 - 2K_1)$$
$$= \frac{d}{2R}(1 - 2K_1) \text{ since } c = d/R.$$

Therefore
$$K_2 = 2K_1.$$

K_1 depends upon temperature gradients, and in certain cases the line of sight can curve upwards rather than downwards; an average value is 0.07 overland. It is the coefficient used in practice, i.e. the coefficient of refraction is defined as the ratio between the angle of refraction and the angle at the centre of the earth.

Now determine the mean vertical angle.

Observed angle of elevation $\alpha = 06°\ 30'\ 59.7''$
angle of depression $\beta = 06°\ 31'\ 37.5''$.

Therefore
$$\alpha + \beta = 13°\ 02'\ 37.2''$$

and the corrected vertical angle $= \dfrac{\alpha + \beta}{2} = \mathbf{06°\ 31'\ 18.6''}$.

Also chord $AC = 2(R + H_A)\sin c/2$ in triangle ABC (Fig. 1.10). But for practical purposes we can say

$$\text{chord } AC = \text{arc } AC = d\,\frac{(R + H_A)}{R}$$

and unless H_A is appreciable, chord AC virtually equals d.

It is possible to compensate further by assuming AC to lie at the mean height of A and B, and to use that height in lieu of H_A in the above expression. This is a matter of judgement, but in this example the horizontal distance between the stations has been specified and we shall write

$$\sin\frac{c}{2} = \frac{1389.396}{2} \times \frac{1}{6\ 383\ 393}$$

i.e.
$$\frac{c}{2} = \mathbf{22.4''},$$

but $(\alpha+c/2-r) = \dfrac{\alpha+\beta}{2}$

$$= 06° \ 31' \ 18.6''$$

and so $\quad c/2-r = 18.9''$

and $\qquad\qquad r = \textbf{3.5}''$,

so $\qquad\qquad \dfrac{r}{c} = \dfrac{3.5}{44.8} = \textbf{0.078}.$

The difference in height between the stations

$$= 1389.396 \tan 06° \ 31' \ 18.6''$$
$$= 158.84 \text{ m}.$$

This example can be solved by the following computer program. The program has been written for the general case, e.g. Problems 13 and 14. For this worked example the heights of instrument, target and level at A should be entered as 0. This version of the program uses the horizontal distance between the stations, but it can easily be converted to use the MSL distance by inserting:

$$135 \ L1 = L1*(R+V1)/R$$

If both observed angles are depressions, e.g. Problem 16, then the three angular components of A1 should be input negative.

Variables

A1	= Elevation angle at A		H1	= Height of instrument at A	A
A2	= Depression angle at B		H2	= Height of instrument at B	B
A3	= Angle subtended at the centre of the earth		H3	= Height of signal at A	
			H4	= Height of signal at B	
A4	= Mean corrected angle		L1	= Horizontal distance AB	
C1	= Correction to angle A1		M	= Input/output, minutes	
C2	= Correction to angle A2		R	= Radius of the earth	
C3	= Correction for curvature		S	= Input/output, seconds	
C4	= Correction for refraction		V1	= Ground level at A	
C5	= Coefficient of refraction		V2	= Ground level at B	
D	= Input/output, degrees				

```
10 REM TRIGONOMETRIC LEVELLING
20 INPUT"HEIGHT OF INSTRUMENT AT A (M)   ";H1
30 INPUT"HEIGHT OF INSTRUMENT AT B (M)   ";H2
40 INPUT"HEIGHT OF SIGNAL AT A (M)       ";H3
50 INPUT"HEIGHT OF SIGNAL AT B (M)       ";H4
60 INPUT"ELEVATION  ANGLE RECORDED AT A (D,M,S) ";D,M,S
70 A1=((3600*D)+(60*M)+S)/206264.8
80 INPUT"DEPRESSION ANGLE RECORDED AT B (D,M,S) ";D,M,S
90 A2=((3600*D)+(60*M)+S)/206264.8
100 INPUT"LEVEL OF GROUND AT A ABOVE MSL. (M) ";V1
110 INPUT"MEAN RADIUS OF THE EARTH (KM)   ";R
120 R=R*1000
130 INPUT"HORIZONTAL DISTANCE AB   (M) ";L1
140 A3=L1/(R+V1)
150 C1=(H4-H1)/L1
160 C2=(H3-H2)/L1
170 A1=A1-C1
180 A2=A2+C2
190 A4=(A1+A2)/2
200 V2=V1+INT((L1*TAN(A4))*1000)/1000
210 C3=INT(A3*2062648/2)/10
220 C4=INT((A1+A3/2-A4)*206264.8*10)/10
```

```
230 C5=ABS(INT(C4*1000/(2*C3))/1000)
240 A4=A4*206264.8
250 D=INT(A4/3600)
260 M=INT((A4-(D*3600))/60)
270 S=INT((A4-(D*3600)-(M*60))*10)/10
280 PRINT"LEVEL AT B                =";V2;"M"
290 PRINT"MEAN OBSERVED ANGLE       =";D;M;S
300 PRINT"CORRECTION FOR CURVATURE  =";C3;"SEC"
310 PRINT"CORRECTION FOR REFRACTION =";C4;"SEC"
320 PRINT"COEFFICIENT OF REFRACTION =";C5
330 END
```

1.7 Levelling over long distances

Two survey stations A and B have been established on opposite sides of a large lake. The level of the station marker at A above water level in the lake is 49.652 m and at B is 176.268 m. The distance between A and B measured at water level is 67 256.4 m.

A theodolite with its trunnion axis 1.475 m above the station at A is used to observe a target mounted 6.237 m above the station at B. Determine the minimum clearance of the line of sight above water level. How far from A does this occur?

Calculate the anticipated vertical angle at A. Will this be elevation or depression?

Take the radius of the earth to be $6.382\,48 \times 10^6$ m and the coefficient of atmospheric refraction to be 0.082.　　　　[Bradford]

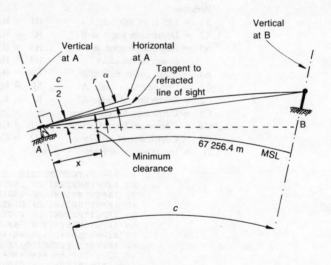

Figure 1.11

Solution.

Height of instrument at A above water level
= 1.475 + 49.652 = 51.127 m.
Height of target at B above water level
= 6.237 + 176.268 = 182.505 m.

Rise of line of sight from A to B
= 131.378 m

Equivalent vertical angle to sight B from A

$$= \frac{131.378}{67\ 256.4} \text{ radian}$$

$$= \frac{131.378}{67\ 256.4} \times 206\ 265$$

$$= 402.9''$$

Over distance x from A curvature correction $\dfrac{c_x}{2}$

$$= \frac{x}{2} \times \frac{206\ 265}{6\ 382\ 480}$$

$$= 0.016\ 159 \times x''.$$

Refraction correction (r)

$$= 0.082 \times c_x$$
$$= 0.082 \times 2 \times 0.016\ 159 \times x$$
$$= 0.002\ 650 \times x''.$$

Therefore combined correction

$$= 0.01351 \times x''.$$

When $x = 67\ 256.4$ m
$c/2 - r = 908.6''.$

If α is the vertical angle anticipated on the theodolite at A when sighting the target at B

$$\alpha + (c/2 - r) = 402.9$$
$$\alpha + 908.6 = 402.9.$$
Therefore $\alpha = -505.7''$
$$= -08'\ 25.7'' \text{ (i.e. depression)}.$$

Elevation of line of sight x from A $= \dfrac{(-505.7 + 0.013\ 51 \times x)x}{206\ 265}$

For a minimum (by differentiation)

$$505.7 = 0.02702 \times x$$
Therefore $x = \mathbf{18\ 716}$ **m.**

Over this distance

the elevation of the line of sight $= \dfrac{(-505.7 + 252.9)}{206\ 265}\ 18\ 716$

$$= -22.938 \text{ m.}$$

Therefore height above water level at lowest point (minimum clearance)

$$= 51.127 - 22.938$$
$$= \mathbf{28.189}\ \textbf{m.}$$

Note that in this example the vertical angle α set off to sight the distant target at B is a depression. Such an eventuality can occur when the height of B above A is small relative to distance AB. The expression

$$H_B - H_A = d \times \tan \frac{\alpha + \beta}{2}$$

was derived on the assumption that α was elevation and β was depression, so in this case it is essential that a negative sign is given to α to acknowledge that both measured vertical angles are depressions.

The expression for height difference in this case is

$$H_B - H_A = d \times \tan \left(\frac{\beta - \alpha}{2} \right).$$

$\alpha + (c/2 - r)$ will be positive because $c/2$ is larger than α. In this example if observations had been taken from both A and B then

$$\alpha + (c/2 - r) = 402.9''$$
$$= \beta - (c/2 - r)$$
$$= \beta - 908.6''.$$

Therefore $\beta = \mathbf{1311.5''}$.

So the two vertical angles recorded would be $-505.7''$ and $-1311.5''$, provided that the instrument heights and target heights are the same. Then

$$H_B - H_A = 67\,256.4 \tan \left(-\frac{505.7'' + 1311.5''}{2} \right)$$
$$= 131.372 \text{ m.}$$

The very small discrepancy between the given height difference and that calculated is due to rounding off during the whole calculation.

1.8 Eye and object correction

Two survey stations A and B are 5126.1 m apart. During the course of taking reciprocal trigonometrical levels between these stations the readings in Table 1.9 have been recorded.

Table 1.9

Instr. at	Height of instr.	Target at	Height of target	Mean observed vertical angle
A	1.5	B	5.0	1° 14′ 24″
B	1.4	A	2.0	—

Assuming that the earth is a sphere with a circumference of 40×10^6 m and that the coefficient of refraction is 0.071 during both sets of observations, compute the value of the missing angle. If the subsequent observations confirm this value what is the difference in level between A and B?   [CEI]

Solution. Determine the 'eye and object corrections'. It will be noted that the height of the instrument at A is not the same as the height of the signal at

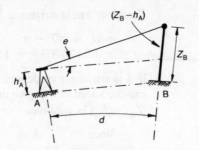

Figure 1.12

B. Thus the observed angle of elevation does not refer to the ground levels at A and B. This difference in height causes the observed vertical angle to be larger than that which would be noted if one could observe directly from those points. A correction (e), termed the 'eye and object' correction, is applied to reduce the observed value to the required value. In its simplest form it can be written as $(Z_B - h_A)/d$ radian. (See Fig. 1.12.)

Hence:

Height of target at B   = 5.00   Height of target at A   = 2.00
Height of instrument at A = 1.50   Height of instrument at B = 1.40

Therefore $Z_B - h_A$ = 3.50   Therefore $Z_A - h_B$ = 0.60

The eye and object correction to be applied to observation at A is

$$\text{Correction} = \frac{3.50}{5126.1} \times 206\,265 = 140.8''$$
$$= 02'\,20.8''.$$

The eye and object correction to be applied to observation at B is

$$\text{Correction} = \frac{0.60}{5126.1} \times 206\,265 = 24.1''.$$

Length subtended by $1''$ of arc at mean sea level

$$= \frac{40 \times 10^6}{360 \times 60 \times 60} = 30.86 \text{ m}.$$

Therefore angle c subtended at earth's centre by AB

$$= \frac{5126.1}{30.86} = 166.1''.$$

Refraction

$$r = 0.071 \times 166.1 = 11.8''.$$
$$\text{Therefore, } \frac{c}{2} - r = 71.2''.$$

Angle of elevation (α_c) from A (corrected for eye and object)

$$= 1°\,14'\,24'' - 2'\,20.8''$$
$$= 1°\,12'\,3.2''.$$

Therefore angle corrected for curvature and refraction

$$= \alpha_c + c/2 - r$$
$$= 1° \ 12' \ 3.2'' + 1' \ 11.2'' = 1° \ 13' \ 14.4''.$$

If β is the angle of depression at B, the relevant eye and object correction must be added to it to give the equivalent ground to ground measurement since $(Z_A - h_B)$ is positive.

Since $\quad\quad \alpha_c + c/2 - r = \beta_c - (c/2 - r)$
$$= 1° \ 13' \ 14.4''.$$
Therefore $\quad \beta_c = \beta + 24.1 = 1° \ 13' \ 14.4'' + 1' \ 11.2''$
$$= 1° \ 14' \ 25.6''$$
and thus $\quad\quad\quad\quad \beta = \mathbf{1° \ 14' \ 1.5''}.$

Now $\quad\quad H_B - H_A = AB \tan \dfrac{\alpha_c + \beta_c}{2}$

where $\quad\quad\quad \alpha_c = 1° \ 12' \ 3'' \text{ and } \beta_c = 1° \ 14' \ 26''.$

Therefore $\quad H_B - H_A = 5126.1 \tan \dfrac{(1° \ 12' \ 3'' + 1° \ 14' \ 26'')}{2}$

$$= \mathbf{109.2 \ m.}$$

Had the slant distance between A and B been measured (as in Problem 15) it should be reduced to its equivalent chord length at the mean height of the stations.

1.9 Parallel plate micrometer

Precise levels are normally fitted with a parallel plate micrometer. Derive an expression relating the displacement of the line of sight d, the thickness of the glass plate t, the angle of rotation of the plate i and the refractive index of the glass n.

Calculate the thickness of a parallel plate micrometer made of glass having a refractive index of 1.7 if it is specified that a 30° rotation of the plate causes a 5.00 mm displacement of the line of sight. State clearly any assumptions made. Check the answer from first principles in order to determine whether or not the assumptions are justified. [Bradford]

Introduction. Fig. 1.13 the parallel plate, fitted in front of the objective lens, is shown rotated through angle i from its vertical position. When vertical, the line of sight to the staff will not be displaced vertically, but when the parallel plate is rotated by means of a micrometer 'knob' that line can be directed upwards or downwards. By adjusting the line to the nearest graduation of the levelling staff the micrometer effectively allows readings to 0.01 mm.

Solution. The derivation of the formula is as follows.
In triangle PQR

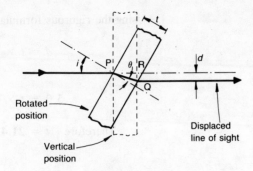

Figure 1.13

$$PQ = t/\cos\theta$$
$$RQ = d.$$

Therefore $\quad d = \dfrac{t}{\cos\theta}\sin(i-\theta)$

$$= t\left(\sin i - \cos i\,\frac{\sin\theta}{\cos\theta}\right)\ \text{where}\ \sin i = n\sin\theta$$

$$= t\sin i\left[1 - \frac{\cos i}{\sin i}\times\frac{\dfrac{\sin i}{n}}{\sqrt{\left(1-\dfrac{\sin^2 i}{n^2}\right)}}\right]$$

$$= t\sin i\left(1 - \frac{\cos i}{\sqrt{n^2-\sin^2 i}}\right)$$

$$= t\sin i\left(1 - \frac{\sqrt{1-\sin^2 i}}{\sqrt{n^2-\sin^2 i}}\right).$$

If we assume that i is small

$$d = t\times i\,\frac{(n-1)}{n}$$

(in which i is expressed in radians).

To determine t, we have

$$d = t\times i\,\frac{(n-1)}{n}.$$

We are given

$$n = 1.7,\ d = 5\ \text{mm and}\ i = 30° = 0.5236\ \text{radian}.$$

Hence $\quad 5.0 = t\times 0.5236\,\dfrac{(1.7-1.0)}{1.7}.$

Therefore $\quad t = \textbf{23.2 mm}.$

Using the rigorous formula

$$d = t \times \sin i \left(1 - \frac{\sqrt{1 - \sin^2 i}}{\sqrt{n^2 - \sin^2 i}} \right)$$

$$5.0 = t \times 0.5 \left(1 - \frac{\sqrt{1 - 0.5^2}}{\sqrt{1.7^2 - 0.5^2}} \right).$$

Therefore $t = \mathbf{21.41}$ **mm.**

Problems

1 Book and reduce the following levels and carry out the necessary checks on the arithmetic. A pair of numbers indicate a change point, the first number being the backsight.

Staff Reading (m)

1.263	OBM, reduced level 26.294 m.
3.279, 0.796	Change point.
0.376	Road level under bridge.
1.627, 0.291	Change point.
−2.162	Soffit of bridge arch.
1.582, 3.526	Change point.
2.014	TBM, reduced level 27.42 m.

Could a lorry 4.1 m high pass under the bridge?   [Salford]
Answer No, clearance under bridge = 3.874 m.

2 Levels are taken to determine the height of two pegs a and b, and to determine the soffit level of an overbridge. Using the values of levels indicated in Table 1.10, and given that the first backsight is taken on a BM at a church (RL 60.270 m), and the final foresight is on a BM at a school (RL 59.960 m), determine the closing error and the height between the underside of the bridge and the ground immediately below it. Use both the collimation level and the rise and fall methods and apply the usual checks.   [Salford]

Table 1.10

Level (m)	Remarks
1.275	Backsight to BM on church 60.270 m OD
2.812	Foresight. Change Point 1
0.655	Backsight. Change Point 1
−3.958	Inverted staff to soffit of bridge
1.515	Ground level beneath centre of bridge
1.138	Foresight. Change Point 2
2.954	Backsight. Change Point 2
2.706	Peg a
2.172	Peg b
1.240	Foresight to BM on school 59.960 m OD

Answer 4 mm; 5.473 m.

3 The rail levels of an existing railway were to be checked and raised as necessary. Points A, B, C, D, E, F and G were marked on the rails at regular 20 m intervals and the following levels were taken (all measurements in metres).

Backsight 2.80 m on OBM 25.10 m.
Intermediate sights on A, B and C: 0.94, 0.76 and 0.57, respectively.
Foresight and backsight at change point D: 0.37 and 1.17, respectively.
Intermediate sights on E and F: 0.96 and 0.75, respectively.
Foresight on G: 0.54.

Book and reduce these readings using the rise and fall method and carry out appropriate checks.

Assuming the levels at A and G were correct, calculate the amount by which the rails would have to be lifted at the intermediate points to give a uniform gradient throughout.   [Salford]
Answer 0.02, 0.03, 0.03, 0.02, 0.01.

4 Complete the extract in Table 1.11 from a level book, applying the usual arithmetic checks.   [Leeds]

Table 1.11

Backsight	Intersight	Foresight	Rise	Fall	Reduced level	Remarks
2.160					220.64	OBM
	1.157					
2.316		0.148				
−0.874		−2.010				Inverted staff
		1.050			225.05	OBM

Answer RL at inverted staff = 226.978 m.

5 A levelling exercise was performed as the first stage of an improvement scheme along a short section of road. The engineer reduced the levels by the height of collimation method whilst on site, and then carelessly dropped his field book in a puddle, obliterating some of the figures. Table 1.12 shows the level book, determine the missing entries and insert them in their appropriate place.

Table 1.12

Backsight	Intersight	Foresight	Height of collimation	Reduced level	Distance	Remarks
			88.41	84.03	0 m	OBM
	4.27				30 m	
	2.65			85.76	60 m	
		0.23	91.80		90 m	cp.
	1.63				120 m	
1.69		0.52			150 m	cp.
	0.97			92.00	180 m	
	0.72			92.25	210 m	
	1.02			91.95	240 m	
	1.10			91.87	270 m	
					300 m	Last RL − First RL
9.69		0.89				= 8.80

The proposed improvement involves regrading the road to a 1 in 20 gradient, rising from chainage 0 to 300, and passing through the existing surface at chainage 180 m. Determine the maximum depths of required excavation and fill. [Salford]

Answer Cut 1.17 m; Fill 5.17 m.

6 Reduce the levels in Table 1.13 by the rise and fall method, applying appropriate checks.

Calculate the headroom of the bridge at A + 80 m and the mean gradient of the ground between the points A and A + 120 m.

[Bradford]

Table 1.13

Backsight	Intersight	Foresight	Rise	Fall	Reduced level	Remarks
2.919					27.113	TBM
3.022		0.461				A
	1.508					A + 20 m
	2.553					A + 40 m
2.298		0.277				A + 60 m
	1.602					A + 80 m
	−3.491					Bridge soffit at A + 80 m
1.782		1.422				A + 100 m
		1.998				A + 120 m

Answer 5.093 m, 1 in 35.2.

7 The values indicated in Table 1.14 refer to levels taken along the centre line of a proposed road at 30 metre intervals.

Determine the reduced levels, and the depth of the earthworks at each point, given that the reduced level at chainage 100 metres is 46.400 metres and formation level is to rise at a gradient of 1 in 50 from a reduced level of 45.000 m at chainage 100 metres.

Table 1.14

Level (m)	Chainage (m)	Remarks
4.230	100	RL 46.400 m
2.870	130	
0.660	160	
0.810		Foresight at change point
3.150		Backsight at change point
4.450	190	

Answer 1.4, 2.16, 3.77, 1.72.

8 A straight section of road XY is to be reconstructed such that it has a constant gradient of 1 in 40, falling from X to Y. The level of the road at X is to remain unaltered. The levels in Table 1.15 were recorded along the centre line of the existing road.

(*i*) Draw up and complete the level book for these readings applying the usual arithmetical checks.

Table 1.15

Backsight	Intermediate sight	Foresight	Remarks
0.738			BM 112.309 AOD
	1.094		Point X
	1.713		30 m from X
	2.265		60 m from X
0.942		2.685	Change point
	1.100		90 m from X
	1.533		120 m from X
	−3.133		Inverted staff on underside of bridge 126.8 m from X
0.741		1.887	Change point
	1.634		150 m from X
	2.472		Point Y (170 m from X)
		2.265	BM 107.895 AOD

(*ii*) Determine the height of the underside of the bridge above the centre line level when the road has been reconstructed.

(*iii*) Calculate the depth of cut or fill at Y when the road has been reconstructed. [Leeds]

Answer (*ii*) 5.654 m; (*iii*) 17 mm fill.

9 A level is set up at point X and readings of 0.219 and 1.674 are taken on to two bench marks A and B respectively. The height of A is 166.84 m above datum and of B is 165.37 m above datum. If the distances XA and XB are 87.6 m and 33.8 m respectively, calculate the collimation error per 100 m. If a further reading of 2.121 is taken from X on to a point Y, 71.6 m from X, calculate the height of Y. [Leeds]

Answer 27.9 mm; 164.934 m.

10 The readings in Table 1.16 were obtained during a levelling operation in which the backsights and foresights were 25 m and 35 m long respectively.

Table 1.16

Station	BS	IS	FS	Distance (m)
A	0.654			0
B		1.755		20
C		3.068		40
D	0.356		3.947	60
E		2.110		80
F		4.085		100
G			3.713	120

All stations and the positions of the level were collinear.

Determine (*i*) the instrument error, and (*ii*) the reduced level of station E given that the reduced levels of A and G were 45.710 m and 39.078 m respectively.

Answer 0.0009 radians; 40.654 m.

11 A series of flying levels was taken up a slope 750 m long, the backsights being half as long again as the foresights. The apparent height of the station at the top of the slope was 89.732 m. Since this did not agree with a previous measurement the level was checked against two

pegs X and Y which were 40 m apart, Y being 0.292 m above X. When the level was set up 40 m from X and 80 m from Y, staff readings of 1.826 m on X and 1.493 m on Y were recorded. Determine the true level of the top station.
Answer 89.886 m.

12 Four points A, B, C and D are pegged out on fairly level ground in a straight line so that the distance AB = BC = CD = 20 m. Readings taken with a level at B gave values of 2.75 m and 1.51 m on staves held vertically at stations A and C, respectively. Readings from station D to stations A and C gave values of 2.27 m and 1.15 m, respectively. State whether the instrument is in or out of adjustment, and if out of adjustment, what staff readings would have been recorded from station D with a correctly adjusted instrument.   [Eng. Council]
Answer 2.45 m; 1.21 m.

13 Reciprocal vertical angles were observed between stations A and B as follows:

Mean observed angle A to B = $+1° 48' 15''$
              B to A = $-1° 48' 02''$.
Height of instrument at A   = 1.35 m
              B   = 1.36 m.
Height of signal at A   = 3.10 m
              B   = 4.50 m.
Reduced level of A   = 185.40 m.
Geodetic distance AB   = 5800 m.
Radius of the earth   = 6370 km.

Calculate (*i*)   the reduced level of B
          (*ii*) the refraction correction.
Answer (*i*) 367.21 m; (*ii*) 13.4''.

14 The horizontal distance between two stations P and Q is 5951.30 m. A theodolite at P is sighted on to a beacon adjacent to station Q at the same time as a theodolite at Q sights on to a beacon adjacent to station P. The following measurements are obtained.

Angle of elevation at P   = $1° 19' 38''$.
Angle of depression at Q   = $1° 21' 01''$.
Height of beacon at P   = 2.85 m
         at Q   = 2.36 m.
Height of instrument at P   = 1.36 m
         at Q   = 1.47 m.

Determine the difference in level between the two stations and the coefficient of atmospheric refraction. Assume that the radius of the earth is 6.37×10^6 m.   [ICE]
Answer 139.27 m; 0.071

15 Vertical angle observations were made between two stations Aga and Beetle whilst a third station, Citation, was also observed from Aga. The results in Table 1.17 were obtained.

Table 1.17

At station	Mean vertical angle	EDM slant range	Inst. ht	Target ht	Time
Aga	− 01° 22′ 09″ to Beetle	6126.45 m	1.65 m	2.27 m	11.55
	− 02° 19′ 45″ to Citation	4321.22 m		2.42 m	12.05
Beetle	+ 01° 20′ 20″ to Aga		1.78 m	3.04 m	11.55

If the height of Aga is 521.46 m AOD and the radius of the earth is 6380 km, calculate the heights of Beetle and Citation.   [Leeds]
Answer 376.24 m; 346.34 m.

16 As part of a hydrographic survey of an estuary across the mouth of which a barrage is to be constructed, a height was transferred by simultaneous reciprocal vertical angles.

Table 1.18

At station	Inst. ht	Target ht	Mean VA	Slant range
A	1.83	2.13	− 00° 01′ 04″	1524.92 m
B	1.72	2.09	− 00° 00′ 16″	

If the height of A is 5.117 m AOD, calculate the height of B.
[Leeds]
Answer 5.014 m.

17 The following simultaneous measurements of zenith distances were made between two stations A and B:

Corrected slope length AB. 8804.8 metres.
ZD A to B. 90° 01′ 22″.
ZD B to A. 90° 02′ 18″.
Height of theodolite at A 1.17 m; target at A 0.89 m.
Height of theodolite at B 1.33 m; target at B 2.58 m.

Find the difference in height between A and B and the coefficient of terrestrial refraction, and explain why both observations are depressions.
[London]
Answer 0.3 m; 0.074.

18 Simultaneous vertical angles are observed between two stations A and B which are 21.8 km apart. The height of station A is 299.65 m above datum. The following observations are recorded:

Instrument at A
Height of instrument above ground = 1.45 m
Height of target at B above ground = 3.10 m
Vertical angle + 00° 11′ 12″

Instrument at B
Height of instrument above ground = 1.52 m
Height of target at A above ground = 3.86 m
Vertical angle − 00° 20′ 52″

If the mean radius of the earth is 6375 km, calculate the coefficient of atmospheric refraction and the height of station B above datum.

[Leeds]

Answer   0.062; 401.67 m.

19   The heights of two stations X and Y above datum are 311.02 m and 315.48 m respectively. There is an intervening hill on the line XY at a height of 310.41 m above datum. A theodolite is set up at the height of 1.44 m above station X, which is 12.6 km from the hill and 17.4 km from Y. If the coefficient of atmospheric refraction is 0.068 and the mean radius of the earth is 6375 km, calculate the minimum height at which a target must be erected above the ground at Y in order that it may be observed from the instrument at X. (Assume that the line of sight must clear the hill by at least 3 m.)   [Leeds]
Answer   4 m.

20   A lighthouse is to be constructed on a cliff top site overlooking the sea. The ground level at this point is 35.260 m above datum. There is an existing illuminated buoy 33.4 km from the lighthouse, the illuminated target on the buoy being 2.4 m above the water level. The highest and lowest tides normally expected are 4.700 m above datum and 0.460 m below datum respectively and the coefficient of atmospheric refraction is found to vary between 0.062 and 0.075. If the mean radius of the earth is 6375 km, calculate the minimum height of the observation point in the lighthouse above the ground such that the buoy is always visible.

[Leeds]

Answer   21.36 m.

21   A gas drilling rig is set up on the sea bed 48 km from each of two survey stations on the coast some distance apart. In order that the exact position of the rig may be obtained, it is necessary to erect a beacon on the rig so that it may be clearly visible from theodolites at the survey stations, each a height of 36 m above high water mark.

Neglecting the effect of refraction, and assuming that the minimum distance between the line of sight and calm water is to be 3 m at high water, calculate the least height of beacon above high water mark at the rig. Prove any formula used.

Calculate the angle of elevation that would be measured by the theodolite when sighted on to this beacon taking refraction into account and assuming that the error due to refraction is one-seventh of the error due to curvature of the earth. Mean radius of the earth = 6370 km.

[ICE]

Answer   62.3 m; −9′ 13″.

22 Two survey stations A and B have reduced levels 193.5 m and 314.0 m respectively; the distance AB is 15 940 m. There is an intervening ridge 238 m high across the line, 5700 m from A. Show that a target 3 m above ground level at B would not be visible from a theodolite 1.5 m above the ground at A.

If the ridge were covered with scrub 10 m high, how high a tower would be required at A in order to observe a signal 5 m high at B, with a clearance of 5 m above the scrub? Assume a value of 0.07 for k and take the radius of the earth as 6380 km.   [London]

Answer 27.39 m.

23 Reciprocal vertical angles were observed between two stations A and B as follows:

Mean observed angle A to B	$= -1'\ 02''$
B to A	$= -2'\ 59''$
Height of instrument at A	$= 1.35$ m
at B	$= 1.33$ m
Height of signal at A	$= 3.10$ m
at B	$= 4.50$ m
Reduced level of A	$= 175.4$ m
Distance AB	$= 11\ 600$ m

Estimate the reduced level of B. What would be the error in that level if only the observation from B to A had been taken and the assumption made that the refraction coefficient was 0.071? A length of 30.95 m at mean sea level may be taken to subtend an angle of $1''$ at the centre of the earth.   [Salford]

Answer 178.0 m; 0.2 m.

24 Given that the refractive index from air to glass is 1.6 determine the angular rotation of a parallel plate, of thickness 10 mm, to give a vertical displacement of the image of the staff through 0.1 mm.

Answer $01°\ 31'\ 40''$

25 A parallel-plate micrometer attached to a level is to provide a displacement of 2.5 mm when rotated through an angle of 15° on either side of the vertical. Using both the precise and approximate equations, calculate the thickness of the plate required if the refractive index of the glass is assumed to be 1.6.   [Eng. Council]

Answer 24.88 mm.

2

Distance measurement

Three methods of distance measurement are reviewed in this chapter, namely by tape, by electromagnetic distance measuring equipment (EDM) and by optical methods.

By tape (or wire)

For many years the very accurate measurement of distance depended upon the careful use of steel tapes or wires. Nowadays, EDM is used almost exclusively for accurate work but the steel tape still is of value for measuring limited lengths and for setting out purposes.

Up to seven corrections may be applied to the measured length, to give the true length, dependent upon the circumstances of the measurement.

Correction for standard

A steel tape will normally be provided with standardizing data, for example it may be designated as 30 m long under a tension of 5 kgf at a temperature of 20 °C when laid on the flat. With use the tape may stretch and it is imperative that the tape is regularly checked against a reference tape kept specifically for this purpose.

Correction for tension

If the tape is of correct length under a standard tension and it is used under a different tension the correction which should be applied is

$$\frac{(P - P_s)L}{A E},$$

where P is the tension applied in the field, P_s is the standard tension, A is the cross-sectional area of the tape, E is Young's modulus for the tape material, and L is the observed length.

The sign of the correction takes that of quantity $(P - P_s)$. Obviously, when P is made equal to P_s the correction is zero and the arithmetic is simplified but the proviso mentioned in the correction for sag should be borne in mind.

Correction for sag

For very accurate work the tape can be allowed to hang in catenary, free of the ground, between suitable supports. In the case of a long tape intermediate supports can be used to reduce the magnitude of the correction (see Fig. 2.1).

If the tape has been standardized on the flat the correction which should be applied to reduce the curved length to the chord length is

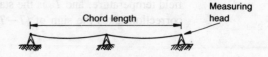

Figure 2.1

$$-\frac{w^2 L^3}{24 \, P^2},$$

where w is the weight of the tape per unit length (P should be larger than $20wL$).

If such a tape is used on a plane surface which can be considered flat then no correction is applicable.

If the tape has been standardized in catenary the equivalent chord length is given by the graduations, provided that the standardizing tension is applied. If such a tape is used on the flat then the correction is *added* to the tape length.

Correction for slope

In surveying it is essential that horizontal lengths are determined. Thus length L measured on the slope must be reduced to its equivalent plan length $L \cos \theta$ (see Fig. 2.2). The correction to be applied is

Figure 2.2

$$(L \cos \theta - L) = -L \, (1 - \cos \theta).$$

Now $\cos \theta = 1 - \dfrac{\theta^2}{2!} + \dfrac{\theta^4}{4!} + \dots$, which is nearly equal to $1 - \dfrac{h^2}{2L^2}$,

where $\theta = h/L$. Thus a simplified form of the correction is $-(h^2/2L)$, provided that h is small. Alternatively, Pythagoras's Theorem can be used to calculate the horizontal distance directly from $\sqrt{(L^2 - h^2)}$ whatever the relative magnitude of h.

It is usual to calculate the corrections for sag and slope separately and then add them together. The change in shape of the catenary has to be considered when h is large, but if it remains below, say, 2 m there should be little loss of accuracy for a tape 30 m long.

Correction for temperature

If a tape is used at a field temperature different from the standardization temperature then the correction is

$$\alpha(T - T_S)L,$$

where α is the coefficient of thermal expansion of the tape material, T is the field temperature, and T_S is the standardization temperature. The sign of the correction takes the sign of $(T - T_S)$.

Correction to mean sea level

In the case of long lines in triangulation surveys the relationship between the length measured on the ground and the equivalent length at mean sea level has to be considered. This is also the case when any distance measurements are being related to National Grid co-ordinates (see Ch. 4). If the measured length is L_m and the height of the line above datum is H then, since both lengths subtend the same angle at the earth's centre (see Fig. 2.3),

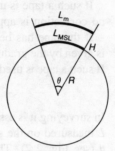

Figure 2.3

$$L_m = (H + R)\theta$$

and

$$L_{ms1} = R\theta,$$

where R is the radius of the earth. Now,

$$\theta = \frac{L_{ms1}}{R} = \frac{L_m}{H + R}$$

and so

$$L_{ms1} = L_m \frac{R}{H + R}.$$

The correction to be applied is

$$L_{ms1} - L_m = L_m \frac{R}{R + H} - L_m = -L_m \frac{H}{R + H}$$

$$= -L_m \frac{H}{R}$$

when H is small compared to R.

Electromagnetic distance measurement (EDM)

Two main types of EDM are encountered in land surveying, namely the electronic or microwave systems and electro-optical instruments. The principle of operation is that a transmitter at the master station sends a modulated continuous wave (carrier wave) to a receiver at the remote station, from which it is returned. The instruments measure slope distance between transmitter and receiver by modulating the continuous carrier wave at different frequencies, and then measuring the phase difference at the master between the outgoing and the incoming signals. Thus an element of double distance is introduced. If ϕ is the measured phase difference and λ the modulation wavelength, the expression for the distance (D) traversed by the wave is

$$2D = n\lambda + \frac{\phi}{2\pi} \lambda + \text{a constant,}$$

where n is the number of complete wavelengths contained within the double distance and will be unknown. The purpose of deploying different modulation frequencies is to evaluate this number by comparing the phase differences of the various outgoing and measuring signals. Note that

$$\lambda = \frac{c_0}{nf},$$

where c_0 is the velocity of the electromagnetic wave in a vacuum, f is the frequency, and n is the refractive index of the medium through which the wave passes.

Nowadays, most local survey work and setting out for civil engineering works will be carried out using infrared based EDM, which falls within the electro-optical group. The infrared carrier wave is transmitted to a passive reflector, usually a retrodioptive prism, from which it is returned to the master at the other end of the line. Ranges of the order of 1–3 km are attainable by standard instruments with an accuracy of ±5 mm and many include slope reduction and calculation functions as part of the electronics.

Electronic or microwave instruments are mainly used over long ranges (up to 100 km). The remote instrument needs an operator acting to instructions from the master at the other end of the line because the signal is transmitted from the master station, received by the remote station and retransmitted to the master station. Lengths of 100 km can be measured to an accuracy of ±50 mm.

There are some instruments available that send modulation pulses along the carrier wave and measure the transit time, but this is not the typical mode of measurement.

Optical distance measurement (tacheometry or tachymetry)

Tacheometry or tachymetry is the method of surveying by which distance and heighting information can be determined from theodolite observations on either a levelling staff (stadia system) or a horizontally mounted bar (subtense system). In practice, measurement by EDM is replacing measurements by optical methods but they are still useful techniques when only conventional equipment is available.

Stadia tacheometry

For stadia tacheometry vertical angles need to be measured when the theodolite's line of sight is inclined. Stadia marks on the instrument diaphragm relate the staff intercept to the slope distance.

For subtense tacheometry the horizontal angle subtended by targets on a special horizontal bar (subtense bar) has to be measured. For heighting, a vertical angle is also required. In Fig. 2.4,

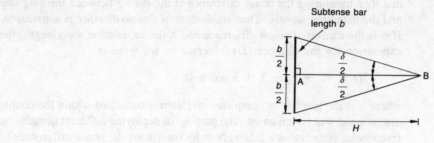

Figure 2.4

$$H = \frac{b/2}{\tan \delta/2} = \frac{b}{2 \tan \delta/2} \simeq \frac{b}{\delta}$$

when δ is very small.

An alternative to the horizontal bar is a vertical bar carrying two targets; in this case vertical angles are measured and the method is termed the tangential system. Examples 7.4 and 8.8 illustrate the use of the subtense bar.

2.1 Basic tape measurements

A survey line was measured with a tape, believed to be 20 m long, and a length of 284.62 m resulted. On checking, the tape was found to measure 19.95 m long.

(a) What was the correct length of the line?
(b) If the line lay on a slope of 1 in 20 what would be the reduced horizontal length used in the plotting of the survey?
(c) What reading is required to produce a horizontal distance of 15.08 m between two site pegs, one being 0.66 m above the other?

Solution. (a) Each time the tape is tensioned under its standardized value a length of 20 m would be booked overall. But actually only a length of 19.95 m has been covered. Hence the tape is reading 'high', the error is positive and any subsequent correction must have a negative effect.

$$\text{Correct length of line} = \frac{19.95}{20.00} \times 284.62$$

$$= \textbf{283.91 m.}$$

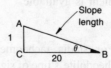

Figure 2.5(a)

(b) A slope of 1 in 20 implies that there is a change in height of 1 m over each 20 m in length horizontally.

In Fig. 2.5(a) angle $\theta = \tan^{-1} 1/20$
$$= 2° 52'.$$

Alternatively, for small angles, $\theta = 1/20$ radian, which gives the same value for θ.

Thus the required length AC = AB cos 2° 52′
$$= 283.91 \times 0.998\ 75$$
$$= \textbf{283.56 m.}$$

Note that horizontal distances are required when plotting surveys.

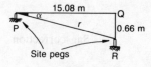

Figure 2.5(b)

(c) In Fig. 2.5(b) rather than estimating the position of Q above R by eye or by spirit level, it is preferable to set out the equivalent slope length between points P and R on top of the site pegs.

Let the slope length required be r.

$$\text{Now} \quad \tan \alpha = \frac{0.66}{15.08}.$$

Therefore $\alpha = 2° 30'$
thus $\quad r = 15.08 \sec 2° 30'$
$$= 15.09 \text{ m}$$

Required tape reading $= 15.09 \times \dfrac{20.00}{19.95} = \textbf{15.13 m.}$

2.2 Tape corrections

In order to determine the horizontal distance between two points A and B set on the floor of a tunnel, measuring heads were set up over those points. A new 30 m steel tape, standardized on the flat under a pull of 49 N at 20°C, is suspended in catenary between the measuring heads, and a pull of 147 N is applied. The mean tape readings at the measuring heads are observed to be 0.422 m and 29.782 m, and the difference in level between the heads is found to be 0.075 m. Determine the corrected horizontal distance between the points A and B.

Mean tape temperature during observations = 26°C
Cross-sectional area of tape = 0.406 cm²
Mass of tape = 0.27 g/cm
Coefficient of linear expansion of steel = 1.15 × 10⁻⁵/°C
Modulus of elasticity of steel = 207 000 N/mm²

[ICE]

Introduction. The tape has been standardized on the flat under a pull of 49 N and is being used in catenary under a pull of 147 N. In addition the measuring heads are not level and the tape temperature is not equal to the standard temperature. Thus corrections for pull, sag, slope and temperature have to be applied.

Solution. The measured length is $29.782 - 0.422 = 29.360$ m.

$$\text{Pull correction} = (P - P_s)\frac{L}{AE}$$

$$= \frac{(147 - 49) \times 29.360}{0.406 \times 10^2 \times 207\,000}$$

$$= +0.0003 \text{ m.}$$

$$\text{Sag correction} = \frac{-w^2 L^3}{24\,P^2}$$

$$= \frac{-(0.027 \times 9.806)^2 \times 29.360^3}{24 \times 147^2}$$

$$= -0.0034 \text{ m.}$$

$$\text{Slope correction} = \frac{-h^2}{2L}$$

$$= \frac{-0.075^2}{2 \times 29.360}$$

$$= -0.0001 \text{ m}$$

$$\text{Temperature correction} = (T - T_s)\alpha L$$
$$= (26 - 20) \times 29.360 \times 1.15 \times 10^{-5}$$
$$= +0.0020 \text{ m.}$$

$$\text{Net correction} = -0.0012 \text{ m}$$

Therefore horizontal length = **29.359 m.**

The following computer program can be used to solve this problem, there is one bay and the height above sea level should be entered as zero. Take care that the data is entered in the correct units.

Variables

A	= Cross-sectional area of tape	L	= Length of any bay
C	= Coefficient of thermal expansion	L1	= Total length of line
		L2	= Corrected length of line
C1	= Correction for tension	L3	= Sum of L^3
C2	= Correction for sag	M	= Tape mass
C3	= Correction for temperature	N	= Number of bays
C4	= Correction for slope	P	= Standard pull
C5	= Correction for sea level	H	= Level differences across any bay
E	= Young's modulus for the tape	I	= Counter

P1	= Pull used at time of survey	T1	= Temperature at time of
S	= Height above sea level		survey
T	= Standard temperature	W	= Weight of tape

```
10 REM TAPE CORRECTION
20 INPUT"INPUT NUMBER OF BAYS ",N
30 FOR I=1 TO N
40 PRINT"BAY";I;
50 INPUT"INPUT LENGTH,LEVEL DIFFERENCE ",L,H
60 L1=L1+L
70 L3=L3+(L↑3)
80 C4=C4-(H*H/(2*L))
90 NEXT I
100 INPUT"TEMPERATURE AT TIME OF SURVEY   ",T1
110 INPUT"PULL USED TO TENSION TAPE IN N ",P1
120 INPUT"HEIGHT ABOVE SEA LEVEL IN M     ",S
130 PRINT"TAPE DETAILS"
140 INPUT"CROSS SECTIONAL AREA IN MM ",A
150 INPUT"COEF.OF THERMAL EXPANSION   ",C
160 INPUT"MASS IN KG/M                ",M
170 INPUT"YOUNGS MODULUS IN N/MM2     ",E
180 INPUT"STANDARD TEMPERATURE        ",T
190 INPUT"STANDARD PULL IN N          ",P
200 W=M*9.806
210 C1=INT(((P1-P)*L1/(A*E))*100000 +0.5)/100
220 C2=INT((L3*W*W*(-1)/(24*P1*P1))*100000 +0.5)/100
230 C3=INT((L1*C*(T1-T))*100000 +0.5)/100
240 C4=INT(C4*100000 +0.5)/100
250 C5=INT((L1*S*(-1)/(6367000 +S))*100000 +0.5)/100
260 L2=L1+(C1+C2+C3+C4+C5)/1000
270 PRINT"APPARENT LENGTH OF BASE LINE =";L1;"M"
280 PRINT"CORRECTION FOR TENSION       =";C1;"MM"
290 PRINT"CORRECTION FOR SAG           =";C2;"MM"
300 PRINT"CORRECTION FOR TEMPERATURE   =";C3;"MM"
310 PRINT"CORRECTION FOR SLOPE         =";C4;"MM"
320 PRINT"CORRECTION FOR SEA LEVEL     =";C5;"MM"
330 PRINT"ACTUAL LENGTH OF BASE LINE   =";L2;"M"
340 END
```

2.3 Comparing two tapes

Tape A of nominal length 30 m, standardized on the flat at 20 °C under a pull of 89 N, was used in catenary under a pull of 410 N at 24 °C. Marks were made at its terminal graduations on the tops of pegs at the same level.

When tape B, also nominally 30 m long, was placed in catenary under a pull of 160 N and a temperature of 25 °C it measured 1.5 mm shorter than tape A. It had been standardized on the flat at 19 °C under a pull of 89 N. Determine the temperature at which it is 30 m long under a pull of 89 N.

Weight of tape A	= 0.455 N/m
Cross-sectional area of tape A	= 6.00 mm^2
Weight of tape B	= 0.255 N/m
Cross-sectional area of tape B	= 3.24 mm^2
Young's modulus	= 207 000 N/mm^2
Coefficient of linear expansion	= 0.000 011/°C

[Salford]

Solution. First determine the horizontal distance between the marks. Tape A was used to establish the marks and the relevant corrections to the nominal length of 30 m are as follows:

$$\text{Pull correction} \quad = \frac{(P - P_s)L}{AE}$$

$$= \frac{(410 - 89) \times 30}{6.00 \times 207\,000}$$

$$= +0.0078 \text{ m.}$$

$$\text{Sag correction} \quad = \frac{-w^2 L^3}{24\,P^2}$$

$$= \frac{0.455^2 \times 30^3}{24 \times 410^2}$$

$$= -0.0014 \text{ m.}$$

$$\text{Temperature correction} = (T - T_s)\alpha L$$
$$= (24 - 20) \times 30 \times 0.000\,011$$
$$= +\ 0.0013 \text{ m.}$$

Therefore net correction $= +0.0077$ m.

The pegs are at the same level and so there is no correction for slope. Therefore the distance between the two marks

$$= 30 + 0.0077$$
$$= 30.0077 \text{ m.}$$

Next determine the length of tape B.

Tape B was hung in catenary between the same marks and the same corrections apply to the measurement. Its nominal length will be used to establish their values.

$$\text{Pull correction} \quad = \frac{(160 - 89) \times 30}{3.24 \times 207\,000}$$

$$= +0.0032 \text{ m.}$$

$$\text{Sag correction} \quad = -\frac{0.255^2 \times 30^3}{24 \times 160^2}$$

$$= -0.0029 \text{ m.}$$

$$\text{Temperature correction} = (25 - 19) \times 30 \times 0.000\,011$$
$$= +0.0020 \text{ m.}$$

Therefore net correction $= +0.0023$ m.

Tape B can now be standardized from the observations.

It was found to be shorter than Tape A by 1.5 mm. Therefore the length between its terminal graduations

$$= 30.0077 - 0.0015 \text{ m}$$
$$= 30.0062 \text{ m.}$$

If L is the correct length of the tape on the flat at 19 °C

$$L + 0.0023 = 30.0062$$

Therefore
$$L = 30.0039 \text{ m}.$$

Let the temperature at which the tape is 30 m long be t_{30}°C

Then $30.0039 = 30 + (19 - t_{30}) \times 30 \times 0.000\,011$

$$(19 - t_{30}) = 11.8$$

Therefore
$$t_{30} = \mathbf{7} \text{ °C (say)}.$$

The basic calculations for this problem can be carried out using the computer program listed with Example 2.3. The program must be run separately for each tape with one bay and the level difference and height above sea level entered as zero. Take care that the data is in the correct units. For tape B the measured length is 30.0062 m and the final calculation for the temperature must be carried out by hand.

2.4 Base line measurement in bays

The information in Table 2.1 was obtained when measuring the length of a line by a tape suspended in catenary, the mean temperature being 16 °C.

Table 2.1

Bay	Length (m)	Difference in level (m)	Tension (N)
1	29.8984	+ 0.382	134
2	29.9498	− 0.234	134
3	29.8826	+ 0.271	134
4	29.9012	− 0.075	134

The length of the line, reduced to mean sea level, was determined as 119.6206 m. Estimate the pull under which the tape was standardized on the flat, assuming a standard temperature of 20 °C.

Cross-sectional area of tape	= 3.24 mm^2
Mass of tape	= 0.026 kg/m
Coefficient of linear expansion	= 0.000 000.9/°C
Young's modulus	= 15.5 × 10^4 MN/m^2
Mean height of line above mean sea level	= 53.78 m
Radius of earth	= 6367 km

[Salford]

Introduction. Assuming that the four bays are collinear, five corrections will have to be evaluated to obtain the reduced length of 119.6206 m. There is sufficient information to compute the corrections for sag, slope, temperature and mean sea level. Thus the unknown factor is the correction for pull which depends upon the standardizing value P_s.

Solution. Tabulate and extend the field measurements. It is convenient to tabulate as in Table 2.2. The measured length of the line is 119.6320 m,

$$\Sigma L^3 = 107\,009.68, \quad \Sigma h^2 / 2L = 0.0046 \text{ m}.$$

Table 2.2

L (m)	L^3	h (m)	h^2	$\dfrac{h^2}{2L}$
29.8984	26726.61	+ 0.382	0.1459	0.0024
29.9498	26864.69	− 0.234	0.0548	0.0009
29.8826	26684.26	+ 0.271	0.0734	0.0012
29.9012	26734.12	− 0.075	0.0056	0.0001
119.6320	107009.68		0.2797	0.0046

$$\text{Sag correction} = -\frac{w^2 L^3}{24\, P^2}$$

$$= -\frac{(0.026 \times 9.806)^2 \times 107\,009.68}{24 \times 134^2}$$

$$= -0.0161 \text{ m.}$$

$$\text{Slope correction} = -\sum \frac{h^2}{2L} = -0.0046 \text{ m.}$$

$$\text{Temperature correction} = \alpha(T - T_s)L$$
$$= 0.000\,000\,9\,(16 - 20) \times 119.6320$$
$$= -0.0004 \text{ m.}$$

$$\text{Mean sea level correction} = -\frac{LH}{R}$$

$$= -\frac{119.6320 \times 53.78}{6\,367\,000}$$

$$= -0.0010 \text{ m.}$$

Next determine the standard pull (P_s). The sum of the four corrections evaluated above

$$= -(0.0161 + 0.0046 + 0.0004 + 0.0010)$$
$$= -0.0221 \text{ m.}$$

Now length reduced to MSL = 119.6206 m
and measured length of line = 119.6320 m,
therefore sum of all corrections = −0.0114 m.
Thus (pull correction − 0.0221) = −0.0114,
therefore pull correction = +0.0107 m.

Hence

$$(P - P_s) \frac{L}{AE} = +0.0107 \text{ m}$$

in which $A = 3.24 \text{ mm}^2$ and $E = 15.5 \times 10^4 \text{ N/mm}^2$. Therefore

$$(134 - P_s) \times \frac{119.6320}{3.24 \times 15.5 \times 10^4} = +0.0107$$

$$(134 - P_s) = 44.9 \text{ N}$$

Therefore $\qquad P_s = 89.1 = \textbf{89 N}$, say.

The computer program listed with Example 2.3 can be used to calculate the background data for this problem by entering the field and standard pull as zero.

2.5 Incorrect measurement of field pull

Derive expressions giving the errors in the pull and sag corrections due to an error of $\pm \delta P$ in the value of the applied tension P.

The length of a baseline was deduced as 1319.774 m when measured by a tape of length 30 m suspended in catenary. Determine the corrected length of the line if the actual field tension was 170 N instead of the intended value of 178 N.

The tape, which was standardized in the flat under a pull of 89 N, had a mass of 0.026 kg/m and a cross-sectional area of 3.25 mm^2. Take Young's modulus as 155 000 MN/m^2 and the acceleration due to gravity as 9.806 m/s^2. [Salford]

Solution. Consider the effects of the error in field pull. Let the error in the nominal applied tension P be $\pm \delta P$. In the case of the pull correction

$$\text{actual pull correction} = \frac{(P \pm \delta P - P_s)L}{AE}$$

$$\text{nominal pull correction} = \frac{(P - P_s)L}{AE}$$

$$\text{error} = \text{actual correction} - \text{nominal correction}$$

$$= \frac{(P \pm \delta P - P_s)L}{AE} - \frac{(P - P_s)L}{AE}$$

$$= \pm \delta P \times \frac{L}{AE}$$

but $\qquad \dfrac{L}{AE} = \dfrac{\text{nominal pull correction}}{P - P_s}.$

Therefore the error in pull correction $= \pm \dfrac{\delta P \times \textbf{nominal correction}}{P - P_s}$, i.e. an increase in P implies that the pull correction increases.

In respect of the sag correction

$$\text{actual sag correction} = \frac{-w^2L^3}{24(P \pm \delta P)^2} = \frac{-w^2L^3}{24P^2\left(1 \pm \dfrac{\delta P}{P}\right)^2}$$

$$\text{nominal sag correction} = -\frac{w^2L^3}{24\,P^2}$$

error in sag correction = actual correction − nominal correction

$$= -\frac{w^2L^3}{24\,P^2}\left(1 \pm \frac{\delta P}{P}\right)^{-2} - \left(-\frac{w^2L^3}{24\,P^2}\right)$$

$$= -\frac{w^2L^3}{24\,P^2}\left(1 \mp \frac{2\delta P}{P}\right) + \frac{w^2L^3}{24\,P^2}$$

neglecting all other terms in $\dfrac{\delta P}{P}$.

$$\text{Therefore error in sag correction} = \mp \frac{2\delta P}{P} \times \textbf{nominal sag correction,}$$

i.e. an increase in P reduces the sag correction.

Next, determine the individual nominal corrections.

$$P_s = 89 \text{ N}, \; P = 178 \text{ N and } \delta P = -8 \text{ N}.$$

$$\text{Nominal pull correction} = (P - P_s)\frac{L}{AE}$$

$$= \frac{(178 - 89)30}{3.25 \times 155\,000}$$

$$= 0.0053 \text{ m per 30 m, say.}$$

$$\text{Nominal sag correction} = -\frac{w^2L^3}{24\,P^2}$$

$$= -\frac{(0.026 \times 9.806)^2 \times 30^2}{24 \times 178^3}$$

$$= -0.0023 \text{ m per 30 m, say.}$$

Finally compute the errors in the total nominal corrections. Any change to the deduced length of the line will depend upon changes in just the pull and sag corrections since these are the corrections which involve the field pull. The deduced length of 1319.744 m implies that there were 44 bays measured by the 30 m tape.

Now $\delta P = -8$ N.

Error due to $-\delta P = \dfrac{-\delta P}{P-P_s} \times$ nominal correction

$$= \dfrac{-8}{178-89} \times 0.0053 \times 44 \text{ for the whole line}$$

$$= -0.0210 \text{ m}.$$

Thus the total pull correction was overestimated by 0.0210 m.

Error due to $-\delta P = +\dfrac{2\delta P}{P} \times$ nominal correction

$$= +\dfrac{2 \times 8}{178} \times (-0.0023) \text{ m per bay}$$

$$= \dfrac{2 \times 8}{178} \times -0.0023 \times 44 \text{ for the whole line}$$

$$= -0.0091 \text{ m}.$$

Thus the total sag correction was too small by 0.0091 m.

The length of the line has therefore been overestimated because the pull correction (too large) would have been added to the measured length, since $P > P_s$, whilst the sag correction (too small) would have been subtracted.

Overestimation $= 0.0210 + 0.0091$
$= 0.0301$ m.
Therefore length of line $= 1319.744 - 0.030$
$= \mathbf{1319.714}$ **m.**

2.6 Principle of EDM

Explain carefully the principles involved in measuring distances using a sinusoidal wave form as the medium of measurement.

A line AB was measured using EDM. The instrument was set up at O in line with AB and on the side of A remote from B. The wavelength of frequency 1 is 10 metres exactly. Frequency 2 is 9/10 frequency 1 and frequency 3 is 99/100 frequency 1.

It is known that AB is less than 200 m.

Calculate the accurate length of AB from the phase difference readings given in metres in Table 2.3.

Table 2.3

	Frequency 1	Frequency 2	Frequency 3
Line OA	4.337	7.670	0.600
Line OB	7.386	1.830	9.911

[Bradford]

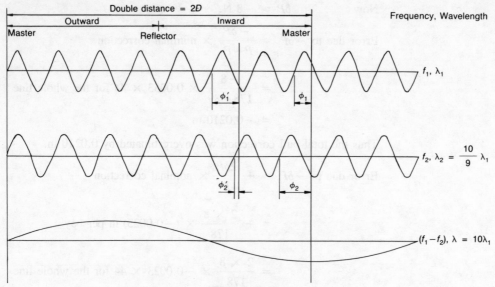

Outward | Inward
Master | Reflector | Master

Double distance = 2D

Frequency, Wavelength

f_1, λ_1

$f_2, \lambda_2 = \dfrac{10}{9} \lambda_1$

$(f_1 - f_2), \lambda = 10\lambda_1$

Figure 2.6

Solution. First, discuss the basic principles. Figure 2.6 shows three different frequencies with their wavelengths enclosed within a distance of 100 m.

Since $\lambda \propto \dfrac{1}{f}$

we have $\dfrac{\lambda_1}{\lambda_2} = \dfrac{f_2}{f_1}$.

Therefore $\lambda_2 = \dfrac{f_1}{f_2} \lambda_1$

$\qquad\qquad = \dfrac{f_1}{9/10\, f_1} \lambda_1$

$\qquad\qquad = \dfrac{10}{9} \lambda_1$.

Now $\lambda_1 = 10$ m, $\lambda_2 = 11.111$ m and accordingly there are 10 whole wavelengths of frequency f_1 and 9 whole wavelengths of frequency f_2 contained within the distance of 100 metres. Note also that one single wave of frequency $(f_1 - f_2)$ has a length of 100 m, since

$$\lambda = \frac{f_1 \lambda_1}{(f_1 - f_2)} = \frac{f_1 \lambda_1}{f_1/10} = 10\lambda_1.$$

At any point within the 100 m length, or stage, the phase of the $(f_1 - f_2)$ frequency wave is equal to the difference in phases of the other two waves. For example, at the 50 m point the phase of f_1 is $5 \times 2\pi$ whilst that of the f_2 frequency is $4.5 \times 2\pi$, giving a difference of π which is the phase of the $(f_1 - f_2)$

frequency. This relationship allows distance to be measured within 100 m stages.

Now, when the third frequency f_3 is considered we have $f_3 = 99/100f_1$, whence $\lambda_3 = 10.101$ m and the wavelength of frequency $(f_1 - f_3) = 1000$ m. The same statement in respect of phase differences applies here as well and further frequencies could be applied to extend the measurements of distance to cover 10 000 m, etc., without any ambiguity. The term 'fine' frequency can be assigned to f_1 which will appear in all the frequency difference values i.e. $(f_1 - f_2)$ whilst the other frequencies needed to make up the stages, or measurements of distance 100 m, 1000 m, etc., are termed 'coarse' frequencies. Lengths for 0 m to 10 m are covered by the f_1 phase difference measured at the 'master' station. In modern EDM the whole procedure is carried out automatically by the electronics.

Next, we determine the lengths OA and OB. The phase differences have been measured in metres in this problem and, as shown above, we have sufficient data to resolve distances from 0 m to 10 m, 0 m to 100 m, and 0 m to 1000 m.

If n_1 is the number of whole wavelengths of frequency f_1, and n_2 is the number of frequency f_2 within OA, then

$$\lambda_1 n_1 + \phi_1 = \lambda_2 n_2 + \phi_2.$$

ϕ_1 and ϕ_2 are measured at the master station. Therefore

$$10n_1 + 4.337 = 11.11n_2 + 7.670.$$

On inspection of Fig. 2.6 it will be seen that two important facts arise.

(a) when $\phi_1 < \phi_2$, $n_1 = n_2 + 1$, and
(b) when $\phi_1 > \phi_2$, $n_1 = n_2$.

These facts are important when evaluating overall phase differences.

Hence $10n_1 + 4.337 = 11.11(n_1 - 1) + 7.670$
and $\qquad n_1 = 7$.

This calculation has removed any ambiguity in the number of complete wavelengths of frequency f_1 lying within the 0 to 100 m stage.

Refer now to the f_3 frequency, which when related to f_1 gives the 0 to 1000 m stage.

Let n_1' be the number of complete waves of f_1 and n_3 be that for frequency f_3, then

$$10n_1' + 4.337 = 10.101n_3 + 0.600.$$

Since the phase difference for f_1 is greater than that for f_3, as measured at the master, $n_1' = n_3$

$$10n_1' + 4.337 = 10.101n_1' + 0.600$$
Therefore $\qquad n_1' = 37$.

Thus there are 37 complete wavelengths within OA; this value confirms the previous computation.

Therefore double length OA $= 4.337 + 370$
$\qquad\qquad\qquad\qquad\qquad = 374.337$ m.

Calculate length OB using the same notation as for OA above within the stages of 100 m and 1000 m, respectively.

$$10n_1 + 7.386 = 11.111n_2 + 1.830.$$

Since $\phi_1 > \phi_2 \; n_1 = n_2$

$$\therefore \; n_1 = 5.$$

Thus there are five whole wavelengths of 10 m length contained within the 0 m to 100 m stage, implying that a distance of 57.386 m is involved. Similarly,

$$10n_1' + 7.386 = 10.101n_3 + 9.911$$

and $\qquad\qquad\qquad n_3 = n_1' - 1.$

Therefore $\qquad\qquad n_1' = 75.$

Thus double length OB = 757.386 m.

Therefore $\quad$ length AB $= \dfrac{757.386 - 374.337}{2}$

$$= \mathbf{191.525 \; m.}$$

2.7 Depth measurement using EDM

Derive the following equation from first principles

$$D = \frac{N\,C_0}{2fn} + \frac{\phi}{2\pi}\,\frac{C_0}{2fn} + k$$

in which: D is the measured distance

N is the number of wavelengths in the distance

f is the modulation frequency

n is the refractive index

ϕ is the phase angle

k is the zero or additive constant

C_0 is the velocity of electromagnetic radiation in free space.

The shaft depth at a mine was recorded by an electromagnetic distance measuring instrument as 750.000 m, the average air temperature t and air pressure p in the shaft being 15 °C and 796.50 mmHg, respectively. The modulation wavelength of the instrument λ_s was 20.000 m, corresponding to a frequency of 14.9854 MHz at a specified combination of air temperature and air pressure of 12 °C and 760 mmHg, respectively. The carrier wavelength λ was 0.875 μm.

Compute the depth of the shaft corrected for atmospheric conditions.

Aide-memoire: $n_g = 1 + [2876.04 + 3(16.288)\lambda^{-2}$
$\qquad\qquad\qquad + 5(0.136)\lambda^{-4}]10^{-7}$

$$n_f = 1 + \frac{n_g - 1}{\left(1 + \dfrac{t}{273.15}\right)}\,\frac{p}{760}$$

$$\lambda_s = \frac{C_0}{fn_s}.$$

The velocity of electromagnetic radiation in free space = 299 792.5 km/s. [CEI]

Introduction. Review the general problem. The general derivation of the formula has been discussed previously with reference to Fig. 2.6. Essentially, double distance is measured but to 'read' out the single distance required we can use an effective wavelength (λ_A) which is half the true wavelength (λ). This is implied in the formula in which

$$\lambda_A = \frac{C_0}{2fn}$$

whilst $\lambda = \dfrac{C_0}{fn}$.

It will be noted that λ is inversely proportional to fn, in which n is the refractive index in the air space. This index is influenced by atmospheric conditions, i.e. air temperature, humidity and pressure, in both microwave and electro-optical systems. It is very important in respect of the longer lines usually measured by the microwave instruments.

In the case of electro-optical instruments the dry-air index is of consequence and this varies through the spectrum. Accordingly a group refractive index n_g is adopted and is then subjected to correction for atmospheric conditions to give n_f.

Solution. Determine shaft depth. λ has been expressed in μm, i.e. m $\times$ 10^{-6} and it is in this form that it is used in the relevant expression.

Modifying the given expression for n_g, we have

$$(n_g - 1) \times 10^7 = 2876.04 + 3(16.288)\lambda^{-2} + 5(0.136)\lambda^{-4}$$

Therefore $n_g = 1.000\,294\,1$ when $\lambda = 0.875$.

At the combination of $t = 12$ °C and $p = 760$ mmHg

$$n_f = 1 + \frac{0.000\,294\,1}{\left(1 + \dfrac{12}{273.15}\right)} \times \frac{760}{760}$$

$$= 1.000\,281\,7.$$

Note that when $f = 14.9854$ MHz

$$\lambda_s = \frac{299\,792\,500}{1.000\,281\,7 \times 14\,985\,400}$$

$$= 20.000 \text{ m.}$$

At the combination of $t = 15$ °C and $p = 796.50$ mmHg

$$n_f = 1 + \frac{0.000\,294\,1}{\left(1 + \dfrac{15}{273.15}\right)} \times \frac{796.50}{760.00}$$

$$= 1.000\,292\,2.$$

Now $D \propto \lambda \propto \dfrac{C_0}{nf}$ in which D is a measured distance. Therefore Dfn is a constant, i.e.

$$D_s fn_s = D_f fn_f,$$

where D_s is the recorded shaft depth and D_f is the depth measured. Therefore

$$750.000 \times 1.000\ 281\ 7 = D_f \times 1.000\ 292\ 2$$
$$D_f = \mathbf{749.992\ m.}$$

2.8 Distance measurement with EDM

A slope distance measured with an electromagnetic distance measuring instrument (EDM) when corrected for meteorological conditions and instrumental constants is 114.652 m. The EDM and its reflector are 1.750 m and 1.922 m above the ground stations, respectively. The vertical angle measured with a theodolite, when corrected for earth curvature and refraction, is $+04° 25' 15''$ from the horizontal. The theodolite and theodolite target are 1.650 m and 1.646 m above ground stations, respectively. Both the EDM instrument and the theodolite were centred over the lower station. What is the horizontal length of the line?

To what precision must the slope angle be measured

(*i*) If the relative precision of the reduced horizontal distance is to be 1/100 000.

(*ii*) If the reduced horizontal distance is to have a standard error of ± 2 mm. [CEI]

Introduction. Slant distance L between the EDM and the reflector (centred over their respective stations) has been measured and the reduced horizontal length, l, is required. To obtain l we require the vertical angle θ which the measured length makes with the horizontal. This incorporates measured angle α together with two corrections β and γ, defined as 'eye and object' corrections, due to the differences in height above the ground stations of the measuring devices and their targets. The 'eye and object' correction and curvature and refraction have been covered in detail in Chapter 1.

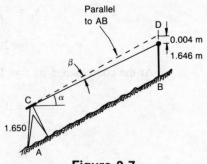

Figure 2.7

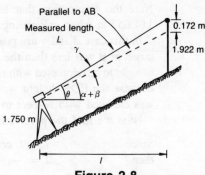

Figure 2.8

It was also mentioned therein that the measured slope length L between two stations can be reduced to its chord length at the mean height of the stations. Accordingly, the equivalent chord length at mean sea level can be computed and a correction then added to determine the spheroidal distance between the stations. Assuming that the slope distance was established by EDM, the relevant corrections are $L^3/43\ R^2$ and $L^3/33\ R^2$ for microwave and infra-red systems, respectively. These include corrections for the curvature of the path of the signal and are of particular importance for long lines. Note that the theoretical difference between spheroidal distance (Fig. 1.10) and chord length is $Rc - 2R \sin c/2 = d^3/24\ R^2$.

Solution. Calculate horizontal length l. The first step is to determine the corrections β and γ. Had the heights of the instruments and their respective 'targets' above the ground stations been equal, the corrections would be zero since the lines of sight would be parallel to AB, and measured angle α would have given the required slope angle. From Fig. 2.7

$$\beta = \frac{1.650 - 1.646}{114.652} \text{ radian}$$

$$= \frac{0.004 \times 206\ 265''}{114.652}$$

$$= 7''.$$

From Fig. 2.8

$$\gamma = \frac{1.922 - 1.750}{114.652} \text{ radian}$$

$$= \frac{0.172 \times 206\ 265''}{114.652}$$

$$= 309''.$$

Therefore   slope angle $\theta = \alpha + \beta + \gamma = 04°\ 25'\ 15'' + 7'' + 5'\ 09''$
$$= 4°\ 30'\ 31''.$$

Therefore horizontal length $l = L \cos \theta$
$$= 114.652 \cos 4°\ 30'\ 31''$$
$$= \mathbf{114.297\ m.}$$

Note the assumption that length CD is equal to the measured length of 114.652 m when calculating β and γ.

In this case β and γ are positive corrections, first because the height of the target at B was less than the theodolite height at A, thereby causing the line of sight to be depressed with respect to CD: thus α was measured low. Second, because the EDM height was less than the reflector height the line of sight was elevated with respect to the parallel to AB.

Next we find the precision of the slope angle.

Since $\qquad\qquad l = L \cos \theta$

then $\qquad\qquad dl = -L \sin \theta \, d\theta$

(an increase $d\theta$ causes a decrease in l).

Relative accuracy $\dfrac{dl}{l} = \dfrac{1}{100\ 000} = \dfrac{L \sin \theta \, d\theta}{L \cos \theta}$

(neglecting the negative sign)

$$= \tan \theta \, d\theta = \tan 4° \, 30' \, 31'' \, d\theta.$$

Therefore $\qquad d\theta = 0.000\ 126\ 8$ radian

$$= \mathbf{26''}.$$

If $\qquad\qquad\qquad dl = \pm \, 2$ mm

$\qquad\qquad 0.002 = L \sin \theta \, d\theta$

$\qquad\qquad\qquad\quad = 114.652 \sin 4° \, 30' \, 31'' \, d\theta.$

$\qquad\qquad\qquad d\theta = 0.000\ 221\ 9$ radian.

Therefore $\qquad d\theta = \pm \, \mathbf{46''}$

for $\qquad\qquad\qquad dl = \pm \, 2$ mm.

Note that the term standard error is defined in Chapter 8.

2.9 Principles of tacheometry

(a) Derive the basic formula for finding horizontal distances and differences in height from tachymetric observations.

(b) The tachymetric observations given in Table 2.4 were made.

Table 2.4

Obs. point	Staff point	ZD	Upper	Lower	Centre
A	1	94° 37′	1.479	0.500	0.990
A	2	119° 25′	1.633	1.000	1.318

The theodolite was 1.57 m above ground level at A. Calculate the horizontal distances and differences in height A−1, A−2. Assume a stadia constant of 100. Assuming that the standard error of a stadia reading is 1.5 mm and of a ZD is 1.5′, calculate the standard errors of the distances and height differences. [London]

Solution. Derive basic formulae. Figure 2.9 shows a simple telescope, whose objective is of focal length f. The image of a staff has been presented

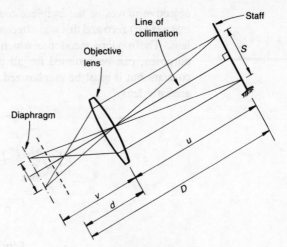

Figure 2.9

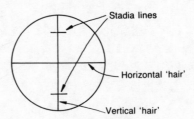

Figure 2.10

at the diaphragm in such a way that the staff intercept s is given by the difference between the upper and lower stadia lines in Fig. 2.10. The staff is normal to the line of sight of the telescope (given by the centre hair reading).

Thus $\dfrac{1}{u} + \dfrac{1}{v} = \dfrac{1}{f}$,

where u is the distance from the staff to the objective and v is the distance from the lens to the diaphragm.

$$\frac{u}{f} = 1 + \frac{u}{v} = 1 + \frac{s}{i}.$$

Therefore $u = f + \dfrac{fs}{i}$.

Distance D is measured to the instrument axis rather than to the objective,

therefore $\quad u+d = (f+d) + \dfrac{fs}{i} = D$.

$(f+d)$ is a constant known as the additive constant and f/i is the multiplying constant (or stadia constant as in the wording of the example). An important point is that the expression was derived for a telescope in which either the objective moves or the eyepiece moves during focusing. It is convenient to make $(f+d)$ zero and this was effected by an internal lens known as an anallatic lens. The modern theodolite which also has an internal lens, for focusing purposes, can be assumed for all practical purposes to have zero additive constant but it must be emphasized that the internal focusing lens is not an anallatic lens.

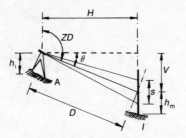

Figure 2.11

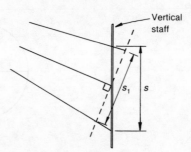

Figure 2.12

In Fig. 2.11 we have an inclined sight being made on to a vertical staff. This is the more usual approach to tacheometric measurement; it also shows the equivalent position of a staff held normal to the line of sight. It is essential that angle θ is measured to determine H and V. ZD (zenith distance) is the angle measured from a vertical pointing such that (on one face) a reading of 90° is given when the line of sight is horizontal. The reading given by the stadia lines on the vertical staff is s, and in Fig. 2.12 we have the reading s_1 which would arise on a staff which is normal to the line of sight. Note that $s_1 \simeq s \cos \theta$.

Now $H = D \cos \theta$

$$= \frac{fs_1}{i} \cos \theta + (f+d) \cos \theta$$

$$= \frac{fs}{i} \cos \theta \times \cos \theta + (f+d) \cos \theta$$

$$= \frac{fs}{i} \cos^2\theta + (f+d) \cos \theta$$

and $V = D \sin \theta$

$$= \frac{fs}{i} \cos \theta \times \sin \theta + (f+d) \sin \theta$$

$$= \frac{fs}{i} \frac{\sin 2\theta}{2} + (f+d) \sin \theta.$$

RL of staff station = RL of instrument station $+ h_i - V - h_m$, where h_i is the height of the instrument axis and h_m is the centre line reading on the staff.

In this example θ is a depression angle but if θ is an elevation angle V in the heighting relationship is additive not subtractive.

Note that if the staff is held normal to the line of sight then

$$H = Ds_1 \cos \theta \pm h_m \sin \theta,$$

where s_1 is the staff intercept and $h_m \sin \theta$ is positive or negative depending upon θ being elevation or depression. Also $V = Ds_1 \sin \theta$ and $h_m \cos \theta$ is the corrected centre hair reading when determining height differences.

(b) Determine distances and height differences. Between A and l, given that

$$\frac{f}{i} = 100 \text{ and } (f+d) = 0,$$

$$H = \frac{fs}{i} \cos^2\theta = 100 (1.479 - 0.500) \cos^2 4° 37'.$$

Since
$$\theta = 94° 37' - 90° 00'$$
$$H = \mathbf{97.27 \ m} = \text{horizontal distance A}-1$$

$$V = \frac{fs}{i} \frac{\sin 2\theta}{2} = 100 \frac{(1.479 - 0.500)}{2} \sin 9° 14'$$

$$= 7.85 \ m.$$

Therefore height difference A$-1 = V + h_m - h_i$ (see Fig. 2.11)

$$= 7.85 + 0.990 - 1.57$$
$$= \mathbf{7.27 \ m.}$$

Between A and 2 $\theta = 119° 25' - 90° 00' = 29° 25'.$
Therefore $H = 100 (1.633 - 1.000) \cos^2 29° 25'$
$$= \mathbf{48.03 \ m} = \text{horizontal distance A}-2$$

$$V = 100 \frac{(1.633 - 1.000)}{2} \sin 58° 50'$$

$$= 27.08 \ m.$$

Therefore height difference A$-2 = 27.08 + 1.318 - 1.57$
$$= \mathbf{26.83 \ m.}$$

The following computer program will carry out stadia tacheometric calculations by either the staff normal or the staff vertical method. The vertical angle input in the program is related to zero as the horizontal and the ZD angles given in this example must have 90° subtracted from them. In lines 20 and 30 the instrument constants are set to $f/i = 100$, $f+d = 0$. For non-standard instruments these lines must be changed before the program is run.

Variables

A = Vertical angle
B = Height of instrument
C = Theodolite multiplying constant (set to 100)
D = Input/output, Degrees
H = Horizontal distance
I = Stadia intercept
K = Theodolite additive constant (set to 0)
L = Difference in level

M = Input/output, Minutes
P = Uphill/downhill indicator
Q$ = Staff normal/vertical indicator
S = Input/output, Seconds
V = Vertical distance
X = Bottom stadia reading
Y = Mid-crosshair reading
Z = Top stadia reading

```
10 REM STAFF TACHEOMETRY
20 C=100
30 K=0
40 INPUT"INPUT BOTTOM STADIA READING",X
50 INPUT"INPUT MID READING          ",Y
60 IF Y<X THEN 190
70 INPUT"INPUT TOP STADIA READING   ",Z
80 IF Z<Y THEN 190
90 I=Z-X
100 INPUT"INPUT VERTICAL ANGLE IN DEG,MIN,SEC ( DEG -VE IF DOWN) ",D,M,S
110 P=1
120 IF D<0 THEN P=-1
130 A=((ABS(D)*3600)+(M*60)+S)/206264.8
140 INPUT"INPUT HEIGHT OF INSTRUMENT ",B
150 INPUT"STAFF NORMAL [N] OR VERTICAL [V] ",Q$
160 IF Q$="N" THEN 250
170 IF Q$="V" THEN 210
180 GOTO 150
190 PRINT"*** INPUT ERROR ***"
200 GOTO 40
210 H=C*I*COS(A)*COS(A)+K*COS(A)
220 V=0.5*C*I*SIN(A*2)+K*SIN(A)
230 L=B+(V*P)-Y
240 GOTO 280
250 H=(C*I+K)*COS(A)+P*Y*SIN(A)
260 V=(C*I+K)*SIN(A)
270 L=B+(V*P)-(Y*COS(A))
280 PRINT"HORIZONTAL DISTANCE   = ";H
290 PRINT"VERTICAL DISTANCE     = ";V
300 PRINT"DIFFERENCE IN GROUND LEVELS = ";L
310 END
```

We now find the standard errors of distances and height differences. This part of the question is covered by Chapter 8 but, for continuity, treatment is given here.

Assuming that the multiplying constant will be unchanged, horizontal distance H is influenced by staff reading errors (ds) and errors in vertical circle reading (dθ).

$$H = \frac{fs}{i} \cos^2\theta,$$

therefore

$$\frac{dH}{ds} = \frac{f}{i} \cos^2\theta$$

and

$$\frac{dH}{d\theta} = -\frac{fs}{i} \sin 2\theta$$

Thus $s_H^2 = dH^2 = \left(\dfrac{dH}{ds}\right)^2 s_s^2 + \left(\dfrac{dH}{d\theta}\right)^2 s_\theta^2$

$$= \left(\dfrac{f}{i}\cos^2\theta\right)^2 s_s^2 + \left(\dfrac{fs}{i}\sin 2\theta\right)^2 s_\theta^2.$$

where s_H, s_s and s_θ are the standard errors in H, s and θ, respectively. Similarly for V

$$\dfrac{dV}{ds} = \dfrac{f}{i}\dfrac{\sin 2\theta}{2}$$

and $\dfrac{dV}{d\theta} = \dfrac{fs}{i}\cos 2\theta$.

Let L be the height difference

$$L = V + h_m - h_i.$$

h_m is a staff reading and will be subject to error (there is no mention of h_i in the question).

$$dL^2 = \left(\dfrac{dL}{dV}\right)^2 s_v^2 + \left(\dfrac{dL}{dh_m}\right)^2 s_{hm}^2,$$

then $s_L^2 = dL^2 = \left(\dfrac{f}{i}\dfrac{\sin 2\theta}{2}\right)^2 \times s_s^2 + \left(\dfrac{fs}{i}\cos 2\theta\right)^2 \times s_\theta^2$

$$+ \left(\dfrac{dl}{dh_m}\right)^2 \times s_{hm}^2,$$

where s_L is the standard error in height difference L. Now the standard error of a single stadia reading is 1.5 mm and there are two such readings involved in s.

Therefore $s_s^2 = 1.5^2 + 1.5^2 = 4.50 \text{ mm}^2$

and $s_\theta^2 = \left(\dfrac{\pi}{180} \times \dfrac{1.5}{360}\right)^2 = (4.3633 \times 10^{-4})^2.$

$$\dfrac{dl}{dh_m} = 1$$

$$s_{hm}^2 = 1.5^2 = 2.25 \text{ mm}^2.$$

Between A and 1

$$\dfrac{f}{i}\cos^2\theta = 100 \cos^2 4° 37' = 99.352$$

$$\dfrac{fs}{i}\sin 2\theta = (100 \times 0.979) \sin 9° 14' \text{m} = 15709 \text{ mm}.$$

Therefore $s_H^2 = (99.352^2 \times 4.5) + 15709^2 \times (4.3633 \times 10^{-4})^2$
$$= 44465.7 \text{ mm}^2$$
$$s_H = 210.9 \text{ mm} = \mathbf{0.21 \text{ m}}.$$

Also $\dfrac{f}{i} \dfrac{\sin 2\theta}{2} = 100 \dfrac{\sin 9° \ 14'}{2} = 8.02$

$$\dfrac{fs}{i} \cos 2\theta = 97.9 \cos 9° \ 14' \text{m}$$
$$= 96 \ 631.52 \ \text{mm}.$$

Therefore
$$s_L^2 = 8.02^2 \times 4.5 + 96 \ 631.52^2 \ (4.3633 \times 10^{-4})^2 + 1 \times 2.25$$
$$= 2069.43 \ \text{mm}^2$$
$$s_L = \textbf{45.5 mm} = \textbf{0.046 m.}$$

Between A and 2, in a similar manner which the reader should check

$$s_H = \textbf{0.19 m,}$$
$$s_L = \textbf{0.09 m.}$$

2.10 Effect of staff verticality

Referring to the previous worked example determine the effects of inclinations of 1° of the staff from the vertical:

(a) towards the instrument for line A−1; and
(b) away from the instrument for line A−2.

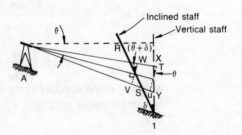

Figure 2.13

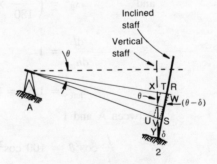

Figure 2.14

Solution. Consider the basic theory. In Fig. 2.13 the staff is inclined through angle δ towards the instrument and so intercept RS would be given by the stadia lines rather than the correct value XY on the vertical staff.

Let WV and TU be perpendicular to the line of sight. Since WV and TU are very near together in practice we can assume that WV = TU. Moreover, WV makes an angle of $(\theta + \delta)$ with RS.

Therefore TU $= XY \cos \theta$

$= WV$

$= RS \cos (\theta + \delta).$

Therefore XY $=$ true staff intercept

$$= \frac{RS \cos (\theta + \delta)}{\cos \theta}.$$

The same relationship holds for elevation θ when the staff is inclined away from the instrument.

In Fig. 2.14 the staff is inclined away from the instrument and so WV makes an angle of $(\theta - \delta)$ with RS.

As above TU $= XY \cos \theta = WV$

$= RS \cos (\theta - \delta).$

Therefore XY $= \dfrac{RS \cos (\theta - \delta)}{\cos \theta}.$

This relationship holds for elevation θ with the staff inclined towards the instrument.

We now find the error in the horizontal distance A−1. In Fig. 2.13 we have $\theta = 4° 37'$, $\delta = 1° 00'$ and RS $= (1.479 - 0.500)$ m.

True horizontal distance $= \dfrac{f}{i} XY \cos^2 4° 37'$

$$= \frac{f}{i} \times \frac{RS \cos (\theta + \delta)}{\cos \theta} \cos^2 4° 37'$$

$$= 100 \times 0.979 \frac{\cos 5° 37'}{\cos 4° 37'} \times \cos^2 4° 37'$$

$$= 97.12 \text{ m}.$$

The apparent distance $= 100 \times 0.979 \cos^2 4° 37'$

$$= 97.27 \text{ m}.$$

Therefore   error $= +0.15$ m or **1 in 647** (approx.).

We next find the error in the horizontal distance A−2. In Fig. 2.14 we have $\theta = 29° 25'$ and $\delta = 1° 00'$, whilst RS $= (1.633 - 1.000)$ m.

True horizontal distance $= \dfrac{f}{i} XY \cos^2 29° 25'$

$$= \frac{f}{i} RS \frac{\cos (\theta - \delta)}{\cos \theta} \cos^2 \theta$$

$$= 100 \times 0.633 \frac{\cos 28° 25'}{\cos 29° 25'} \times \cos^2 29° 25'$$

$$= 48.50 \text{ m}.$$

The apparent distance $= 100 \times 0.633 \cos^2 29° 25'$
$$= 48.03 \text{ m.}$$

Therefore $\quad$ error $= -0.47$ m or **1 in 102** (approx.).

When the staff has been held normal to the line of sight θ is of no influence. However δ causes an increase in the staff reading and the error in horizontal distance is positive. Accordingly, this method of staff holding is to be preferred when θ exceeds, say, $10°$.

2.11 Subtense tacheometry

A horizontal subtense bar of length 2 m is set up at station A. Its targets subtend a mean angle of $1° 58' 46.8''$ at a theodolite set up at station B. Determine:

(a) the length of line AB; and
(b) the fractional error in that length if the angle was measured with an accuracy of $\pm 1''$.

Introduction. The bar is set up horizontally by means of a levelling head, and by means of a sighting device it is arranged to be virtually perpendicular to AB. Invar steel is used in construction of the bar to ensure that temperature effects are of no consequence.

θ is measured in the horizontal plane by the theodolite, and so H is the reduced horizontal distance. Moreover, since the bar is horizontal refraction has no effect on the readings and pointings may be made using a single face.

Solution. Finding the length of line AB, from Fig. 2.4

$$H = \frac{b/2}{\tan \theta/2} = \frac{b}{2 \tan \theta/2}$$

$$= \frac{2.00}{2 \tan \left(\dfrac{1° 58' 46.8''}{2} \right)}$$

$$= \textbf{57.88 m.}$$

We now find the fractional error in H.

In $\quad H = \dfrac{b}{2 \tan \theta/2},$

since θ is small we can write $2\tan \theta/2 = \theta$ radian.

Therefore $\quad H = b/\theta$

$$dH = \frac{-b}{\theta^2} \, d\theta$$

$$= \frac{-H^2}{b} \, d\theta,$$

i.e. increase in θ gives decrease in H. Given that the accuracy of measurement is ± 1 second of arc,

$$d\theta = \frac{1}{206\ 265} \text{ radian}$$

Therefore $dH = \dfrac{-57.88^2}{2} \times \dfrac{1}{206\ 265}$

$$= 0.008 \text{ m.}$$

Fractional error = **1 in 7234.**

Note that dH depends upon H^2 for a given value $d\theta$, and consequently accuracy is reduced markedly over ranges exceeding hundreds of metres. Measures taken to improve the fractional errors include:

(*a*) increasing the number of readings of θ to enhance the accuracy $d\theta$;

(*b*) subdividing the line into separately measured bays; and

(*c*) using an auxiliary base.

Further mention of these will be found in Chapter 8.

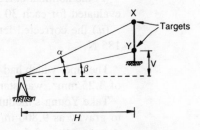

Figure 2.15

In the tangential system the bar is in a vertical position and the angles α and β are measured in the vertical circle. Figure 2.15 shows the principle

$$
\begin{aligned}
\text{XY} &= \text{distance between targets} \\
&= H \tan \alpha - H \tan \beta
\end{aligned}
$$

and $\quad V = H \tan \beta.$

Problems

1 During the measurement of a base line a 30 m steel tape was used in catenary to obtain the following individual bay lengths: 29.9016 m, 29.8834 m, 29.9502 m, 29.9782 m, 29.9218 m. If the total sag correction was 0.0149 m, estimate the total pull correction.

The tape, having a mass of 0.026 kg/m and cross-sectional area of 3.24 mm², had been standardized on the flat under a pull of 89 N. Take Young's modulus as 155 000 N/mm² and the acceleration due to gravity as 9.806 m/s². [Salford]

Answer +0.0200 m.

2 A steel tape, 50 m long on the flat at 20 °C under a pull of 89 N, has a cross-sectional area of 6 mm^2 and a mass of 2.32 kg. It is to be used to measure lengths when supported at mid-span.

Assuming that the ends of the tape are at the same level and the field temperature is 20 °C determine the tension to the nearest newton to be applied to ensure that errors greater than ±2.5 mm will not occur when measuring a length of 50 m.

Take Young's modulus to be 207 000 N/mm^2 and g to be 9.81 m/s^2.

[Salford]

Answer 198 N; 255 N.

3 A tape, which was standardized on the flat under a tension P_s, was used in catenary to measure the length of a baseline. Show that the nominal corrections for pull and sag must be modified by factors of $\pm \delta P/(P - P_s)$ and $\mp 2\delta P/P$ respectively, if an error of $\pm \delta P$ occurred in the applied field tension P.

The length of the line was deduced as 659.870 m, the apparent field tension being 178 N. Determine

(*i*) the nominal corrections for pull and sag which would have been evaluated for each 30 m tape length, and

(*ii*) the corrected length of the line, if the actual field tension was 185 N.

The tape, which had a mass of 0.026 kg/m and a cross-sectional area of 3.25 mm^2, was standardized on the flat under a pull of 89 N.

Take Young's modulus as 155 000 MN/m^2 and the acceleration due to gravity as 9.806 m/s^2. [Salford]

Answer (*i*) +5.3 mm, −2.3 mm; (*ii*) 659.883 m.

4 A surveyor measured a base line in five bays using a tape in catenary under a pull of 10 kg f. He recorded the data given in Table 2.5.

Table 2.5

Length of span (m)	Rise/fall between end of span (m)
29.913	+0.31
29.935	−0.22
29.872	+0.81
29.914	+1.05
23.721	−0.49

Unfortunately, he poorly recorded the field temperature and could not decide whether he had written down 15 °C or 18 °C. He returned to the site with an EDM and with the prism axis set at the same height as the instrument collimation. The EDM gave a reading of 143.212 ± 5 mm.

If the tape was standardized on the flat under a pull of 7 kg f at 20 °C, which of the field temperature readings is correct?

$\alpha = 0.000011/°C$
$E = 207000 \text{ MN/m}^2$
Tape Area = 6 mm^2
Tape Density = 7700 kg/m^3 [Salford]
Answer 15 °C

5 Five bays of base line AB were measured under a tension of 149 N
and the data given in Table 2.6 was recorded. The tape was standardized
on the flat under a pull of 89 N and at temperature 20 °C.

Table 2.6

Length of span (m)	Rise/fall between ends of span (m)	
29.940	−0.258	End A
29.096	0.118	
29.101	0.292	
29.891	0.325	
29.846	0.424	End B

Field temperature = 17 °C

Tape details

Coefficient of thermal expansion	0.000 011/°C
Young's Modulus	207 kN/mm^2
Density	7700 kg/m^3
Cross-sectional area	6 mm^2

An EDM was set up 1.37 m above end A of the base line and a prism
was set up 1.56 m above end B. The distance between them was read
as 147.859 ± 5 mm. What is the maximum and minimum possible length
of the 30 m division on the tape?   [Salford]
Answer 30.008; 30.006 m.

6 Explain carefully how a transmission having a sinusoidal wave form
can be used to measure distances. Hence derive an expression for N_1
which is the whole number of complete waves beyond multiples of ten
wavelengths. Consider both $dx_1 > dx_2$ and $dx_1 < dx_2$, where dx_1 and
dx_2 are phase differences on frequencies 1 and 2, respectively.

An EDM instrument produces a wavelength of exactly 10 m when
transmitting on frequency 1. Frequency 2 = 9/10 frequency 1 and
frequency 3 = 99/100 frequency 1.

This instrument was set up at O in line with P and Q in order to
determine the length PQ. Station O is on the side of P remote from Q.
Determine the shortest possible value for the length of the line from the
observations given in Table 2.7.

Explain how this method eliminates the effects of zero error in the
instrument.   [Bradford]
Answer 281.601 m

Table 2.7

Length	Phase differences		
	dx_1	dx_2	dx_3
OP	2.561	1.450	0.440
OQ	5.763	9.096	8.086

7 The length of a certain line was measured as 1873.574 m by an EDM instrument which had a design refractive index of 1.000 274, modulation frequency of 14.985 52 MHz and carrier wavelength of 0.874 μm.

Air temperature and pressure were 15 °C and 740 mmHg, respectively. Determine

(i) the refractive index of the air at the time of measurement,

(ii) the corrected length of the line.

Answer (i) 1.000271; (ii) 1873.580 m

8 (a) With one type of EDM instrument the determination is based on the formula $(x_2 - x_1) \dfrac{\lambda_1 \lambda_2}{\lambda_1 - \lambda_2}$ where x_1 and x_2 are fractions of two different measuring patterns with wave lengths λ_1 and λ_2 respectively. Answer the following questions relating to this formula and its use.

(i)   What distance is the formula intended to represent?

(ii)   How is the fomula derived?

(iii)   What relationship must exist between λ_1 and λ_2?

(iv)   When does the formula fail to produce the required distance even though x_1 and x_2 have been correctly measured and what must then be done to obtain the distance? Explain why.

(b) If an instrument capable of displaying the fractions referred to in part (a) to 0.001 cycle is to be designed, propose with reasons wave lengths for its measuring patterns which will allow distances up to 10 km to be calculated by elementary arithmetic, i.e. requiring no electronic or mechanical aids, with a resolution of 0.01 m.   [London]

Answer 20 m, 22.222 m, 20.202 m, 20.020 m. (Assuming that $C_0 = 300 \times 10^6$ m/s)

9 The formula given in a manufacturer's instruction manual for computing the atmospheric correction (c_m) to measured electro-optical distance measurements is

$$c_m = \frac{1.000\ 281\ 95}{\left[\dfrac{0.000\ 294\ 335}{1 + 0.003\ 660\ 86t} \times \dfrac{P}{1013} + 1 \right]}$$

t = ambient atmospheric temperature (°C)

P = ambient atmospheric pressure (mb)

corrected slope distance = measured slope distance $\times c_m$

The modulated wavelength of the instrument (λ_s) is 20.00 000 m corresponding to a frequency of 14.985 400 MHz at specified meteorological reference data of 12 °C (t) and 1013 mb (P) and carrier wavelength (λ) of 0.860 μm.

A survey line forming part of a precise test network was measured with the instrument and a mean value of 2999.097 m recorded. The mean ambient temperature t and pressure P were 13.4 °C and 978.00 mb, respectively.

Compute the atmospheric correction using the formula given in the instruction manual and from first principles, and compare the results. Assume the velocity of electromagnetic radiation in free space to be 299 792.5 km/s.

It was later discovered that the field barometer was in error by +24 mb. Compute the correction in the distance due to this error. What conclusions can be drawn from these calculations?

Aide-mémoire

$$n_a = 1 + \frac{n_g - 1}{\alpha T} \times \frac{P}{1013.25},$$

$$n_g = 1 + \left[28\,760.4 + 3 \times \frac{162.88}{\lambda^2} + 5 \times \frac{1.36}{\lambda^4} \right] \times 10^{-8},$$

$$\lambda_s = \frac{c_0}{f n_s},$$

where

n_a = group refractive index of atmosphere,
n_g = group refractive index of white light (1.000 294),
n_s = group refractive index for standard conditions,
α = 3.661 × 10^{-3} K^{-1},
T = ambient temperature (K),
P = ambient pressure (mb),
λ_s = modulated wavelength (20.000 000 m),
λ = carrier wavelength (0.860 μm),
c_0 = velocity of electromagnetic radiation in free space (299 792.5 km/s),
f = modulation frequency (14.985 40 MHz). [Eng. Council]
Answer +0.054 m

10 The data in Table 2.8 has been abstracted from the field notes of

Table 2.8

From	To	Theodolite height (m)	Target height (m)	Measured vertical angle	EDM height (m)	Prism height (m)	Measured slope distance (m)
BH	LB	1.68	1.56	−03° 00′ 56″	1.63	1.65	1253.687
LB	BH	1.65	1.68	+03° 00′ 03″			

a mine surface trilateration survey. Calculate the horizontal distance between BH and LB. [CEI]

Answer 1251.963 m

11 An electromagnetic distance-measuring instrument is to be used to measure a slope distance of 3 km inclined at 10° to the horizontal. Compute the accuracy of the vertical angle measurement required in order to determine (*a*) slope corrections and (*b*) differences in level with a standard error of ±10 mm. What conclusions can be drawn from the results of this computation and what instructions should be given to the field party regarding the measurement of the vertical angles? [CEI]

Answer (*a*) ±4.0″; (*b*) ±0.7″

12 A theodolite was set up over station P and observations were taken on to a subtense bar set up over Q and R in turn. The results in Table 2.9 were recorded.

Table 2.9

Instrument station	Subtense bar station	Subtense angle	Horizontal circle
P	Q	01° 05′ 26″	116° 37′ 52″
P	R	00° 58′ 56″	179° 12′ 05″

Calculate the area of the triangle PQR. [Leeds]

Answer 5440 m^2

13 A theodolite with constants 100 and 0 was set up 1.38 m above Station A and the tacheometric readings given in Table 2.10 were recorded on a staff held normal to the line of sight.

Table 2.10

Station	Horizontal circle	Vertical circle	Stadia
B	123° 10′ 20″	−11° 2′ 20″	1.000/1.200/1.400
C	33° 10′ 20″	+ 4° 1′ 20″	1.891/2.321/2.751

Boreholes at A, B and C showed the depth of bedrock below the surface to be 7.65 m, 2.46 m, and 11.27 m respectively.

If the level of the ground at station A was 126.026 m AOD, find the bearing of the line of steepest rock slope relative to line AB. [Salford]

Answer 16° 28′ 34″

14 Three survey stations, F, A and B have been set out in line on steeply inclined ground. A target at B, 1.219 m above the ground, is sighted from the two instrument stations A and F. The angles of elevation are 45° 30′ from A and 30° 20′ from F. The height of the instrument axis at A above the ground is 1.554 m and at F, 1.451 m. The horizontal distance from A to F is 60.961 m. A levelling staff is held at F, and a reading of 2.585 m obtained from the instrument at A, the telescope being set to the horizontal.

Assuming station F to be 24.384 m above Ordnance Datum, calculate (*i*) the horizontal distance from A to B, and (*ii*) the reduced level of station B to Ordnance Datum. [Eng. Council]
Answer (*i*) 79.852 m; (*ii*) 107.509 m

15 A theodolite with a multiplying constant of 100 and no additive constant is set up over station X, which has coordinates of 2241.03 m E, 2106.58 m N and is 224.260 m above datum. The height of the instrument above the station is 1.320 m. A station Y with coordinates of 2368.56 m E, 2002.40 m N is sighted. Readings are then taken successively on a staff held vertically on stations A and B and the observations in Table 2.11 are recorded.

Table 2.11

Instrument station	Station sighted	Stadia hair readings	Centre hair readings	Vertical circle	Horizontal circle
X	Y	—	—	—	15° 20′ 31″
X	A	2.684 1.505	2.095	+1° 15′ 20″	22° 28′ 57″
X	B	2.745 1.188	1.967	−2° 04′ 50″	141° 15′ 36″

If the positive and negative signs in the vertical circle column denote angles of elevation and depression respectively, calculate:

(*i*) the horizontal length of AB
(*ii*) the bearing of AB
(*iii*) the heights of points A and B above datum. [Leeds]
Answer (*i*) 236.02 m; (*ii*) 281° 06′ 58″; (*iii*) 226.06 m, 217.96 m

16 Describe the method of tangential tacheometry. Two points B and C were observed from A. Field notes are given in Table 2.12.

Table 2.12

From	To	Bearing	Vertical angles	Staff (m)
A	B	165° 36′	−3° 18′; −3° 00′	1.00; 3.00
	C	216° 06′	−2° 54′; −2° 06′	1.00; 3.00

(*a*) Calculate the coordinates of B if the coordinates of C are 70.00 mE, 120.00 mN.

(*b*) Calculate the ground level at B if the ground level at C is 59.130 m AOD. [Bradford]
Answer (*a*) 261.17 mE, 124.25 mS; (*b*) 44.41 m

17 A theodolite, with a multiplying constant of 100 and no additive constant, is set up at A and sighted on a level staff at B. The upper stadia hair, centre hair and lower stadia hair readings are 2.975, 1.890 and 0.805 respectively and the angle of elevation of the telescope is 4° 36′. If the height of the instrument above the ground at A is 1.250 m, calculate

(a) the horizontal distance AB and

(b) the difference in height between A and B.   [Leeds]

Answer   (a) 215.60 m; (b) 16.71 m

18 A tacheometer with a multiplying constant of 100 and no additive constant is set up over station X and sighted on a level staff held vertically on a bench mark, B, and then on two points Y and Z. The height of B above datum is 278.36 m. The results are in Table 2.13.

Table 2.13

Instrument station	Station sighted	Stadia hair readings	Centre hair reading	Vertical circle	Horizontal circle
X	B	2.310 1.464	1.887	+2° 15′ 10″	—
X	Y	2.466 0.732	1.600	+3° 36′ 00″	28° 12′ 15″
X	Z	2.522 0.388	1.455	−2° 42′ 30″	93° 51′ 55″

The positive and negative signs in the vertical circle column denote angles of elevation and depression respectively. Calculate:

(a) the horizontal distance YZ,

(b) the heights of points Y and Z above datum and

(c) the error (in mm) which would occur if the slope distance rather than the horizontal distance of YZ was plotted on a 1/500 scale plan.

[Leeds]

Answer   (a) 211.79 m; (b) 286.19 m, 265.39 m; (c) 2.0 mm

19 The following values were recorded during a theodolite tacheometric survey

Stadia readings: 3.33 m (top)   2.20 m (middle)
                          1.07 m (bottom)
Vertical angle: +11° 40′
Instrument height: 1.48 m
Height of collimation: 269.01 m

Find the horizontal distance between the staff and instrument station, and the reduced level of the staff station. Assume that the telescope is anallactic, the multiplying constant 100, and the staff vertical.

Determine the error in the horizontal and vertical distances due to an error of ±5 minutes of arc in the measurement of the vertical angle.

[CEI]

Answer   216.76 m; 222.05 m; ±0.130 m; ±0.30 m

20 The gradient of the line joining two stations A and B was known to be 1 in 18. A tacheometer having a multiplying constant of 100 and zero additive constant was set up at A and, with an angle of depression of 3°, observations were taken on a staff held vertically at B. If the height of the instrument axis was 1.190 m and the reading of the lower stadia line was 1.000 m, estimate the other staff readings, and deduce the horizontal distance between A and B.

What other method of staff positioning may be adopted? Compare the two methods with particular reference to the effects of staff tilt.

[CEI]

Answer 1.511; 2.023; 102.05 m

21 Use the slope correction expression $- (h^2/2L)$ to show that the difference between the actual correction and the nominal correction has a magnitude of

$$2 \, \frac{\delta h}{h} \times \text{nominal correction}$$

due to an error of δh in h.

The slope length of a line was measured as 29.8984 m and h was booked as 1.382 m instead of the measured height 1.392 m. What error will arise in the corrected horizontal length of the line?

Answer error $= +0.0005$ m (correction $= -0.0005$ m)

3

Theodolite and traverse surveying

The theodolite

There are three important lines or axes in a theodolite, namely the line of sight, the horizontal axis and the vertical axis. The line of sight has to be perpendicular to the horizontal axis (trunnion axis), and their point of intersection has to lie on the vertical axis (see Fig. 3.1). The line of sight then coincides with the line of collimation which, in a correctly adjusted theodolite, describes a vertical plane when rotated about the horizontal axis. In use, the vertical axis has to be centred as accurately as possible over the station at which angles are being measured and that axis has to be truly vertical.

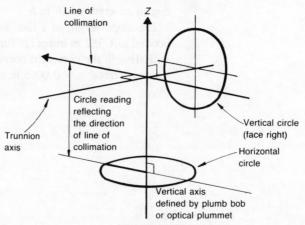

Figure 3.1

Errors in horizontal circle reading

Errors in horizontal circle readings due to certain maladjustments of the theodolite are given below, h being the altitude of the signal observed.

Maladjustment	*Error*
(a) Line of collimation making an angle $(90-c)$ with the trunnion axis, i.e. vertical hair of diaphragm displaced laterally.	$c \sec h$
(b) Trunnion axis inclined at $(90-i)$ to the vertical axis, i.e. tilted from the horizontal.	$i \tan h$

Errors in vertical circle reading

It can be taken that errors in the vertical circle readings due to (a) and (b) are negligible and that, in each case, the mean of horizontal circle observations taken face left and face right will be free from error. This is not the case when the vertical axis is not truly vertical. The error in the horizontal circle reading is of the form given for (*b*) above but *i* is now variable, depending upon the pointing direction of the telescope. Its maximum value occurs when the trunnion axis lies in that plane containing the vertical axis of the instrument and the true vertical.

Traverse

By means of a traverse survey a framework of stations or control points can be established, their positions being determined by measuring the distances between the stations and the angles subtended at the various stations by their adjacent stations. For accurate work the theodolite is used to measure the angles, with distances measured by methods discussed in Chapter 2, although stadia tacheometry would only be used in low-order work. Modern EDM instruments can be mounted on, or incorporated within, theodolites and this allows the two to be used in combination in an efficient manner, particularly with data recorders.

Co-ordinates

Normally, plane rectangular co-ordinates are used to identify the stations on a traverse. A specific point is defined by its perpendicular distances from each of two co-ordinate axes which are based on north−south and east−west directions. The former is the reference axis and it can be:

(*a*) true north;
(*b*) magnetic north;
(*c*) National Grid north; and
(*d*) a chosen arbitrary direction, which could be one of the traverse lines if so wished.

The intersection of the axes gives the origin for the survey and usually it is to the south and west to ensure that all points have positive co-ordinates.

Easting/northing

The co-ordinates given by the perpendicular distances from the two main axis are termed:

eastings (distance from the north−south axis)
northings (distance from the east−west axis)

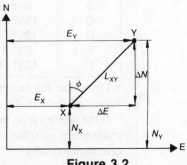

Figure 3.2

as indicated in Fig. 3.2 for the points X and Y. Relative positions are given by co-ordinate differences $\Delta E = E_Y - E_x$
and $$\Delta N = N_Y - N_x.$$

Bearings

The position of a point may also be referenced by stating length XY and bearing ϕ of line XY, and these are referred to as polar co-ordinates. The bearing ϕ is termed the whole circle bearing (WCB) of XY. It is measured clockwise from 0° to 360° at X between the north–south reference direction and the direction of Y from X. In Fig. 3.3 the whole circle bearing of YZ is θ, and the whole circle bearing of ZY is $(\theta - 180°)$.

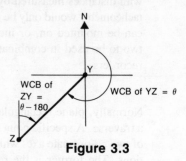

Figure 3.3

Similarly for YX in Fig. 3.2 the bearing is $(\phi + 180°)$ and, in general, bearing of line 1–2 = bearing of line 2–1 ± 180°, 1 and 2 being points within the system.

3.1 Calculation of bearings

(a) Briefly explain the difference between the following pairs of terms

(i) true north and magnetic north

(ii) whole circle bearings.

(b) The included angles given in Table 3.1 are recorded at stations forming a closed traverse survey around the perimeter of a field.

Table 3.1

Station	Included angle
A	122° 42′ 20″
B	87° 16′ 40″
C	133° 08′ 20″
D	125° 55′ 20″
E	92° 47′ 40″
F	158° 06′ 40″

Determine the amount of angular error in the survey and adjust the values of the included angles.

If the whole circle bearing of the line BC is 45° calculate the whole circle bearings of the traverse lines and the corresponding values in the centesimal system. [Salford]

Introduction. True north refers to the north geographical pole. The true or geographical meridian through a point is the trace of the plane through the north and south poles and the point in question.

Magnetic north does not coincide with geographical north: the magnetic meridian is the direction revealed by a freely floating magnetic needle. The angle between it and the true meridian is termed declination.

The whole circle bearing of a line has been defined previously as the angle, lying between 0° and 360°, between the direction of north and the direction of the line, measured clockwise.

Solution. Determine the angular error and apply corrections.

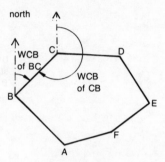

Figure 3.4

Figure 3.4 shows the traverse survey, the orientation of line BC being 45° from the magnetic meridian. This form of traverse is known as a closed-loop traverse since it begins and ends at the same point. The sum of the internal angles of a polygon is $(2n-4)$ right angles, where n is the number of angles.

Thus the sum of the six angles of this example must be eight right angles, or 720° 00′ 00″, whereas by measurement it is 719° 57′ 00″. The total error is therefore −3′ 00″ or −180″, and hence a total correction of +180″ has to be applied. Note that this is of some magnitude and implies a relatively low order of work, for example fourth order implies an error of the order of $60\sqrt{N}$ sec, i.e. $60\sqrt{6} = 147″$.

Table 3.2

Angle	Observed value	Correction	Adjusted value
A	122° 42′ 20″	+30″	122° 42′ 50″
B	87° 16′ 40″	+30″	87° 17′ 10″
C	133° 08′ 20″	+30″	133° 08′ 50″
D	125° 55′ 20″	+30″	125° 55′ 50″
E	92° 47′ 40″	+30″	92° 48′ 10″
F	158° 06′ 40″	+30″	158° 07′ 10″
Total	719° 57′ 00″	Total	720° 00′ 00″

The corrections can be applied equally to each angle on the assumption that conditions were constant at the time of measurement and that the angles had been measured with the same accuracy. Hence a correction of $(+180''/6) = +30''$ is given to each angle in this example.

Next we calculate whole circle bearings.

It is usual to proceed in an anti-clockwise manner round the traverse when internal angles have been measured. To determine the whole circle bearing of the line to the forward station it is necessary to add the whole circle bearing of the previous line, i.e. that from the back station, to the internal angle at the station, and then to add or deduct 180° depending upon whether that sum is less or greater than 180°. For instance, at A we require the whole circle bearing of AF knowing that of BA.

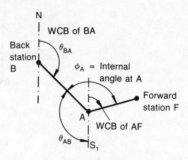

Figure 3.5

Therefore from Fig. 3.5, WCB of AF $= \theta_{BA} + \phi_A - 180°$
since $\qquad N\hat{B}A = \theta_{BA} = S_1\hat{A}B.$

In this example we are given the bearing of BC but to move round the traverse in the anti-clockwise direction B, A, F, E, D, C, B we need the bearing of CB instead.

Now whole circle bearing of BC $= 45°$
therefore whole circle bearing of CB $= 45° + 180°$
$\qquad\qquad\qquad\qquad\qquad\qquad\qquad = \mathbf{225°}.$

We can now proceed, using the quoted rule, to determine the whole circle bearings of the six traverse lines as follows:

	WCB of CB =	225° 00′ 00″
Add	angle B =	87° 17′ 10″
		312° 17′ 10″ (Exceeds 180°)
Deduct	180° =	180° 00′ 00″
Therefore	WCB of BA =	**132° 17′ 10″**
Add	angle A =	122° 42′ 50″
		255° 00′ 00″

Deduct	180°	180° 00′ 00″
Therefore	WCB of AF =	**75° 00′ 00″**
Add	angle F =	158° 07′ 10″
		233° 07′ 10″
Deduct	180°	180° 00′ 00″
	WCB of FE =	**53° 07′ 10″**
Add	angle E =	92° 48′ 10″
		145° 55′ 20″ (Less than 180°)
Add	180°	180° 00′ 00″
Therefore	WCB of ED =	**325° 55′ 20″**
Add	angle D =	125° 55′ 50″
		451° 51′ 10″
Deduct	180°	180° 00′ 00″
Therefore	WCB of DC =	**271° 51′ 10″**
Add	angle C =	133° 08′ 50″
		405° 00′ 00″
Deduct	180°	180° 00′ 00″
Therefore	WCB of CB =	225° 00′ 00″ (Check)

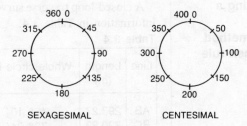

SEXAGESIMAL   CENTESIMAL

Figure 3.6

Calculate centesimal values.

In the centesimal system the major graduations of the instrument range from zero to 400 gon, as against zero to 360° in the sexagesimal system (Fig. 3.6). Subdivision in the centesimal system is carried out in steps of ten and readings may be made to 0.0001 gon. It will be evident that:

1 degree is equivalent to $\frac{10}{9}$ gon

1 minute is equivalent to $\frac{1}{54}$ gon

1 second is equivalent to $\frac{1}{3240}$ gon.

Hence in the case of line FE, 53° 07′ 10″ in the sexagesimal system is equivalent to

$$53 \times \tfrac{10}{9} = 58.8889$$

$$7 \times \tfrac{1}{54} = 0.1296$$

$$10 \times \tfrac{1}{3240} = \underline{0.0031}$$

$$59.0216 \text{ gon}$$

Similar calculations are used to complete Table 3.3.

Table 3.3

Line	Whole circle bearing	
	Sexagesimal system	Centesimal system
AF	75° 00′ 00″	83.3333 gon
FE	53° 07′ 10″	59.0216 gon
ED	325° 55′ 20″	362.1358 gon
DC	271° 51′ 10″	302.0586 gon
CB	225° 00′ 00″	250.0000 gon
BA	132° 17′ 10″	146.9846 gon

3.2 Correcting a traverse by Bowditch's method and the Transit rule

A closed-loop traverse survey ABCDEA, shown in Fig. 3.7, gave the information in Table 3.4.

Table 3.4

Line	Length (m)	Whole circle bearing
AB	293.27	45° 10′ 10″
BC	720.83	72° 04′ 55″
CD	497.12	161° 51′ 45″
DE	523.34	228° 43′ 10″
EA	761.87	300° 41′ 50″

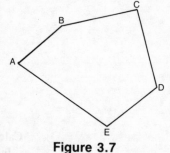

Figure 3.7

Determine the closing error and hence, after adjustment by Bowditch's method, determine the co-ordinates of the traverse stations given that the co-ordinates of A are 1200.00 mE, 1200.00 mN.

Compare the corrections given by Bowditch's method with those given by the Transit rule.

It may be assumed that the whole circle bearings do not need an adjustment.

Introduction. The previous example covered the first stage in traverse computation in that the measured angles were assessed for error, and then duly corrected. This was then followed by the determination of the whole circle bearings of the lines. This example allows us to follow the remainder of the procedure to ultimately establish the co-ordinates of the stations, starting and finishing at A.

Solution. First determine the easting and northing differences. In the introduction it was pointed out that

(a) easting difference $\Delta E = E_Y - E_x$
(b) northing difference $\Delta N = N_Y - N_x$

for points X and Y, which can be the stations at each end of a line as in Fig. 3.8.

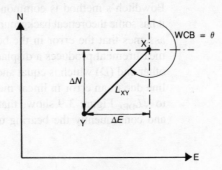

Figure 3.8

It will be seen that $\Delta E = l_{XY} \sin(WCB)$
and $\Delta N = l_{XY} \cos(WCB)$.

The signs of ΔE and ΔN automatically follow the trigonometrical terms. Since the whole circle bearing lies between 180° and 270° both sin (WCB) and cos (WCB) are negative. This causes ΔN and ΔE to be negative in respect of the positive directions of N and E.

The resultant easting differences and northing differences can now be computed for the traverse lines, as in Table 3.5.

Table 3.5

Line	Length (m)	WCB	ΔE (m)	ΔN (m)
AB	293.27	45° 10′ 10″	+ 207.99	+ 206.76
BC	720.83	72° 04′ 55″	+ 685.87	+ 221.77
CD	497.12	161° 51′ 45″	+ 154.75	− 472.42
DE	523.34	228° 43′ 10″	− 393.28	− 345.27
EA	761.87	300° 41′ 50″	− 655.11	+ 388.94
Totals	2796.43		+ 0.22	− 0.22

Next, the closing error is determined. The algebraic sums of the easting differences and northing differences should be zero because the traverse starts and ends at A, but in fact we have total errors in so far as this traverse is concerned of $\Delta E = +0.22$ m and $\Delta N = -0.22$ m.

$$\begin{aligned} \text{Closing error} &= \sqrt{(\Delta E^2 + \Delta N^2)} \\ &= \sqrt{(0.22^2 + 0.22^2)} \\ &= \textbf{0.31 m.} \end{aligned}$$

Expressed fractionally in terms of the total length of the traverse the linear error is 0.31 m in 2796.43 m or 1 in 9021, say. This lies between third-order accuracy of 1 in 5000 and second-order accuracy of 1 in 10 000.

Finally, traverse is adjusted and the co-ordinates are determined. There are a number of methods of correcting the error in easting and northing differences. Bowditch's method is commonly adopted in civil-engineering surveys since it has some theoretical background and is relatively simple to apply. The method assumes that the error in the bearing of a line caused by inaccurate angular measurement produces a displacement at one end of a line (E) relative to the other end (D) which is equal and perpendicular to the displacement along that line due to an error in linear measurement, which is taken to be proportional to $\sqrt{L_{DE}}$. Figure 3.9 shows that the method causes E to be displaced to E', and consequently the bearing of DE changes.

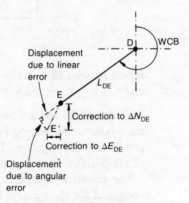

Figure 3.9

For individual lines Bowditch's method states:

$$\text{correction to easting difference } \Delta E_{DE} = \frac{dE \times \text{length of line DE}}{\text{total length of traverse}},$$

$$\text{correction to northing difference } \Delta N_{DE} = \frac{dN \times \text{length of line DE}}{\text{total length of traverse}}.$$

In which dE and dN are the total corrections required for the easting differences and northing differences, respectively.

In this example $dE = -0.22$ m and $dN = +0.22$ m. Hence the corrections for line DE are

$$\text{correction to easting difference} = -0.22 \times \frac{523.34}{2796.43}$$

$$= -0.04 \text{ m}$$

$$\text{correction to northing difference} = +0.22 \times \frac{523.34}{2796.43}$$

$$= +0.04 \text{ m}.$$

Note also that in Fig. 3.9

$$\text{the change in bearing of the line} = \tan^{-1} \frac{\text{correction to } \Delta E_{\text{DE}}}{\text{correction to } \Delta N_{\text{DE}}}$$

$$= \tan^{-1} \frac{\text{d}E}{\text{d}N}.$$

This is the bearing of the closing error and the correction in bearing applies throughout the traverse for all lines.

The corrections can be tabulated as in Table 3.6.

Table 3.6

Line	ΔE	Correction to ΔE	Corrected ΔE	ΔN	Correction to ΔN	Corrected ΔN
AB	+ 207.99	− 0.02	+ 207.97	+ 206.76	+ 0.02	+ 206.78
BC	+ 685.87	− 0.06	+ 685.81	+ 221.77	+ 0.06	+ 221.83
CD	+ 154.75	− 0.04	+ 154.71	− 472.42	+ 0.04	− 472.38
DE	− 393.28	− 0.04	− 393.32	− 345.27	+ 0.04	− 345.23
EA	− 655.11	− 0.06	− 655.17	+ 388.94	+ 0.06	+ 389.00
		− 0.22	0.00		+ 0.22	0.00

To determine the co-ordinates of the stations we apply the corrected difference ΔE and ΔN to the previous station co-ordinates, i.e. for B

$$\text{easting co-ordinate of B} = \text{easting co-ordinate of A} + \Delta E_{\text{AB}}$$
$$\text{northing co-ordinate of B} = \text{northing co-ordinate of A} + \Delta N_{\text{AB}}$$

Hence we obtain the values in Table 3.7.

There is an alternative method of obtaining corrections, by the Transit rule. This method has no theoretical background, but it is such that if a line has no easting difference it will not be given an easting correction. This is not so with the Bowditch approach. The rule states

$$\text{correction to } \Delta E_{\text{DE}} = \frac{\text{d}E \times \Delta E_{\text{DE}}}{\Sigma \, \Delta E}$$

$$\text{correction to } \Delta N_{\text{DE}} = \frac{\text{d}N \times \Delta N_{\text{DE}}}{\Sigma \, \Delta N}$$

Table 3.7

Station	Co-ordinates	
	E (m)	N (m)
A (line AB)	1200.00 + 207.97	1200.00 + 206.78
B (line BC)	**1407.97** + 685.81	**1406.78** + 221.83
C (line CD)	**2093.78** + 154.71	**1628.61** − 472.38
D (line DE)	**2248.49** − 393.32	**1156.23** − 345.23
E (line EA)	**1855.17** − 655.17	**811.00** + 389.00
A	1200.00	1200.00 (Check)

In the example $\Sigma \, \Delta E = 2097.00$ m
and $\Sigma \, \Delta N = 1635.16$ m.

Magnitudes of the differences are considered but the signs are ignored.
Thus for line DE we have

$$\text{correction to } \Delta E_{DE} = -0.22 \times \frac{393.28}{2097.00}$$

$$= -0.04$$

$$\text{correction to } \Delta N_{DE} = +0.22 \times \frac{345.27}{1635.16}$$

$$= +0.05 \text{ m.}$$

For the complete traverse we obtain the values in Table 3.8.

Table 3.8

Line	ΔE	Correction to ΔE	Corrected ΔE	ΔN	Correction to ΔN	Corrected ΔN
AB	+ 207.99	− 0.02	+ 207.97	+ 206.76	+ 0.03	+ 206.79
BC	+ 685.87	− 0.07	+ 685.80	+ 221.77	+ 0.03	+ 221.80
CD	+ 154.75	− 0.02	+ 154.73	− 472.42	+ 0.06	− 472.36
DE	− 393.28	− 0.04	− 393.32	− 345.27	+ 0.05	− 345.22
EA	− 655.11	− 0.07	− 655.18	+ 388.94	+ 0.05	+ 389.99
		− 0.22	0.00		+ 0.22	0.00

The following computer program will correct a closed-loop traverse by Bowditch's method. Stations must be numbered clockwise around the traverse, the co-ordinates of point 1 being fixed and the whole circle bearing of the line joining station 1 and station N must be known. The internal angles at the stations should have been measured and the traverse lines must not cross. Since whole circle bearings are given in this exercise and internal angles are required by the program it is necessary to convert the data to: A = 75° 31′ 40″, B = 153° 5′ 15″, C = 90° 13′ 10″, D = 113° 8′ 35″, E = 108° 01′ 20″ and WCB_{AE} = 120° 41′ 50″.

Line 230 in the program compares the actual angular misclosure to the allowable one and calculation stops if this is exceeded, this check can be removed by deleting lines 230–250.

For readers with programmable calculators the DIM statement in line 20 has been set for a 10-sided traverse, minimum values are $L(N)$, $E(N+1)$, $N(N+1)$, $A(N)$, $B(N)$. The data is output in a table and lines 430, 500, 540 and 560 will need to be reformatted for machines with a limited display.

Variables

A(i)	= Angle i		P1	=	Actual angular misclosure
B	= WCB of line 1 to N		S	=	Input/output, Seconds
B(i)	= WCB of line i to i+1		S1	=	Minimum theodolite
C	= 360° in seconds				division (least count)
D	= Input/output, Degrees		T	=	Sum of measured angles
E(i)	= Easting of station i		T1	=	Expected sum of angles
I	= Loop counter		U1	=	Easting misclosure
K	= Traverse accuracy factor		U2	=	Northing misclosure
	(1−3)		U3	=	Linear error
L	= Sum of side lengths		U4	=	Fractional error
L(i)	= Length of side i to i+1		U5	=	Cumulative easting
M	= Input/output, Minutes				correction
N	= Number of sides		U6	=	Cumulative northing
N(i)	= Northing of station i				correction
P	= Allowable angular		X	=	I+1
	misclosure				

```
10 REM CLOSED LOOP TRAVERSE CORRECTED BY BOWDITCH'S METHOD
20 DIM L(10),E(11),N(11),A(11),B(10)
30 INPUT"NUMBER OF SIDES ",N
40 FOR I=1 TO N
50 PRINT"INPUT ANGLE";I;"IN DEG,MIN,SEC"
60 INPUT D,M,S
70 A(I)=(D*3600)+(M*60)+S
80 T=T+A(I)
90 X=I+1
100 IF I=N THEN X=1
110 PRINT"INPUT LENGTH";I;"TO";X
120 INPUT L(I)
130 L=L+L(I)
140 NEXT I
150 INPUT"MINIMUM THEODOLITE DIVISION IN SEC (NORMALLY 1 OR 20) ",S1
160 INPUT"TRAVERSE ACCURACY FACTOR FROM 1 TO 3        ",K
170 IF K<1 OR K>3 THEN 160
180 P=INT(K*S1*SQR(N)+0.5)
190 T1=(2*N-4)*90*3600
200 P1=T-T1
210 PRINT"ANGULAR MISCLOSURE =";P1;"SEC"
220 PRINT"ALLOWED MISCLOSURE =";P;"SEC"
230 IF ABS(P1)<P THEN 260
240 PRINT"DATA ERROR - CALCULATION STOPPED"
250 STOP
260 C=360*3600
270 INPUT"INPUT CO-ORDINATES OF POINT 1 ",E(1),N(1)
```

```
280 PRINT"INPUT W.C.BEARING OF LINE 1 TO";N;"IN DEG,MIN,SEC"
290 INPUT D,M,S
300 B=(D*3600)+(M*60)+S
310 FOR I=1 TO N
320 B(I)=B-A(I)+(P1/N)
330 IF B(I)<0 THEN B(I)=B(I)+C
340 B=B(I)+(180*3600)
350 IF B>C THEN B=B-C
360 N(I+1)=N(I)+L(I)*COS(B(I)/206264.8)
370 E(I+1)=E(I)+L(I)*SIN(B(I)/206264.8)
380 NEXT I
390 U1=INT((E(N+1)-E(1))*1000+0.5)/1000
400 U2=INT((N(N+1)-N(1))*1000+0.5)/1000
410 U3=INT((SQR((U1*U1)+(U2*U2)))*1000+0.5)/1000
420 U4=INT((L/U3)+0.5)
430 PRINT"STATION   BEARING      LENGTH       CO-ORDINATES"
440 PRINT
450 FOR I=1 TO N
460 U5=U5+(U1*L(I)/L)
470 U6=U6+(U2*L(I)/L)
480 E(I+1)=INT((E(I+1)-U5)*1000+0.5)/1000
490 N(I+1)=INT((N(I+1)-U6)*1000+0.5)/1000
500 PRINT I;TAB(30);E(I);TAB(40);N(I)
510 D=INT(B(I)/3600)
520 M=INT((B(I)-(D*3600))/60)
530 S=INT(B(I)-(D*3600)-(M*60))
540 PRINT TAB(4);D;TAB(10);M;TAB(15)S;TAB(20);L(I)
550 NEXT I
560 PRINT" 1";TAB(30);E(N+1);TAB(40);N(N+1)
570 PRINT"EASTING ERROR    =";U1
580 PRINT"NORTHING ERROR   =";U2
590 PRINT"LINEAR ERROR     =";U3
600 PRINT"FRACTIONAL ERROR = 1 IN";U4
610 END
```

3.3 Vector misclosure of a traverse

A five-sided loop traverse (whose angles have an accepted misclosure) has been computed giving the co-ordinate differences in Table 3.9 for each leg.

Table 3.9

Leg	ΔE (m)	ΔN (m)
AB	− 43.62	− 61.39
BC	+ 70.45	− 34.71
CD	+ 50.85	+ 48.10
DE	− 23.01	+ 73.37
EA	− 53.73	− 25.86

(*i*) Determine the easting, northing and vector misclosure of the traverse.

(*ii*) The vector misclosure indicates a mistake of 1 m in the length of one of the sides of the traverse. Find which side contains the mistake and, after eliminating its effect, recompute the easting, northing and vector misclosures.   [London]

Solution.   (*i*) Calculate the vector misclosure of the traverse. The easting misclosure is +0.94 m and the northing misclosure is −0.49 m, as indicated in Fig. 3.10.

Table 3.10

Leg	E	N
AB	−43.62	−61.39
BC	+70.45	−34.71
CD	+50.85	+48.10
DE	−23.01	+73.37
EA	−53.73	−25.86
Misclosures	+ 0.94	− 0.49

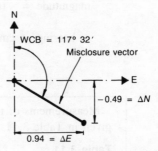

Figure 3.10

Whence vector misclosure
$$= \sqrt{((+0.94)^2 + (-0.49)^2)}$$
$$= \mathbf{1.06\ m}$$

with a bearing of $\tan^{-1} \dfrac{0.94}{-0.49} = \mathbf{117°\ 32'}$, say.

(*ii*) Now find the side that is in error. We are given that the angles as measured produced an acceptable misclosure, and accordingly the magnitude of the misclosure cannot be attributed to their measurement. In such an eventuality we have to search for a side in the traverse which has the same bearing, approximately, as that of the closing error. Bearings have not been given in this example, but scrutiny of the data reveals that line BC is the only one whose easting difference and northing difference bear some proportional relationship with the corresponding differences of the closing error, i.e. +70.45 m for ΔE and −34.71 m for ΔN against +0.94 m and −0.49 m, respectively, in the closing error. Thus we can assume that BC is in error by the amount stated.

The apparent length of BC $= \sqrt{((70.45)^2 + (-34.71)^2)}$
$= 78.54$ m.

Also, for BC, $\qquad \dfrac{\Delta E}{\Delta N} = \dfrac{70.45}{-34.71} = -2.03$

and the closing error $\dfrac{\Delta E}{\Delta N} = \dfrac{0.94}{-0.49} = -1.92.$

Since the other four sides of the traverse will have some influence on the closing error we can accept that the length of BC should have been booked as 77.54 m (given that it has been subjected to an error of 1 m) so as to reduce the magnitudes of the misclosures.

$$\text{Therefore corrected } \Delta E \text{ for BC} = \frac{+77.54}{78.54} \times 70.45 = +69.55 \text{ m}$$

$$\text{corrected } \Delta N \text{ for BC} = \frac{-77.54}{78.54} \times 34.71 = -34.27 \text{ m.}$$

Thus for the amended closing error, $\Delta E = +0.04$ m $\quad \Delta N = -0.05$ m

$$\text{magnitude} = \sqrt{((0.04)^2 + (0.05)^2)} = 0.06 \text{ m}$$

$$\text{bearing} = \tan^{-1} \frac{0.04}{-0.05} = 141° \ 19'.$$

3.4 Closed-link traverse

Measurements of the traverse ABCDE, as shown in Fig. 3.11, are given in Table 3.11.

Table 3.11

Station	Clockwise angle	Length (m)
A	260° 31′ 18″	129.352
B	123° 50′ 42″	81.700
C	233° 00′ 06″	101.112
D	158° 22′ 48″	94.273
E	283° 00′ 18″	

WCB of EY = 282° 03′ 00″.

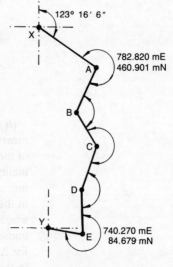

Figure 3.11

The measured angles are as shown in the figure. Keeping the bearings XA and EY and also the co-ordinates of A and E fixed, obtain the adjusted co-ordinates for B, C and D using an equal shifts angular adjustment and Bowditch linear adjustment. [Bradford]

Introduction. Although in an extended form, rather than starting and finishing at one point, this traverse is still a closed traverse, since it runs between two points whose co-ordinates are fixed and two lines whose bearings are fixed. It is known as a closed-link traverse and it can be readily adjusted.

Solution. Determine whole circle bearings. Starting with the bearing of XA we should end with the given bearing of EY; if not then there will be some angular error to balance.

Using measured angles:

WCB of XA =	123° 16′ 06″
Â =	260° 31′ 18″
Add =	383° 47′ 24″
Deduct	180° 00′ 00″
WCB of AB =	203° 47′ 24″
B̂ =	123° 50′ 42″
Add =	327° 38′ 06″
Deduct	180° 00′ 00″
WCB of BC =	147° 38′ 06″
Ĉ =	233° 00′ 06″
Add =	380° 38′ 12″
Deduct	180° 00′ 00″
WCB of CD =	200° 38′ 12″
D̂ =	158° 22′ 48″
Add =	359° 01′ 00″
Deduct	180° 00′ 00″
WCB of DE =	179° 01′ 00″
Ê =	283° 00′ 18″
Add =	462° 01′ 18″
Deduct	180° 00′ 00″
WCB of EY =	282° 01′ 18″
But given WCB of EY =	282° 03′ 00″

Using corrected angles:

WCB of XA =	123° 16′ 06″
Corrected Â =	260° 31′ 39″
Add =	383° 47′ 45″
Deduct	180° 00′ 00″
WCB of AB =	203° 47′ 45″
Corrected B̂ =	123° 51′ 02″
Add =	327° 38′ 47″
Deduct	180° 00′ 00″
WCB of BC =	147° 38′ 47″
Corrected Ĉ =	233° 00′ 26″
Add =	380° 39′ 13″
Deduct	180° 00′ 00″
WCB of CD =	200° 39′ 13″
Corrected D̂ =	158° 23′ 08″
Add =	359° 02′ 21″
Deduct	180° 00′ 00″
WCB of DE =	179° 02′ 21″
Corrected Ê =	283° 00′ 39″
Add =	462° 03′ 00″
Deduct	180° 00′ 00″
WCB of EY =	282° 03′ 00″ Check

therefore angular error = −01′ 42″
and total correction = +01′ 42″

This has to be shared out to five angles, i.e. 21″ to A and E and 20″ to B, C and D.

Next determine the easting and northing differences. Having adjusted the observed angles and deduced the whole circle bearings of the lines, the easting and northing differences are calculated for each line on the basis of

$$\Delta E = l \sin (WCB) \text{ and } \Delta N = l \cos (WCB).$$

From Table 3.12 it will be seen that the total difference is −42.553 mE and −376.247 mN. The fixed co-ordinates for A and E are as in Table 3.13. Thus

Table 3.12

Line	Length (m)	WCB	ΔE (m)	ΔN (m)
AB	129.352	203° 47′ 45″	− 52.191	− 118.356
BC	81.700	147° 38′ 47″	+ 43.721	− 69.017
CD	101.112	200° 39′ 13″	− 35.664	− 94.614
DE	94.273	179° 02′ 21″	+ 1.581	− 94.260
Totals	406.437		− 42.553	− 376.247

Table 3.13

Point	E (m)	N (m)
E	740.270	84.679
A	782.820	460.901
Differences	− 42.550	− 376.222

the easting differences and northing differences given by the actual measurements are in error by

$$\text{eastings } -42.553 - (-42.550) = -0.003 \text{ m}$$
$$\text{northings } -376.247 - (-376.222) = -0.025 \text{ m}.$$

Therefore corrections of +0.003 m and +0.025 m are required for the easting differences and northing distances, respectively.

Make the Bowditch adjustment of the easting and northing differences. As stated previously the Bowditch linear adjustment for a particular line is

$$\text{correction to easting difference } = \frac{\text{length of line} \times dE}{\text{total length of traverse}}$$

$$\text{correction to northing difference } = \frac{\text{length of line} \times dN}{\text{total length of traverse}},$$

in which $dE = +0.003$ m and $dN = +0.025$ m for this traverse. Hence the corrections to be applied are as in Tables 3.14 and 3.15.

Table 3.14

	AB	BC	CD	DE	Totals
Length (m)	129.352	81.700	101.112	94.273	406.437 m
ΔE correction (m)	+ 0.001	+ 0.000	+ 0.001	+ 0.001	+ 0.003 m
ΔN correction (m)	+ 0.008	+ 0.005	+ 0.006	+ 0.006	+ 0.025 m

Table 3.15

Line	Correction to ΔE	Corrected ΔE	Correction to ΔN	Corrected ΔN
AB	+ 0.001	− 52.190	+ 0.008	− 118.348
BC	0.000	+ 43.721	+ 0.005	− 69.012
CD	+ 0.001	− 35.663	+ 0.006	− 94.608
DE	+ 0.001	+ 1.582	+ 0.006	− 94.254
Totals		− 42.550 m		− 376.222 m (Check)

The computation of co-ordinates is given in Table 3.16.

Table 3.16

Station		E (m)		N (m)
Co-ordinates A		782.820		460.901
ΔE_{AB}		− 52.190	ΔN_{AB}	− 118.348
Co-ordinates B		**730.630**		**342.553**
ΔE_{BC}		+ 43.721	ΔN_{BC}	− 69.012
Co-ordinates C		**774.351**		**273.541**
ΔE_{CD}		− 35.663	ΔN_{CD}	− 94.608
Co-ordinates D		**738.688**		**178.933**
ΔE_{DE}		+ 1.582	ΔN_{DE}	− 94.254
Co-ordinates E		**740.270**		**84.679** (Check)

3.5 Open traverse

A theodolite traverse PabQ was carried round a hillside to connect stations P and Q.

The horizontal lengths were: Pa = 175.29 m, ab = 316.78 m, bQ = 98.15 m.

The horizontal angles, measured clockwise from the back station were: Pab = 117° 48′ 20″, abQ = 132° 21′ 40″ and the bearing of Pa = 255° 34′ 00″. Calculate the length and bearing of PQ.   [Salford]

Introduction. This example relates to a type of traverse known as an open traverse. Although like the previous example in so far as general shape is concerned, there are no co-ordinate data, etc., to allow checking of either angular measurement or length measurement. Hence all field measurements have to be accepted as observed. The technique should thus only be used as a last resort or where checks can be applied by intersecting prominent features (that lie off the line of the traverse) from intermediate stations along the traverse.

Solution. First determine the whole circle bearings and easting and northing differences for the various lines. Effectively, the traverse has been carried out in an anti-clockwise direction measuring internal angles. Accordingly a whole circle bearing can be evaluated by adding the internal angle at a station to the whole circle bearing of the previous line and applying 180°.

$$WCB \text{ line } Pa = 255° \ 34' \ 00''$$
$$\text{Angle } P\hat{a}b = \underline{117° \ 48' \ 20''}$$

$$\text{Add} = 373° \ 22' \ 20''$$
$$\text{Deduct} = 180° \ 00' \ 00''$$

$$WCB \text{ line } ab = 193° \ 22' \ 20''$$
$$\text{Angle } a\hat{b}Q = \underline{132° \ 21' \ 40''}$$

$$\text{Add} = 325° \ 44' \ 00''$$
$$\text{Deduct} = 180° \ 00' \ 00''$$

$$WCB \text{ line } bQ = 145° \ 44' \ 00'' \ \text{(no check)}$$

The easting and northing differences may now be calculated as in previous examples to give the values tabulated in Table 3.17.

Table 3.17

Line	Length (m)	WCB	ΔE (m)	ΔN (m)
Pa	175.29	255° 34′ 00″	− 169.76	− 43.69
ab	316.78	193° 22′ 20″	− 73.26	− 308.19
bQ	98.15	145° 44′ 00″	+ 55.26	− 81.11
Totals			− 187.76	− 432.99

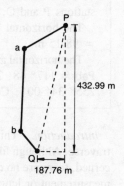

Figure 3.12

Finally, determine the length and bearing of PQ. The differences in easting and northing coordinates of P and Q have been found as -187.76 mE and -432.99 mN, respectively. Hence in Fig. 3.12, Q is south and west of P, line PQ lying in the third quadrant with respect to the N−S reference axis and the E−W axis.

$$\text{Length PQ} = \sqrt{((-187.76)^2 + (-432.99)^2)}$$
$$= \mathbf{471.95\ m.}$$

$$\text{Bearing PQ} = \tan^{-1} \frac{-187.76}{-432.99}$$

$$= \tan^{-1} (+0.433\ 64)$$
$$\text{Therefore} \quad \text{WCB of PQ} = \mathbf{203°\ 26'\ 36''.}$$

3.6 Errors introduced when centring the theodolite

The magnitude of the errors in centring a theodolite over a station by various methods has been investigated and the following results quoted

(*i*)   plumb bob ± 3 mm
(*ii*) centring rod ± 2 mm
(*iii*) optical plummet ± 1 mm
(*iv*) constrained centring ± 0.1 mm.

Compute the maximum errors in the measurement of a horizontal traverse angle of approximately 120° resulting from each of the above centring errors, assuming that the lengths of the adjacent drafts are 15 m and 25 m.

[Eng. Council]

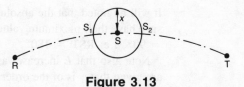

Figure 3.13

Introduction.   In Fig. 3.13 let the true centring position of the theodolite be S. It is possible for the vertical axis of the theodolite to lie anywhere within a circle of radius x from that point, x being one of the centring errors quoted above. However, there will be two points on the perimeter of that circle at which the true horizontal angle RST will be subtended. These are S_1 and S_2 which lie on the circumference of the circle containing R, S, and T. Accordingly $R\hat{S}_1T = R\hat{S}T = R\hat{S}_2T$ because all three angles stand on chord RT.

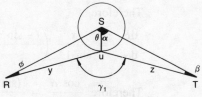

Figure 3.14

Solution. First determine the maximum angular error due to a centring error of ± 3 mm. In Fig. 3.14 the theodolite has been centred at U, distance x from S. The measured horizontal angle $\hat{RUT} = \gamma_1$,

$$\text{whilst the correct angle } \hat{RST} = \gamma$$
$$= \theta + \alpha$$
$$\text{Now } \gamma_1 = \phi + \theta + \alpha + \beta$$

Therefore the error in measurement $E = \phi + \beta$

But
$$\frac{\sin \beta}{x} = \frac{\sin \alpha}{z}$$

in which
$$z = TU$$

and
$$\frac{\sin \phi}{x} = \frac{\sin \theta}{y}$$

in which
$$y = UR.$$

For small angles
$$\sin \beta = \beta'' \times \sin 1''$$
and
$$\sin \phi = \phi'' \times \sin 1''$$
Therefore
$$E = \phi + \beta \text{ seconds of arc}$$

$$= \frac{x \sin \theta}{y \sin 1''} + \frac{x \sin \alpha}{z \sin 1''}$$

$$= \frac{x}{\sin 1''} \left(\frac{\sin \theta}{y} + \frac{\sin \alpha}{z} \right).$$

It will be noted that the absolute maximum error E is given when $\sin \alpha$ and $\sin \theta$ have their maximum values. This occurs when $\alpha = 90°$ and $\theta = (\gamma - \alpha) = 90°$, i.e. $\hat{RST} = 180°$.

Note also that E increases as y and z decrease. However, in this case we are given that γ is of the order of $120°$, and for the maximum error arrising in this case $\dfrac{dE}{d\alpha} = 0$.

Now

$$E = \frac{x}{\sin 1''} \left(\frac{\sin(\gamma - \alpha)}{y} + \frac{\sin \alpha}{z} \right)$$

$$= \frac{x}{\sin 1''} \left(\frac{(\sin \gamma \times \cos \alpha) - (\cos \gamma \times \sin \alpha)}{y} + \frac{\sin \alpha}{z} \right).$$

Therefore

$$\frac{dE}{d\alpha} = \frac{x}{\sin 1''} \left(\frac{(-\sin \gamma \times \sin \alpha) - (\cos \gamma \times \cos \alpha)}{y} + \frac{\cos \alpha}{z} \right) = 0.$$

Therefore $\dfrac{\cos \alpha}{z} = \dfrac{(\sin \gamma \times \sin \alpha) + (\cos \gamma \times \cos \alpha)}{y}$

or
$$y = z (\sin \gamma \times \tan \alpha + \cos \gamma)$$

$$\tan \alpha = \frac{y - z \cos \gamma}{z \sin \gamma}$$

$$= \frac{15 - 25 \cos 120}{25 \sin 120} = \frac{15 + 12.5}{21.65}$$

$$= 1.270\ 207\ 8.$$

Therefore $\alpha = 51° 47'$, say.

The direction of displacement of U is significant.

Taking case (i), $x = 3$ mm $= 0.003$ m, thus

$$E = \frac{0.003}{\sin 1''} \left(\frac{\sin(120 - 51° 47')}{15} + \frac{\sin 51° 47'}{25} \right)$$

$$= \frac{0.003}{\sin 1''} \left(\frac{0.928\ 59}{15} + \frac{0.785\ 68}{25} \right)$$

in which $\dfrac{1}{\sin 1''} = 206\ 265$ and so $E =$ **57.8 seconds.**

3.7 Errors due to maladjustment of the theodolite

Derive expressions for the error in horizontal circle readings taken using a theodolite having the following maladjustments:

(i) the line of collimation not perpendicular to the trunnion axis by a small amount c

(ii) the trunnion axis not perpendicular to the vertical axis by a small amount i.

Hence prove that the effect of these maladjustments is eliminated by using the mean of face left and face right readings.

Calculate the true value of angle PRQ from the readings given in Table 3.18 which were taken using a theodolite at R on face left only for which instrument c is known to be 23 seconds left and i is known to be 15 seconds down at the right.

Table 3.18

Station	Horizontal circle	Vertical circle	
P	27° 15′ 27″	42° 15′ 12″	depression
Q	112° 27′ 53″	28° 12′ 34″	elevation

[Bradford]

Introduction.   In Fig. 3.15 the line of sight is shown to make a small angle c with the perpendicular to the trunnion axis (which would be its correct position). It sweeps along circle Z_1QT_1 when the telescope is rotated about the trunnion axis for the pointing on Q. The reading of the horizontal circle,

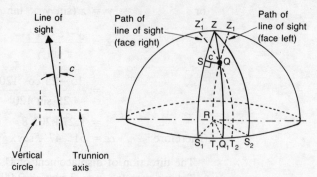

Figure 3.15

however, is as if Q were in vertical circle ZS_1 (to which Z_1QT_1 is parallel) whereas it is actually in vertical circle ZQ_1. Consequently, the error in the horizontal circle reading is S_1Q_1 for this sighting, and it is positive on a clockwise reading circle.

Let SQ be at right angles to ZS_1, i.e. S = 90°. Then in spherical triangle ZSQ

$$\frac{\sin Z}{\sin SQ} = \frac{\sin S}{\sin ZQ}.$$

Therefore $\sin Z = \dfrac{\sin SQ \sin S}{\sin ZQ}$.

For small angles we can write

$$Z = \frac{SQ \times \sin S}{\sin(90-h)}$$

in which $h = QQ_1$ (the altitude of Q).

Therefore $Z = \dfrac{SQ}{\cos h}$ (since $\sin S = 1$)

$$= SQ \sec h$$
$$= c \sec h,$$

writing SQ $= c$.

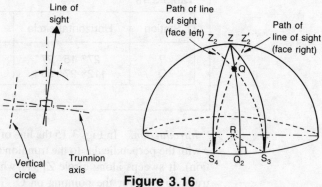

Figure 3.16

In Fig. 3.16 the left-hand support of the trunnion axis is higher than the right-hand support and consequently the line of sight sweeps out Z_2QS_3, making angle i with the vertical circle ZS_3. Q appears to be on that circle but is in fact on vertical circle ZQ_2. Thus the error in the horizontal circle reading for this particular case is S_3Q_2 and it is negative.

Consider spherical triangle QQ_2S_3 in which $Q_2 = 90°$

$$\sin Q_2S_3 = \tan i \times \tan QQ_2.$$

Therefore for small angles

$$Q_2S_3 = i \tan h,$$

in which h is the altitude of Q.

The same expressions hold when depression angles are observed. In the case of collimation error c the senses of the errors are the same since path Z_1QT_1 is parallel to ZS_1 throughout (see Fig. 3.15). However, in Fig. 3.16 Z_2QS_3 is inclined to ZS_3, the two effectively crossing at S_3 when moving from elevation to depression. Thus there is a change in sense and P, in this particular worked example, will have a positive error.

When face right observations are made the paths of the lines of sight change direction Z_1, moving to Z_1' in Fig. 3.15, thus giving an error of Q_1S_2 in the horizontal circle reading. This is of similar magnitude but of opposite sense to S_1Q_1. Similarly, in Fig. 3.16, Z_2 moves to Z_2' and S_3 to S_4, giving error S_4Q_2 which is equal in magnitude but of opposite sense to error Q_2S_3.

Thus in each case the means of the face left and face right observations will give the true value of the horizontal circle reading.

Solution. Determine angle $P\hat{R}Q$. Tabulating the errors we have the values in Table 3.19.

Table 3.19

Sighting	$c \sec h$	$i \tan h$
RP RQ	$+23''\ \sec 42°\ 15'\ 12'' = +31.1''$ $+23''\ \sec 28°\ 12'\ 34'' = +26.1''$	$+15''\ \tan 42°\ 15'\ 12'' = +13.6''$ $-15''\ \tan 28°\ 12'\ 34'' = -8.0''$

Applying corrections equal and opposite in sense to the above errors we obtain the values given in Table 3.20.

Table 3.20

Sighting	Observed angle	Net correction	Corrected angle
RQ	112° 27′ 53″	−18.1″	112° 27′ 34.9″
RP	27° 15′ 27″	−44.7″	27° 14′ 42.3″
			85° 12′ 52.6″

Therefore the corrected angle $P\hat{R}Q = \mathbf{85°\ 12'\ 53''}$ to the nearest second.

3.8 Striding level observations

The readings in Table 3.21 were recorded on a theodolite, known to be in adjustment, at Station A.

Table 3.21

Pointing	Striding level on trunnion axis				Circle readings	
	L	R	L	R	Vertical	Horizontal
R	11.3	3.3	9.7	4.9	+15° 46′ 40″	108° 19′ 46″
P	10.5	4.1	8.9	5.7	+8° 14′ 00″	76° 24′ 18″

The value of one division of the striding level is 12″. Determine:

(a) the corrected value of angle PAR;

(b) the maximum inclination of the trunnion axis as the telescope is rotated about the vertical axis of the theodolite.

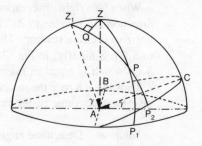

Figure 3.17

Introduction. In Fig. 3.17 the vertical axis AZ_1 is not truly vertical, being inclined at γ to the true vertical AZ. This causes the trunnion axis, in the main, to be tilted.

Z_1PP_2 is the path swept out when the line of sight is directed to P and ZQ, perpendicular to Z_1P, gives the inclination of trunnion axis AB for this pointing.

The maximum inclination γ of the trunnion axis occurs when that axis lies in the plane containing the true vertical axis AZ and the inclined axis AZ_1. Zero inclination occurs when it is at right angles to that plane. By means of striding level observations inclination ZQ can be determined, as indeed it can by observations of the horizontal plate bubble.

The horizontal circle readings for the pointing on P are at P_2 on the inclined horizontal circle, and at P_1 on the correctly positioned circle. Thus the error in the horizontal circle reading due to the tilt of the vertical axis is $(Z - Z_1)$. As in Example 3.7, this error is related to the altitude of P by the expression

$$Z - Z_1 = ZQ \tan h.$$

The error in measured altitude is $(P_1P - P_2P)$ which can be taken to be Z_1Q, i.e. $\gamma \cos Z_1$.

Hence the maximum error in measured altitude is γ. In practice, the mean of the observed altitudes, measured face left and face right with the altitude bubble brought to the centre of its run using the clip screw, will give the required altitude.

Solution. First determine the corrected horizontal angle PAR. The instrument is in adjustment and therefore the trunnion axis is at right angles to the vertical axis. Thus the vertical axis is not truly vertical since the readings of the striding level bubble indicate that the left-hand support of the trunnion axis is higher than the right-hand support.

The inclination of the trunnion axis is given by the expression

$$ZQ = \frac{\theta}{4} [(l_1 + l_2) - (r_1 + r_2)]$$

in which l_1 and l_2 are the readings of the left-hand end of the striding level bubble, r_1 and r_2 refer to the right-hand end and θ is the angular value of one division of the bubble. From this we obtain the values in Table 3.22.

Whence

Table 3.22

Pointing	Striding level values	Inclination ZQ	$(Z - Z_1)$
R	$\frac{12}{4} [(11.3 + 9.7) - (3.3 + 4.9)]$	$+38.4''$	$+38.4 \tan 15° 46' 40'' = +10.8''$
P	$\frac{12}{4} [(10.5 + 8.9) - (4.1 + 5.7)]$	$+28.8''$	$+28.8 \tan 08° 14' 00'' = +4.2''$

Since the left-hand support is high the correction must be added to the observed values as follows in Table 3.23.

Table 3.23

Pointing	Observed horizontal circle reading	Correction $(Z - Z_1)$	Corrected circle reading
R	108° 19′ 46″	+11″	108° 19′ 57″
P	76° 24′ 18″	+ 4″	76° 24′ 22″
Therefore corrected angle PAR =			31° 55′ 35″

Now determine the maximum inclination γ of the trunnion axis. In the spherical triangle ZQZ_1 in which $Q = 90°$

$$\frac{\sin ZZ_1}{\sin Q} = \frac{\sin ZQ}{\sin Z_1}.$$

For small angles we can write $ZQ = ZZ_1 \sin Z_1$, and so ZQ is a maximum when $Z_1 = 90°$.

For the pointing on P

$$ZQ = 28.8'' = \gamma \sin Z_1$$

and for the pointing on R, which is related to Z_1 by the difference in horizontal circle readings

$$38.4'' = \gamma \sin (Z_1 + 108°\ 19'\ 46'' - 76°\ 24'\ 18'')$$
$$= \gamma \sin (Z_1 + 31°\ 55'\ 28'').$$

Therefore $\dfrac{38.4}{28.8} = \dfrac{\sin Z_1 \cos 31°\ 55'\ 28'' + \cos Z_1 \sin 31°\ 55'\ 28''}{\sin Z_1}$

$$\cot Z_1 = 0.916\ 389\ 3,$$

i.e. $\quad Z_1 = 47°\ 29'\ 53''.$

Thus $\quad 28.8'' = \gamma \sin 47°\ 29'\ 53''$ and $\gamma = \mathbf{39.1''}.$

Problems

1 A closed traverse survey involved the measurement of angles at and distances between five stations, A, B, C, D and E.

From the information given in the table below, calculate the closing error in eastings and northings using the Bowditch method and hence determine the corrected values of station co-ordinates if the known co-ordinates of station A are 1500 N and 650 E.

Assume that all the angles, as given in Table 3.24, are correct and therefore do not require any adjustment.

Table 3.24

Station	Line	Length (m)	Whole-circle bearing
A	AB	293	45° 10′ 00″
B	BC	721	72° 05′ 00″
C	CD	496	161° 52′ 00″
D	DE	522	228° 43′ 00″
E	EA	762	300° 42′ 00″
A			

[Salford]

Answer Co-ordinates of C 1543.5 mE, 1927.8 mN

2 A, B, C, D are the stations of a four sided loop traverse; angles are measured with a Wild T1A and distances with a 50 m steel band, with the results in Table 3.25.

Table 3.25

Angle A	42° 47′ 55″	AB	329.88 m
B	135° 37′ 30″	BC	181.60 m
C	137° 31′ 50″	CD	265.15 m
D	44° 02′ 30″	DA	650.14 m

(*i*) Compute the traverse, and say whether you would consider the misclosure satisfactory.

(*ii*) If you consider the misclosure to be unsatisfactory, say where the bad observation responsible is most likely to have occurred.

(*iii*) On the basis that your assumption is correct, obtain adjusted co-ordinates for B, C, D on the basis of:

> Co-ordinates of A: 1000 mE, 2000 mN
> Bearing of AB: 30°

[London]

Answer Co-ordinates of C 1339.84 mE, 2334.59 mN

3 The measured internal angles of a looped traverse ABCDEA are

$$
\begin{aligned}
\hat{A} &= 51° 37' 40'' \\
\hat{B} &= 192° 08' 55'' \\
\hat{C} &= 101° 51' 55'' \\
\hat{D} &= 87° 29' 35'' \\
\hat{E} &= 106° 53' 35''
\end{aligned}
$$

The lengths of the sides are measured in metres as AB = 88.355, BC = 65.205, CD = 76.405, DE = 112.960, EA = 125.400.

Using an equal shifts angular adjustment and a transit linear adjustment, calculate the co-ordinates of the points B, C, D and E if the co-ordinates of A are 50.235 mE, 75.170 mN and the bearing of the line AB is 56° 56' 40''. [Bradford]

Answer Co-ordinates of C 170.325 mE, 169.515 mN

4 Three control points A, B and C have been set out on a construction site. A traverse was run between A and B. It was lettered APQRB. Field notes were:

Measured clockwise angle: CAP = 123° 36' 00''
Measured deflection angles at: P = −28° 17', Q = −43° 23'
 R = −41° 53'
 B (to C) = −57° 05'
Measured lengths (m): AP = 42.67, PQ = 55.13
 QR = 89.15, RB = 24.50
Co-ordinates (m): E_A = 162.970, N_A = 34.160
 E_B = 251.865, N_B = 181.815
 E_C = 27.305, N_C = 186.520

Using an equal shifts angular adjustment and a transit linear adjustment calculate the co-ordinates of the points P, Q and R.

(Deflection angles are negative when measured in an anticlockwise sense.) [Bradford]

Answer Co-ordinates of Q, 249.11 mE, 72.98 mN

5 A traverse is carried out between two points P and Q, the co-ordinates of which are known to be 1268.49 mE, 1836.88 mN and 1375.64 mE,

Table 3.26

Line	Length	Easting +	Easting −	Northing +	Northing −
PA	104.65	26.44		101.26	
AB	208.96	136.41		158.29	
BC	212.45	203.88			59.74
CD	215.98		146.62		158.59
DQ	131.18		112.04	68.23	

1947.05 mN, respectively. Table 3.26 gives the values of the eastings and northings of the sides as calculated from the field observations.

Adjust the closing error using Bowditch's Method and hence calculate the adjusted co-ordinates of A, B, C and D.   [Leeds]
Answer   Co-ordinates of C 1634.67 mE, 2037.13 mN

6   From an established point P a point R is to be set out so that PR is 350 m and $WCB_{PR} = 100° 00'$. The estimated position of R is not visible from P so a traverse PSTU is established such that R is visible from U. Field notes were:

$WCB_{PS} = 115° 00'$, PS = 174.34 m
deflection angle at S = $-83° 00'$, ST = 94.74 m
deflection angle at T = $+92° 00'$, TU = 157.92 m

Calculate the distance UR and the angle TUR. If ST intersects PR at X calculate the distance PX.   [Bradford]
Answer   21.59; 70° 54′; 186.63 m

7   In the looped traverse ABCDEA station D could not be occupied, but was observed from C and E. The internal angles were: at A 95° 18′, at B 105° 45′, at C 86° 09′ and at E 120° 54′. The lengths (m) were AB = 86.61, BC = 79.48 and EA = 52.19. If the co-ordinates of A are 127.50 mE, 146.25 mN and the bearing AB is 39° 54′ calculate the co-ordinates of D.   [Bradford]
Answer   225.97 mE, 124.48 mN

8   A closed traverse survey ABCDEA in which lines CD and AE cross, was carried out in a confined area. Various obstructions prevented distances BC and CD from being measured although their whole circle bearings were able to be recorded. Determine the lengths of BC and CD from the values given in Table 3.27.

Table 3.27

Side	Length (m)	Whole circle bearing
AB	141.2	325° 40′ 00″
BC	?	69° 10′ 00″
CD	?	41° 00′ 00″
DE	58.6	305° 55′ 00″
EA	347.0	194° 50′ 00″

Assume that A has co-ordinates 500.00 N, 500.00 E.   [Salford]
Answer 88.9 m, 202.5 m

9 Derive an expression for the error in the horizontal circle reading of a theodolite (h) caused by the line of collimation not being perpendicular to the trunnion axis by a small amount c.

A theodolite under test for error in collimation and alignment of the trunnion axis is set with its axis truly vertical. Exactly 20.000 metres away is a vertical wire carrying two targets at different levels. An accurate scale perpendicular to the line of sight is graduated from − 100 mm to + 100 mm and is mounted just touching the wire and in the same plane as the trunnion axis. The zero graduation coincides with the wire.

The theodolite is first pointed at a target, the telescope is then lowered to read the scale with the results given in Table 3.28. Determine the magnitude and sense of the error in collimation of the theodolite and the inclination of the trunnion axis.

Table 3.28

Target	Vertical angle	Scale reading
A	65° 27′ 15″	+ 4.11 mm
B	30° 43′ 27″	− 6.24 mm

(The error in horizontal circle reading caused by a trunnion axis misalignment t is $t \tan \alpha$ where α is the altitude.)   [Bradford]
Answer Collimation 148.8″ left; Trunnion 182.9″ left support down

10 Horizontal angle PQ̂R was measured, face left and face right, R being clockwise of P. The mean value of the angle was computed as 75° 30′ 30″, the vertical angles being booked as 22° 00′ elevation on P and 35° 00′ depression on R.

When measuring the angle it was noted that the horizontal plate bubble (mounted parallel to the trunnion axis) had moved off centre during the observations, in Table 3.29.

Table 3.29

Pointing	Face left	Face right
P	+ 20″	+ 30″
R	0″	+ 10″

What value should be assigned to angle PQ̂R? The positive signs imply bubble movement towards the left.
Answer 75° 30′ 16″

11 The readings in Table 3.30 were obtained whilst measuring angle AB̂C, the theodolite known to be in adjustment.

The value of one division of the striding level is 15 seconds. Determine the corrected reading of angle AB̂C.
Answer 59° 55′ 20″

Table 3.30

Pointing	Striding level				Circle readings	
	L	R	L	R	Horizontal	Vertical
A	8.3	11.1	8.5	10.9	126° 20′ 20″	+22° 45′ 00″
C	8.6	10.8	8.8	10.6	186° 15′ 35″	+12° 16′ 10″

12 Calculate the maximum error in the measurement of a horizontal angle of magnitude 135° resulting from a centring error of ±2 mm; the lengths of the adjacent legs of the traverse are 35.50 m and 26.26 m.
Answer 25.3″

13 The readings in Table 3.31 were taken by theodolite.

Table 3.31

Instrument station	Pointing	Vertical circle	Horizontal circle	
			Face left	Face right
A	P	+63° 22′ 00″	27° 24′ 14″	246° 18′ 53″
	R	+12° 16′ 20″	119° 47′ 11″	338° 41′ 28″

Determine the corrected horizontal circle readings given that the line of sight was not at right angles to the trunnion axis, this being the only maladjustment.
Answer 27° 24′ 34″; 246° 18′ 33″; 119° 47′ 20″; 338° 41′ 19″

14 In a certain theodolite the left-hand support of the trunnion axis is higher than the right-hand support when the telescope is 'face left', the inclination of the trunnion axis being 30″ to the horizontal.

Pointings were made on two targets and the readings in Table 3.32 booked.

Table 3.32

Pointing	Horizontal circle reading (face right)	Vertical circle reading
P	246° 18′ 53″	+63° 22′ 00″
R	338° 41′ 28″	+12° 16′ 20″

Determine the angle subtended at the instrument, assuming that there is no other maladjustment.
Answer 92° 23′ 28″

15 If the horizontal axis of a theodolite makes an angle of $(90° + \alpha)$ with the vertical axis and if the instrument is otherwise in adjustment, show that the difference between circle left and circle right measurement of the horizontal angle subtended by two targets, whose elevations are θ and ϕ above horizontal is $2\alpha(\tan \theta - \tan \phi)$.

In a certain theodolite, the horizontal axis is 0.025 mm out in 100 mm and the instrument is otherwise in correct adjustment. Find the difference, to the nearest second, between circle left and circle right values of the horizontal angle subtended by two targets whose elevations are 55° 30′ and 22° 00′. [London]

Answer 01′ 48″

16 An angle of elevation was measured by vernier theodolite and it was noted that the altitude bubble was not in the centre of its run in either the face-left or face-right positions. Deduce the value of that angle from the data in Table 3.33.

Table 3.33

Face	Vernier readings		Altitude level	
			O	E
Left	25° 20′ 40″	25° 21′ 00″	3.5 div.	2.5 div.
Right	25° 21′ 00″	25° 21′ 20″	4.5 div.	1.5 div.

O and E refer to the objective and eyepiece end respectively of the bubble, and one division of the altitude level is equivalent to 20 seconds.

[I. Struct. E.]

Answer 25° 21′ 20″

4

Triangulation and the National Grid

Uses of triangulation

Triangulation surveys can be used for the accurate location of control points for plane surveys, aerial surveys and geodetic surveys covering appreciable areas. They can also be used for the setting out of civil engineering works such as the piers and abutments of long span bridges, and the measurement of the deformation of dams.

Control points

The control points lie at the corners of geometrical figures such as triangles, quadrilaterals with diagonals and polygons with central points. All angles of these figures are measured and their most probable values are determined as illustrated in Chapter 8.

Base line

Providing one side length has been measured all others can be calculated by trigonometry; this length is referred to as the base line. At least one other length should be measured, particularly over large areas, so that comparison may be made between their lengths, calculated and measured. The accuracy of the work can then be assessed.

If EDM systems are available, and all the sides are measured, a trilateration survey is established. However, angular measurements define the shape of the survey network better than wholly linear measurements and so it is likely that a number of angles will be included in a trilateration survey.

Survey of Great Britain

In the geodetic triangulation of Great Britain, commenced in 1936, the primary triangles had side lengths generally lying between 40 km and 70 km. These triangles supported a secondary system which, in turn, divided into tertiary and fourth-order systems. In the latter case stations were at 1 km to 2 km intervals in urban areas and were often fixed by the techniques of intersection or resection.

National Grid

Triangulation stations in Great Britain have co-ordinates expressed to 0.001 m within the National Grid. This is composed of lines parallel to a central meridian (through 2°W) and lines perpendicular thereto, producing a square grid of side 10 km. The grid was derived from a Transverse Mercator projection whose origin lies on the central meridian at a latitude of 49°N. The scale is constant at every point in all directions, but during the projection some adjustments were made. Consequently, the scale changes with distance from the central meridian, being correct on lines nearly parallel to the central meridian about two-thirds of the way towards the boundaries.

The reader will appreciate that if a 'local' engineering survey can be tied into Ordnance Survey points it should be 'strengthened'.

Convergence

The direction of Grid North is that of the central meridian through 2°W but elsewhere a meridian does not align with Grid North. Thus, in general, the grid bearing of a line will not equal the true bearing of that line if measured at a station by an astronomical method or by gyro-theodolite. To convert the former to the latter a convergence factor has to be applied.

4.1 Sine and cosine rule

In a triangulation survey for a certain civil engineering project line AC was measured and found to be 1210.46 m long. Two stations B and D were established on opposite sides of AC and the following angles were observed.

ABD = 44° 40′ 59″
DBC = 67° 43′ 55″
ADB = 63° 19′ 28″
BDC = 29° 38′ 50″

Calculate the length BD.

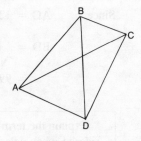

Figure 4.1

Introduction. In Fig. 4.1, ABCD is a quadrilateral with diagonals, referred to as a braced quadrilateral, and it is one of the basic figures employed in triangulation. Further examples of this type of 'three-point problem are covered in Chapter 7.

Solution. First calculate angles BÂD and BĈD.

In triangle ABD   In triangle BCD

$\hat{ABD}$ = 44° 40′ 59″ $\hat{DBC}$ = 67° 43′ 55″
$\hat{ADB}$ = 63° 19′ 28″ $\hat{BDC}$ = 29° 38′ 50″
                 ―――――――   ―――――――
          108° 00′ 27″   97° 22′ 45″

Therefore $\hat{BAD} = 180° \ 00' \ 00'' - 108° \ 00' \ 27'' = 71° \ 59' \ 33''$

$\hat{BCD} = 180° \ 00' \ 00'' - \ 97° \ 22' \ 45'' = 82° \ 37' \ 15''$

Next use sine rule to establish lengths AB and BC in terms of BD in triangle ABD.

$$\frac{BD}{\sin 71° \ 59' \ 33''} = \frac{AB}{\sin 63° \ 19' \ 28''}$$

$$AB = \frac{\sin 63° \ 19' \ 28''}{\sin 71° \ 59' \ 33''} \times BD = 0.939\ 588 \times BD.$$

In triangle BCD

$$\frac{BD}{\sin 82° \ 37' \ 15''} = \frac{BC}{\sin 29° \ 38' \ 50''}$$

$$BC = \frac{\sin 29° \ 38' \ 50''}{\sin 82° \ 37' \ 15''} \times BD = 0.498\ 789 \times BD.$$

Finally, use cosine rule in triangle ABC to obtain BD.

$$AC^2 = AB^2 + BC^2 - (2AB \times BC \times \cos ABC)$$

in which $\hat{ABC} = (44° \ 40' \ 59'' + 67° \ 43' \ 55'')$.

Therefore $AC^2 = (0.939\ 588 \times BD)^2 + (0.498\ 789 \times BD)^2$
$+ (0.357\ 408\ 8 \times BD^2)$
$= 1.489\ 025\ BD^2.$

Since $AC = 1210.46$ m

$$BD = \sqrt{\left(\frac{1210.46^2}{1.489\ 025}\right)}$$

$$= \mathbf{991.97\ m.}$$

4.2 Satellite station

Explain the term triangulation survey and state the factors to be considered when selecting stations for such surveys.

In triangle ABC, C could not be occupied and a satellite station S was established north of C. The angles in Table 4.1 were then registered by a theodolite set up at S.

Table 4.1

Pointing on	Horizontal circle reading
A	14° 43′ 27″
B	74° 30′ 35″
C	227° 18′ 12″

The lengths of AC and BC were estimated to be 17 495 m and 13 672 m, respectively, and the angle $\hat{ACB}$ was deduced to be $59° \ 44' \ 53''$. Calculate the distance of S from C. [Salford]

Introduction. The term triangulation survey has been discussed at the beginning of this chapter. In so far as selection factors are concerned:

(a) stations must be visible from each other;

(b) the triangles should be well conditioned, that is with no angle less than 30° and all sides tending to the same length;

(c) the sides should be as long as possible; and

(d) when selecting stations on the edge of an area, connection to adjoining surveys should be considered.

A satellite station (or eccentric station) may be used when it is very difficult, or impossible, to centre the theodolite over a particular station, even though it is clearly visible from other stations in the network. In addition, a satellite station can be included in the observation programme if there are line-of-sight difficulties on to one station but not on to the nearby satellite station.

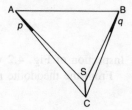

Figure 4.2

Solution. Establish the relative location of A, B, C and S. Assuming that the theodolite is graduated in a clockwise manner, the observations from S imply that B and C lie in a clockwise direction from A with angle $A\hat{S}B = 59° 47' 08''$ and angle $A\hat{S}C = 212° 34' 45''$. Since S is stated to be north of C then the general disposition of the four stations is shown in Fig. 4.2 with S lying within triangle ABC. Lengths AC and CB could have been estimated since the length of AB would have been determined from previous observations, and during the observation programme angles $A\hat{B}C$ and $B\hat{A}C$ would have been measured. Taking $A\hat{C}B$ as $(180 - B\hat{A}C - A\hat{B}C)$ then the approximate lengths of AC and BC are readily obtained.

First we determine distance SC. In Fig. 4.2 let angles $C\hat{A}S$ and $C\hat{B}S$ be *p* and *q*, respectively. SC will be a short distance in practice (less than AC/1000, preferably) and so *p* and *q* will be relatively small angles. From Fig. 4.2

$$\frac{SC}{\sin p} = \frac{AC}{\sin A\hat{S}C}$$

and

$$\frac{SC}{\sin q} = \frac{BC}{\sin B\hat{S}C}.$$

Now

$$A\hat{S}C = 360° - 212° 34' 45''$$
$$= 147° 25' 15''$$

and

$$B\hat{S}C = 227° 18' 12'' - 74° 30' 35''$$
$$= 152° 47' 37''.$$

Thus
$$\sin p = \frac{SC}{AC} \sin 147° \, 25' \, 15''$$

and
$$\sin q = \frac{SC}{BC} \sin 152° \, 47' \, 37''.$$

For small angles
$$\sin p = p'' \sin 1''$$
and
$$\sin q = q'' \sin 1''.$$
Now
$$\sin 1'' = 1/206\,265.$$

Therefore
$$p'' = \frac{SC}{17\,495} \times \frac{\sin 147° \, 25' \, 15''}{\sin 1''}$$
$$= 6.348 \times SC.$$

$$q'' = \frac{SC}{13\,672} \times \frac{\sin 152° \, 47' \, 37''}{\sin 1''}$$
$$= 6.898 \times SC.$$

Inspection of Fig. 4.2 will show that $A\hat{S}B = A\hat{C}B + p + q$. From the theodolite readings at S

$$A\hat{S}B = 74° \, 30' \, 35'' - 14° \, 43' \, 27''$$
$$= 59° \, 47' \, 08''$$
also $\quad\quad\quad A\hat{C}B = 59° \, 44' \, 53''$ from the data.
Therefore $p+q+59° \, 44' \, 53'' = 59° \, 47' \, 08''$
$$p+q = 02' \, 15'' = 135'',$$
i.e. $(6.348 \times SC) + (6.898 \times SC) = 135''$ and SC = **10.192 m.**

4.3 Spherical triangles

In a geodetic survey the mean angles in Table 4.2 were observed in one triangle, each having been observed the same number of times under similar conditions.

Table 4.2

Station	Mean value
A	62° 24′ 18.4″
B	64° 56′ 09.9″
C	52° 39′ 34.4″

Side AB was known to be 37 269.280 m long. Estimate the corrected values of the three angles. Take the radius of the earth to be 6383.393 km.

Introduction. The theodolite measures horizontal angles in the horizontal plane but in the case of large triangles, i.e. primary triangles in the National Triangulation, the curvature of the earth means that such planes are not parallel at the apices, Fig. 4.3. Accordingly, the three angles of a large triangle do

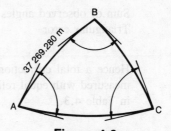

Figure 4.3

not total 180°, as is the case for a plane triangle, but to 180° + spherical excess. The spherical excess depends upon the area of the triangle, and triangular error is given by

Σ observed angles − (180° + spherical excess).

Solution. First determine the spherical excess. The sum of the three mean angles at A, B and C is 180° 00′ 02.7″. In order to estimate the spherical excess it is necessary to estimate the area of triangle ABC. For this purpose it is sufficiently accurate to assume that the triangle is plane. The three angles should thus sum to 180° and we can deduct 2.7″/3 = 0.9″ from each to give

A = 62° 24′ 17.5″
B = 64° 56′ 09.0″
C = 52° 39′ 33.5″

180° 00′ 00″

Now $\dfrac{AB}{\sin C} = \dfrac{BC}{\sin A}$

$\dfrac{37\ 269.280}{\sin 52° 39′ 33.5″} = \dfrac{BC}{\sin 62° 24′ 17.5″}$

Therefore   BC = 41 544.469 m

Area of triangle = $\frac{1}{2}$ × AB × BC × sin 64° 56′ 9.0″

= 701.3 km².

Spherical excess can be calculated from the expression,

$$\frac{\text{Area of triangle}}{R^2 \sin 1″}.$$

where R is the radius of the earth.

Whence the spherical excess of triangle ABC = $\dfrac{701.3}{6383.393^2 \times \sin 1″}$

= 3.6″.

Now calculate the triangular error and the angle corrections.

Theoretical sum of the angles = 180° + spherical excess
= 180° 00′ 03.6″.

Sum of observed angles = 180° 00' 02.7".
Triangular error = 180° 00' 02.7" − 180° 00' 03.6"
= −0.9".

Hence a total correction of +0.9" is required and since the angles were measured with equal reliability a correction of +0.3" is applied to each as in Table 4.3.

Table 4.3

Station	Corrected angle
A	62° 24' 18.4" + 0.3" = **62° 24' 18.7"**
B	64° 56' 09.9" + 0.3" = **64° 56' 10.2"**
C	52° 39' 34.4" + 0.3" = **52° 39' 34.7"**
	Total = 180° 00' 03.6"

4.4 Effect of the earth's curvature

A traverse was run as in Table 4.4

Table 4.4

Station	Length (m)	Clockwise angle from rear station
X	4215.65	
Q	3778.46	205° 36' 12"
R	5237.28	144° 23' 20"
Y		

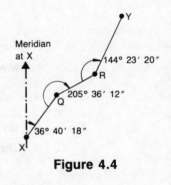

Figure 4.4

The bearing of XQ is 36° 40' 18" and the latitude of X is 52° 20' 45" N. Determine the bearings of YX and YR at Y and estimate the latitude of Y.

	Length of 1" of latitude	Length of 1" of longitude
52° 20'	30.9022 m	18.9364 m
52° 25'	30.9107 m	18.9008 m

Introduction. The curvature of the earth causes the azimuth or bearing of a 'straight' survey line to constantly change so that the bearing of X from Y is not equal to the bearing of Y from X ±180°. In Fig. 4.5 let XY be the

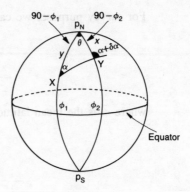

Figure 4.5

line in question, X and Y having latitudes of ϕ_1 and ϕ_2 respectively, p_n and p_s are the terrestrial poles so that the meridians of X and Y are $p_s X p_n$ and $p_s Y p_n$. These meridians are parallel at the equator but this changes with progress towards the poles. At the poles the angle between the meridians is θ, which is also the difference in longitude of X and Y. The bearings of XY at X and XY at Y are shown as α and $\alpha + \delta\alpha$, respectively, $\delta\alpha$ indicating the convergence of the meridians, i.e. the angle between the meridian at X and the meridian at Y. $\delta\alpha$ can be readily derived from spherical triangle Xp_nY (Fig. 4.5), in which

$$Xp_n = 90 - \phi_1$$
$$Yp_n = 90 - \phi_2$$

$$\tan\left(\frac{X+Y}{2}\right) = \frac{\cos\left(\dfrac{x-y}{2}\right)}{\cos\left(\dfrac{x+y}{2}\right)} \times \cot\frac{p_n}{2}$$

$$\tan\left[\frac{\alpha+(180-\alpha-\delta\alpha)}{2}\right] = \frac{\cos\left[\dfrac{(90-\phi_2)-(90-\phi_1)}{2}\right]}{\cos\left[\dfrac{(90-\phi_2)+(90-\phi_1)}{2}\right]} \times \cot\frac{\theta}{2}$$

$$\tan\left(90-\frac{\delta\alpha}{2}\right) = \frac{\cos\left(\dfrac{\phi_1-\phi_2}{2}\right)}{\sin\left(\dfrac{\phi_1+\phi_2}{2}\right)} \times \cot\frac{\theta}{2}.$$

Whence
$$\tan\frac{\delta\alpha}{2} = \frac{\sin\left(\dfrac{\phi_1+\phi_2}{2}\right)}{\cos\left(\dfrac{\phi_1-\phi_2}{2}\right)} \times \tan\frac{\theta}{2}.$$

For survey purposes we can take $(\phi_1 - \phi_2)$ and θ to be small, so that

$$\delta\alpha = \theta \, \sin\left(\frac{\phi_1 + \phi_2}{2}\right)$$

$$= \theta \, \sin \bar\phi$$

where $\bar\phi$ is the mean latitude of X and Y.

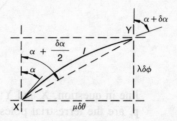

Figure 4.6

For calculation purposes it is quite adequate to assume that the meridians at X and Y are parallel when the two points are relatively close together, say less than 40 km apart. This allows a rectangular grid to be established, as in Fig. 4.6, the lines being spaced apart at mid-latitude distances. Thus λ represents the length of $1''$ of latitude, whilst μ represents the length of $1''$ of longitude, at the mean latitude, $\bar\phi$, of X and Y in each case. l is the length of XY, which is actually part of a great circle, and the average bearing of

XY is $\left(\alpha + \dfrac{\delta\alpha}{2}\right)$. Thus we write

$$\lambda \times \delta\phi = l \cos\left(\alpha + \frac{\delta\alpha}{2}\right)$$

$$\mu \times \delta\theta = l \sin\left(\alpha + \frac{\delta\alpha}{2}\right)$$

and $\qquad \delta\alpha = \delta\theta \times \sin \bar\phi.$

Note that had X and Y been widely separated, triangle Xp_nY could be solved for angles $\hat{X}$ and $\hat{Y}$ using the standard expressions for $\tan[(X+Y)/2]$ and $\tan[(X-Y)/2]$.

Solution. First calculate the traverse details. In the first instance the relationship between X and Y is established as if the surface of the earth is plane.

(*a*) Whole circle bearings (meridian at X serving as reference):

$$\text{WCB of XQ} = \quad 36°\ 40'\ 18''$$
$$\text{add} \quad \underline{205°\ 36'\ 12''}$$

$$242° \ 16' \ 30''$$

$$\text{deduct} \quad 180° \ 00' \ 00''$$

$$\text{WCB of QR} = \quad 62° \ 16' \ 30''$$
$$\text{add} \quad 144° \ 23' \ 20''$$

$$206° \ 39' \ 50''$$
$$\text{deduct} \quad 180° \ 00' \ 00''$$

$$\text{WCB of RY} = \quad 26° \ 39' \ 50''$$

(*b*) The easting and northing differences are given in Table 4.6.

Table 4.6

Line	Length (m)	WCB	Easting difference (m)	Northing difference (m)
XQ	4215.65	36° 40′ 18″	2517.71	3381.25
QR	3778.46	62° 16′ 30″	3344.66	1757.85
RY	5237.28	26° 39′ 50″	2350.26	4680.32
			8212.63	9819.42

$$\text{Length of XY} = \sqrt{(8212.63^2 + 9819.42^2)}$$
$$= 12\ 801.11 \text{ m} = l \text{ (Fig. 4.6)}.$$

$$\text{Bearing of XY} = \tan^{-1}\left(\frac{8212.63}{9819.42}\right)$$

$$= 39° \ 54' \ 28.7'' = \alpha.$$

Next determine the convergence effect. Had the latitude and longitude of X and Y been given, $\delta\phi$, $\delta\theta$ and ϕ would be known and the expressions derived previously could have been used directly. However, the latitude of Y and the longitude difference $\delta\theta$ are unknown and accordingly the mid-latitude of X and Y has to be estimated and $\delta\phi$, $\delta\theta$ and $\delta\alpha$ evaluated by successive approximations.

For the first trial we assume that the mid-latitude of XY is that of X, i.e. 52° 20′ 45″ N.

$$\text{Therefore } \lambda = 30.9022 + \frac{45}{300}(30.9107 - 30.9022)$$

$$= 30.9035 \text{ m}.$$

The approximate difference in latitude between Y and X

$$= \frac{l \cos \alpha}{30.9035} = \frac{\text{northing difference}}{30.9035} = \frac{9819.42}{30.9035}$$

$$= 317.74''.$$

Whence the approximate mid-lattitude

$$= 52° \ 20' \ 45'' + \frac{317.74''}{2}$$

$$= 52° \ 23' \ 23.9''.$$

For this latitude

$$\mu = 18.9364 - \frac{203.9}{300} \times 0.0356$$

$$= 18.9122 \text{ m}$$

and $\quad \delta\theta = \dfrac{l \sin \alpha}{18.9122} = \dfrac{\text{easting difference}}{18.9122}$

$$= 434.25''.$$

Therefore $\quad \delta\alpha = \delta\theta \times \sin \bar{\phi}$

$$= 434.25 \sin 52° \ 23' \ 23.9''$$

$$= 344.01'' = 5' \ 44.01''.$$

For the second trial we take the mid-latitude to be $52° \ 23' \ 23.9''$

$$\lambda = 30.9022 + \frac{203.9}{300} (30.9107 - 30.9022)$$

$$= 30.9080 \text{ m}.$$

Now $\qquad \delta\alpha = 344.01''$

$$\frac{\delta\alpha}{2} = 172.00''.$$

Hence $\qquad \delta\phi = \dfrac{l}{\lambda} \cos \left(\alpha + \dfrac{\delta\alpha}{2} \right)$

$$= \frac{12\ 801.11}{30.9080} \cos 39° \ 57' \ 20.7''$$

$$= 317.48''.$$

Revised mid-latitude $= 52° \ 20' \ 45'' + \dfrac{317.48''}{2}$

$$= 52° \ 23' \ 23.7''.$$

Whence $\qquad \mu = 18.9364 - \left(\dfrac{203.7}{300} \times 0.0356 \right)$

$$= 18.9122 \text{ m}.$$

Revised $\qquad \delta\theta = \dfrac{l}{\mu} \sin \left(\alpha + \dfrac{\delta\alpha}{2} \right)$

$$= \frac{12\ 801.11}{18.9122} \sin 39° \ 57' \ 20.7''$$

$$= 434.68''$$

and
$$\delta\alpha = \delta\theta \sin \bar{\phi}$$
$$= 434.68 \sin 52° 23' 23.7''$$
$$= 344.35''.$$

It will be seen that only small changes have occurred in the values of $\delta\phi$ (decrease), $\delta\theta$ (increase) and $\delta\alpha$ (increase). A further trial is not warranted, particularly in view of the fact that the data were expressed to an accuracy of a single second.

Whence
$$\delta\alpha = 344.35$$
$$= 05' 44'', \text{ say.}$$
Therefore bearing of RY at R $= 26° 39' 50'' + 05' 44''$
$$= 26° 45' 34''.$$
Therefore bearing of YR at Y $= 26° 45' 34'' + 180°$
$$= \mathbf{206° 45' 34''.}$$

Similarly,
bearing of YX at Y $= 39° 54' 29'' + 05' 44'' + 180°$
$$= \mathbf{220° 00' 13''.}$$
latitude of Y $=$ latitude of X $+ \delta\phi$
$$= 52° 20' 45'' + 317.48''$$
$$= \mathbf{52° 26' 02'' N.}$$

4.5 Co-ordinates

Two survey stations A and B have the National Grid co-ordinates and latitudes given in Table 4.7.

Table 4.7

	Easting	Northing	Latitude
A	460 257.664	350 258.130	N 53° 02′ 45″
B	460 476.691	350 340.727	N 53° 02′ 48″

Calculate

(i) the grid length of the base line
(ii) the National Grid bearing of the line A to B
(iii) the National Grid bearing of the line B to A
(iv) the convergence at station A
(v) the convergence at station B
(vi) the geographical bearing of the line A to B
(vii) the geographical bearing of the line B to A.

Assume the local radius of the earth to be 6384.100 km, and since the two points are close to the central meridian an approximate formula may be used in the calculation of convergence.   [Eng. Council]

Introduction. The earth can be idealized as being spheroidal, and for geodetic purposes a spheroid of reference may be adopted to satisfy a particular area. In the calculation of the Transverse Mercator projection referred

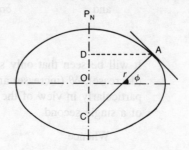

Figure 4.7

to at the beginning of the chapter the Airy values equivalent to 6 377 563.394 m
and 6 356 256.908 m were used for the lengths of the major semi-axis and
the minor semi-axis, respectively, of the spheroid of reference.

Figure 4.7 shows a spheroid with point A on its surface. The normal at A
meets the minor axis at C and AC represents the radius of curvature (r) perpendicular to the meridian. ϕ is the geographical latitude of A, and AD is the
radius of the parallel of latitude through A, having a value of $r \cos \phi$.

Hence the length of $1''$ of longitude at this latitude is $r \cos \phi \times \sin 1''$.

Solution. (*i*) Find grid length AB. The co-ordinates are given in Table 4.8.

Table 4.8

	Easting	Northing
B	460 476.691	350 340.727
A	460 257.664	350 258.130
Differences	219.027 m	82.597 m

Therefore grid length AB = $\sqrt{(219.027^2 + 82.597^2)}$
= **234.084 m.**

(*ii*) Find the National Grid bearing of line A to B. Let α_{AB} be the required
bearing.

$$\tan \alpha_{AB} = \frac{E_B - E_A}{N_B - N_A}$$

$$= \frac{219.027}{82.597}$$

$$= 2.651\ 754\ 9.$$

Therefore $\alpha_{AB} = $ **69° 20′ 17″.**

(*iii*) Find the National Grid bearing of line B to A.
Let α_{BA} be the required bearing. Since BA is a straight line within the grid

$$\alpha_{BA} = \alpha_{AB} + 180°$$

$$= \textbf{249° 20′ 17″.}$$

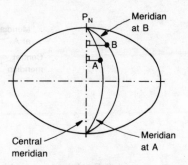

Figure 4.8

(*iv*) Find the convergence at station A. All north–south lines on the grid are parallel to the central meridian running through 2°W, the true origin lying at 49°N thereon. However, to ensure that all eastings are positive and all northings are less than 1000 km a false origin 400 km west of and 100 km north of the true origin was adopted. Hence the grid easting of the central meridian is 400 km, so that A is at a perpendicular distance of (460 257.664 − 400 000) m = 60 257.664 m from the central meridian. Referring to Fig. 4.8, let L be the perpendicular distance from the central meridian to A.

The difference in longitude between that meridian and A is

$$\delta\theta = \frac{L}{r \cos \phi \times \sin 1''}.$$

But the change in bearing over that length is

$$\delta\alpha = \delta\theta \sin \phi$$

$$= \frac{L}{r \cos \phi \times \sin 1''} \times \sin \phi$$

$$= 206\ 265 \frac{L}{r} \tan \phi''.$$

Since the north–south grid lines are parallel to the central meridian, $\delta\alpha$ is the convergence, C_A, at A.

Therefore $C_A = 206\ 265 \times \dfrac{60\ 257.664}{6\ 384\ 100} \tan 53°\ 02'\ 45''$

$$= 2587.9''$$

$$= \mathbf{43'\ 07.9''}.$$

(*v*) Find the convergence at station B. The perpendicular distance from the central meridian to B is 460 476.691 − 400 000 = 60 476.691 m. Therefore the convergence

$$C_B = 206\ 265 \times \frac{60\ 476.691}{6\ 384\ 100} \times \tan 53°\ 02'\ 48''$$

$$= 2597.38''$$

$$= \mathbf{43'\ 17.38''}.$$

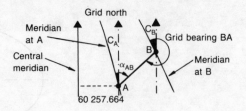

Figure 4.9

(*vi*) Find the geographical bearing of the line A to B. The geographical bearing of line AB is the angle between the meridian at A and the line. Meridians of points to the east of the central meridian are inclined to the direction of Grid North as shown in Figs 4.8 and 4.9.

Hence the geographical bearing of line AB at A is $\alpha_{AB} + C_A$

$$= 69° \, 20' \, 17.15'' + 43' \, 07.9''$$
$$= \textbf{70° 03' 25.0''}, \text{ say.}$$

(*vii*) Find the geographical bearing of the line BA at B.
The bearing is $\alpha_{BA} + C_B$

$$= 249° \, 20' \, 17.15'' + 43' \, 17.38''$$
$$= \textbf{250° 03' 34.5''}, \text{ say.}$$

Note that when the station is to the west of the central meridian, i.e. its easting co-ordinate is less than 400 000 m, its meridian will be inclined in the other direction with respect to the central meridian. Convergence (*C*) takes a negative sign and is deducted from grid bearing to give geographical bearing. A further point in respect of long lines is discussed in Example 4.6.

4.6 (*t* − *T*) correction

Calculate the $(t-T)$ direction corrections to obtain straight-line directions at two stations A and B having the National Grid co-ordinates

A	E 300 120 m	N 385 920 m
B	E 310 970 m	N 380 230 m

Aide-mémoire

$$(t_A - T_A)'' = (2y_A + y_B)(N_A - N_B) \times 0.845 \times 10^{-9}$$
$$(t_B - T_B)'' = (2y_B + y_A)(N_B - N_A) \times 0.845 \times 10^{-9} \qquad \text{[CEI]}$$

Introduction. The direct line of sight between A and B becomes effectively curved in the projection. This is of particular importance for long-sighting distances and correction $(t-T)$ has to be applied when deducing the geographical bearing of lines from the National Grid co-ordinates. Convergence, *C*, has to be taken into account also as indicated in Fig. 4.10 so that

True bearing = Grid bearing + $C - (t-T)$.

Note that the curved line is always concave to the meridian.

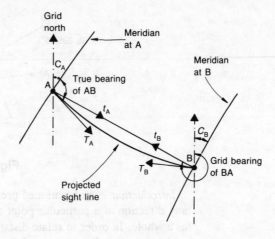

Figure 4.10

Solution. Determine the $(t - T)$ corrections. In the quoted expressions

$$y = (\text{Easting} - 400\,000) \text{ m.}$$

Therefore
$$y_A = (300\,120 - 400\,000)$$
$$= -99\,880 \text{ m}$$

and
$$y_B = (310\,970 - 400\,000)$$
$$= -89\,030 \text{ m.}$$

Thus
$$2y_A + y_B = -288\,790 \text{ m}$$
$$2y_B + y_A = -277\,940 \text{ m}$$

and
$$N_A - N_B = +5690 \text{ m}$$
$$N_B - N_A = -5690 \text{ m.}$$

$$(t_A - T_A)'' = (2y_A + y_B)\,(N_A - N_B) \times 0.845 \times 10^{-9}$$
$$= -288\,790 \times 5690 \times 0.845 \times 10^{-9}$$
$$= \mathbf{-1.39''.}$$

$$(t_B - T_B)'' = -277\,940 \times -5690 \times 0.845 \times 10^{-9}$$
$$= \mathbf{+1.34''.}$$

4.7 Local scale factor

The measured slope distances from a survey station A situated on a colliery headgear 101.15 m above Ordnance Datum, to two Ordnance Survey stations B and C, are AB 1923.400 m and AC 1398.446 m. The National Grid co-ordinates and levels of stations B and C in metres are as follows:

Station B: E 323 679.35   N 340 431.32   Level 385.13 AOD.
Station C: E 324 022.07   N 342 846.89   Level 259.99 AOD.

Assuming the local radius of the earth and the local scale factor to be 6383.391 km and 0.999 672, respectively, compute the National Grid co-ordinates of station A. Station A is to the east generally of B and C.

[CEI]

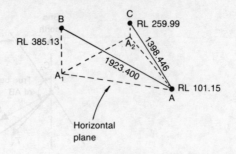

Figure 4.11

Introduction. As mentioned previously, although the scale is constant in any direction at a particular point there are variations across the projection as a whole. In order to relate distance (S) on the spheroid at mean sea level to the corresponding distance (s) in the projection, the expression $s = F \times S$ is used. F is known as the local scale factor. Thus we can either deduce the spheroid distance from a known grid distance or transform a measured distance (reduced to mean sea level) to grid distance. On the central meridian the value of F is 0.999 60, whilst at eastings of 220 km and 580 km the value is 1.000 00. The values of local scale factor are symmetrical about the central meridian and can be closely approximated by the expression $F = F_0(1 + k\,y^2)$ in which F_0 is the scale factor on the central meridian and y is defined in Example 4.6.

At the easting of 220 km

$$1.000\ 00 = 0.999\ 60\ (1 + k \times 180^2),$$
$$\text{since } y = (200 - 400) \text{ km.}$$
$$\text{Therefore } k = 1.235\ 06 \times 10^{-8}.$$

For the easting of 670 km

$$y = (670 - 400) = 270 \text{ km}$$
$$\text{and so} \quad F = 0.999\ 60\ (1 + (270^2 \times 1.235\ 06 \times 10^{-8}))$$
$$= 1.000\ 50.$$

The reader could check the easting satisfying the given F.

Solution. First determine horizontal distances AB and AC. The difference in height between A and B is $385.13 - 101.15$

$$= 283.98 \text{ m.}$$
Slope distance from A to B = 1923.400 m.

Therefore horizontal distance

$$AA_1 = \sqrt{(1923.400^2 - 283.98^2)}$$
$$= 1902.320 \text{ m.}$$

The difference in height between A and C is $259.99 - 101.15$

$$= 158.84 \text{ m.}$$
Slope distance from A to C = 1398.446 m.

Therefore horizontal distance

$$AA_2 = \sqrt{(1398.446^2 - 158.84^2)}$$
$$= 1389.396 \text{ m}.$$

Note that since the height differences are appreciable when compared to the slope distances Pythagoras's theorem has been used to determine the horizontal distances, rather than the slope correction $-h^2/2L$ referred to in Chapter 2.

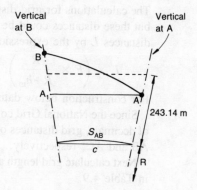

Figure 4.12

Grid distances can now be determined. In order to calculate the corresponding grid distances between A and B, and A and C we must initially establish the distances at mean sea level. The horizontal distances AA_1, etc., are positioned at their mean heights, h_m, as shown by the dotted line in Fig. 4.12. Chord lengths, c, are then calculated by proportion. For A to B

$$\frac{c}{1902.320} = \frac{6\,383\,391}{6\,383\,391 + 243.14}$$

since the mean height of AB is $(101.15 + 385.13)/2$ and R is given as 6383.391 km.

Therefore $c = 1902.248$ m.

Similarly, for A to C

$$\frac{c}{1389.396} = \frac{6\,383\,391}{6\,383\,391 + \dfrac{(101.15 + 259.99)}{2}}$$

Therefore $c = 1389.357$ m.

The same lengths would be obtained by applying the mean sea level correction, $-Lh_m/R$, referred to in Chapter 2, to the horizontal distances. Over the distances in this example the chord lengths are the same as the curved lengths at mean sea level, but the conversion needs to be considered when long lengths are involved.

Thus the distances at mean sea level are

$$S_{AB} = 1902.248 \text{ m}$$
and $\quad S_{AC} = 1389.357 \text{ m.}$

Now $\quad FS = s.$

Therefore $\quad s_{AB} = 0.999\ 672 \times 1902.248$
$$= 1901.624 \text{ m}$$
and $\quad s_{AC} = 0.999\ 672 \times 1389.357$
$$= 1388.901 \text{ m.}$$

The calculations for grid distances have been presented in stages for clarity, but these distances could be readily determined directly from the horizontal distances L by the expression

$$L \times \left(F \times \frac{R}{R+h_{m}} \right).$$

For construction below datum h_{m} takes a negative sign.

Since the National Grid co-ordinates of B and C are expressed to two places of decimals, grid distances of 1901.62 m and 1388.90 m will be accepted for AB and AC, respectively.

Next calculate grid length and grid bearing of BC. The co-ordinates are given in Table 4.9.

Table 4.9

Station	E (m)	N (m)
C	324 022.07	342 846.89
B	323 679.35	340 431.32
Difference	+ 342.72	2 415.57

Grid lengths of BC $= \sqrt{(342.72^2 + 2415.57^2)}$
$$= 2439.76 \text{ m.}$$

Grid bearing of BC $= \tan^{-1} \dfrac{E_C - E_B}{N_C - N_B} = \tan^{-1} \dfrac{342.72}{2415.57}$

$$= \tan^{-1} 0.141\ 879\ 6$$
$$= 08^\circ\ 04'\ 30.7''.$$

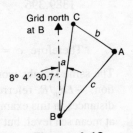

Figure 4.13

Finally calculate the co-ordinates of A. In Fig. 4.13

$$b^2 = a^2 + c^2 - 2\,ac\cos B,$$

whence $2\,ac\cos B = a^2 + c^2 - b^2$

$$a^2 = 2439.76^2 = 5\ 952\ 428.858$$
$$c^2 = 1901.62^2 = \underline{3\ 616\ 158.624}$$

$$\text{add} = 9\ 568\ 587.482$$
$$b^2 = 1388.90^2 = \underline{1\ 929\ 043.210}$$

$$\text{deduct} = 7\ 639\ 544.272$$

Therefore $\quad \cos B = \dfrac{7\ 639\ 544.272}{2 \times 2439.76 \times 1901.62}$

$$= 0.823\ 316\ 1.$$

Therefore $\quad \hat{B} = 34°\ 34'\ 54.7''$.

The grid bearing of BA is $34°\ 34'\ 54.7'' + 08°\ 04'\ 30.7'' = 42°\ 39'\ 25.4''$.
Whence the easting difference between B and A is $1901.62 \sin 42°\ 39'\ 25.4''$
$= +1288.55$ m, and the northing difference between B and A is
$1901.62 \cos 42°\ 39'\ 25.4'' = +1398.49$ m. Thus

	E	N
Station B	323 679.35	340 431.32
+	1 288.55	+ 1 398.49
Station A	324 967.90	341 829.81

Grid co-ordinates of Station A are: **E 324 967.90 m   N 341 829.81 m.**

4.8 Gyro-theodolite observations— reversal method

Describe carefully the 'reversal' method of observation using a gyro-theodolite.

Using the Schuler mean values calculate the bearing of reference object B from the following observations taken at Station A using a gyro-theodolite.

Horizontal circle readings to B: face left 45° 20.8′, face right 225° 20.2′. The turning-point readings on the gyro-theodolite were as follows:

Left turning-point	Right turning-point
146° 27.2′	151° 46.3′
146° 28.8′	151° 44.8′
146° 30.2′	151° 43.5′

The calibration constant of the instrument is +2.4′:   [Bradford]

Introduction. In its operating mode the spinner of the suspended gyroscope is driven at a high speed about its spin axis, which lies in a horizontal plane. Although the earth's rotation causes the spin axis to move into the meridian, it does not settle there but tends to oscillate about it in a sinusoidal manner. The mid-position of the oscillations gives the orientation to north.

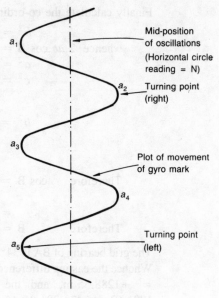

Mid-position
of oscillations
(Horizontal circle
reading = N)

Turning point
(right)

Plot of movement
of gyro mark

Turning point
(left)

Figure 4.14

Solution. First, describe the reversal method. Essentially, the theodolite and gyroscope should be directed within a few degrees of north. The spinner is activated and an index mark, indicating the behaviour of the spin axis, is observed through a viewing system against a horizontal scale and is kept centred thereon using the horizontal motion system of the theodolite. The spin axis and the line of sight of the theodolite should now be in the same vertical plane.

In the middle of an oscillation the gyromark moves quickly but it stops momentarily at a reversal, or turning point, before changing its direction. The observer must read the horizontal circle at this position, and then carry on tracking the gyro mark towards the next turning point. At least four such observations are made at successive turning points, as illustrated in Fig. 4.14, in which a_1, etc., are the horizontal circle readings. The mid-position of oscillation can be estimated as $(a_1+a_2)/2$ and in fact this approach can be used to orient the line of sight of the theodolite before beginning the series of observations.

Calculate the bearing of reference object B. The Schuler mean of three turning-point observations is given by the expression $(a_1+2a_2+a_3)/4$. Hence for the given set of observations we have the following Schuler mean values

$$\frac{146°\ 27.2' + (2 \times 151°\ 46.3') + 146°\ 28.8'}{4} = 149°\ 7.15'$$

$$\frac{151°\ 46.3' + (2 \times 146°\ 28.8') + 151°\ 44.8'}{4} = 149°\ 7.175'$$

$$\frac{146° \ 28.8' + (2 \times 151° \ 44.8') + 146° \ 30.2'}{4} = 149° \ 7.15'$$

$$\frac{151° \ 44.8' + (2 \times 146° \ 30.2') + 151° \ 43.5'}{4} = 149° \ 7.175'.$$

Mean value = 149° 7.2'.

Thus the mean horizontal circle reading implied by the turning-point observations is 149° 7.2'.

The bearing of the reference object is given by the expression $M-N+E$, in which M is the mean circle reading for the pointing on the reference object (the face-left reading being adopted) and N is the deduced circle reading for the mid-position of oscillations. C is the calibration constant of the instrument, being the horizontal angle between the telescope's line of sight at the mid-position of oscillation and the meridian at a station. It should be checked regularly against a known meridian.

Assuming that the horizontal circle is graduated in a clockwise manner, the readings indicate that reference object B is 'west' of north since the relevant reading, M, was lower than N. This means that 360° must be added to that reading.

Thus $M = 45° \ 20.5'$ (+360° 00')
$\quad \ \ N = 149° \ 7.2'$
$\quad \ \ C = +2.4'$

Therefore the bearing of AB

$= 45° \ 20.5'$ (+360° 0.0') $- 149° \ 7.2' + 2.4'$
$= 256° \ 15.7'$ clockwise from north.

4.9 Gyro-theodolite observations — transit method

Horizontal circle readings on a reference object were recorded as 297° 32.8' and 117° 32.6' face left and face right, respectively, and an approximate gyro orientation of 159° 37.0' was then established. Successive transit times and amplitudes were recorded as follows:

Times of transit:
   00 min 00.0 s   03 min 21.1 s   07 min 13.8 s   10 min 34.7 s
Amplitudes:
       +9.3   −11.6   +9.2

Determine the azimuth of the reference object given that the proportionality factor was 0.046'/s and the instrument calibration factor was −2.5'. The gyro attachment was fixed with the telescope in the face left position and the gyro scale is marked positive and negative left and right respectively of the zero mark.

Introduction. This method of observation, the transit method, requires that the telescope pointing remains unaltered, once it has been placed in a pointing within $\pm 15'$ of true north. The gyro oscillations are damped so that the gyromark remains within 15 graduations either side of the scale zero. The horizontal circle reading is recorded for the fixed pointing of the telescope, and the moving gyro mark is followed, recording the times of its transit through the zero of the graduated gyro scale and the amplitude of swing, i.e. maximum scale readings (Fig. 4.15).

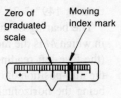

Zero of graduated scale   Moving index mark

Figure 4.15

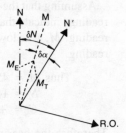

Figure 4.16

Solution. The relationship between the true and estimated meridians is needed. In this method the torques due to the earth's rotation and tape torsion play a part, the mid-position of oscillation being referenced by the resultant of the two torques, M_E and M_T, as shown in Fig. 4.16. It can be shown that

$$\delta N = \delta \alpha \, \frac{(M_E + M_T)}{M_E}.$$

Furthermore

$$\delta N = ca' \, \delta t,$$

in which c is the proportionality factor, a' is the amplitude measured in scale units, and δt is the time difference derived from the times of three successive transits. The circle reading, N', is corrected by δN to achieve the true meridian pointing, N.

Determine the bearing of the reference object. The data can be tabulated, and reduced, as in Table 4.10. We can see that the gyro mark was moving to the left at the transit time of 00 min 00.0 s since the sign of the amplitude reading between the first two transits was positive, i.e. to the left. The second transit occurred with the gyro mark moving through the scale zero towards

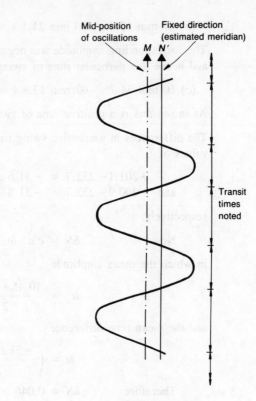

Mid-position of oscillations — M

Fixed direction (estimated meridian) — N′

Transit times noted

Figure 4.17

Table 4.10

Time of transit	Time of swing left + right (s)	Time difference ST (s)	Amplitude reading	Mean amplitude reading
00 min 00.0 s				
	+ 201.1	− 31.6	+ 9.3	
03 min 21.1 s				10.45
	− 232.7		− 11.6	
07 min 13.8 s		− 31.8		10.40
	+ 200.9		+ 9.2	
10 min 34.7 s				

the right, recording a negative amplitude. Accordingly, the times of swing have been determined as follows:

(a) 03 min 21.1 s − 00 min 00.0 s = 03 min 21.1 s = 201.1 s.

This has been booked as a positive time of swing: the gyro mark was to the left of the scale zero during this period of time.

(b) 07 min 13.8 s − 03 min 21.1 s = 03 min 52.7 s = 232.7 s.

The corresponding amplitude was negative, i.e. to the right of the scale zero and hence this particular time of swing has a negative sign.

(c) 10 min 34.7 s − 07 min 13.8 s = 03 min 20.9 s = 200.9 s.

As in (a) this is a positive time of swing.

The differences in successive swing times can now be calculated to give δt values of

$$+201.1 - 232.7 = -31.6 \text{ s}$$
$$\text{and } +200.9 - 232.7 = -31.8 \text{ s,}$$

respectively.

Now $$\delta N = c\, a'\, \delta t,$$

in which the mean amplitude

$$a' = \frac{10.45 + 10.40}{2} = 10.42,$$

and the mean time difference

$$\delta t = \left(\frac{-31.6 - 31.8}{2} \right) = -31.7.$$

Therefore $$\delta N = 0.046 \times 10.42 \times (-31.7)$$
$$= -15.2'.$$

The sign of δN depends upon the sign of δt, and the negative sign indicates that δN must be deducted from the horizontal circle reading for the fixed direction N' to determine the circle reading for the True North pointing.

Now the bearing of the reference object $= M - N + E$

$$= 297° \; 32.7' - (159° \; 37.0' - 15.2') - 2.5'$$
$$= \mathbf{138° \; 08.4' \text{ clockwise from north.}}$$

Note that since the gyro attachment was positioned with the telescope face left, the horizontal circle reading for the fixed pointing N' was observed on that face and the corresponding face reading had to be taken for the mean pointing on the reference object.

Problems

1 Two stations A and B are 773.58 m apart. From theodolite stations P and Q on opposite sides of AB the following angles were observed:

$$\text{A}\hat{\text{P}}\text{Q} = 61° \; 12' \qquad \text{B}\hat{\text{Q}}\text{P} = 53° \; 28'$$
$$\text{Q}\hat{\text{P}}\text{B} = 44° \; 11' \qquad \text{P}\hat{\text{Q}}\text{A} = 41° \; 29'$$

Calculate the distance between stations P and Q. [London]
Answer 651.36 m

2 Clockwise angles were observed in three stations A, B, and C by a theodolite set up at an eccentric station S, distant 5.204 m from, and south of, A. Angles, $\hat{ASB} = 56° \, 14' \, 10''$ and $\hat{ASC} = 142° \, 52' \, 50''$ were measured. The approximate lengths of AB and AC were estimated as 7014.4 m and 6201.2 m, respectively. Calculate the value of angle BAC.

Answer 86° 38′ 17″

3 Explain the use of a satellite station.

The following clockwise angles were observed on to three stations A, B and C with a theodolite set up at a satellite station S, distant 4.550 m from A. A, zero; B, 57° 10′ 36″; C, 131° 27′ 40″. The approximate lengths of AB and AC were estimated as 2460.0 m and 3090.0 m respectively. Calculate the value of $\hat{BAC}$.   [London]

Answer 74° 15′ 31″

4 In a triangulation network stations were established at A, B and C. The measured angle ACB was 81° 52′ 17″. The distances AC and BC determined by electronic distance measurement were 5136.25 m and 4242.73 m respectively.

At a later date the station marker at C had to be removed because of a dispute over land ownership. It was replaced by a new station marker at D 7.32 m from C such that the angle ACD was 119° 40′ 23″.

Using the theory of satellite stations calculate the angle at D subtended by A and B. State clearly any assumptions made.

Show from first principles how this answer would still be correct to the nearest second even though the angle ACD was only determined to the nearest minute.   [Bradford]

Answer 81° 52′ 54″

5 In setting out for a nuclear reactor it is necessary to establish four control points A, B, C and D (listed clockwise) which form an exact square with sides precisely 100 m long. The line from A to B runs due north.

During construction a check is required on these control points but because of site obstructions a theodolite cannot be mounted on station A although observations can be made towards it. A satellite station P is therefore established at P 315 mm away from A on a bearing 319° 33′ 15″.

Using normal satellite theory, calculate the exact anticipated theodolite readings at P when observing A, B and D. Station C is assumed to be the reference object with a reading 00° 00′ 00″.

What percentage error can be tolerated in length AB without introducing an error greater than 1″ in the angle BPA?   [Bradford]

Answer $\hat{CPB}$ = 314° 59′ 23.5″

6 In a triangulation a tall pointed spire, referred to as station C, is clearly visible from stations A and B. The horizontal distances CA and CB have been calculated from the unadjusted triangulation to be 3911 m and 3034 m respectively. A fourth station S is set up close to C so that A, B and C can be seen from it. The horizontal distance between S and a further station T is measured and found to be 22.10 m. The mean horizontal circle readings in Table 4.11 are obtained from a theodolite set up at S and T.

Table 4.11

Instrument at	Station sighted	Mean horizontal circle reading
S	C	038° 08′ 46″
	A	174° 52′ 20″
	B	232° 01′ 06″
	T	347° 23′ 34″
T	C	190° 13′ 53″
	S	255° 40′ 18″

Calculate the magnitude of the angle ACB.   [ICE]
Answer 56° 49′ 14″

7 In triangle ABC, C could not be occupied and a satellite station S was established east of C.

The angles in Table 4.12 were then registered by theodolite set up at S.

Table 4.12

Pointing	Horizontal circle reading
A	9° 27′ 15″
B	71° 04′ 42″
C	307° 38′ 15″

The lengths of AC and BC were found to be 10 566 m and 11 525 m respectively.

If the angle ACB was deduced to be 61° 37′ 05″ calculate the distance of S from C.

Estimate the length of AB and the values of angles ABC and BAC.
    [Salford]

Answer SC = 9.679 m

8 A line, of length L, following a parallel of latitude through station X, is to be located by means of offsets from a direction at 90° to the meridian at X. Show that the offsets may be determined from the expression,

$$\frac{L^2 \tan \phi}{2R},$$

where ϕ is the latitude of X and R is the local radius of the earth.

Such a line, 30 km long, is to be set out along the 53° parallel of latitude from a station. Compute data for locating points at 10 km intervals assuming that $R = 6384$ km.

Answer At 30 km, 93.54 m (at 89° 38′ 34″)

9 A line AB of length 9946 m has an azimuth of 286° 21′ 40″ at A in latitude 53° 21′ 00″ N and longitude 02° 16′ 00″ W. Line BC makes an angle of 30° 40′ 15″ clockwise from BA at B. Calculate the azimuth of BC.

Table 4.13

Latitude	Length of 1′ of latitude	Length of 1″ of longitude
53° 20′	30.9155 m	18.5065 m
53° 25′	30.9160 m	18.4704 m

[Salford]

Answer 137° 08′ 49″

10 If line AB in Problem 9 had a bearing of 73° 38′ 20″, estimate the latitude and longitude of B and the bearing of AB at B.
Answer 53° 22′ 31″, 2° 7′ 24″ W, 253° 45′ 14″

11 Derive an approximate formula for the convergence of meridians at a survey station in terms of the latitude distance from the central meridian of the station, and the local radius of the earth.

Two survey stations A and B have the following National Grid co-ordinates and latitudes:

A: E 448 315.186 N 347 987.763 Latitude N 53° 01′ 36″
B: E 448 026.224 N 347 768.921 Latitude N 53° 01′ 29″

The mean of a series of gyro theodolite observations from station A to station B gave the geographical bearing of the line AB as 233° 26′ 20″. Assuming the local radius of the earth to be 6384.100 km, compare the geographical bearing determined from National Grid co-ordinates with that obtained from the gyro theodolite observations, and state the difference in seconds of arc. Note that since the two points are close to the central meridian, an approximate formula may be used in the calculation of convergence. [CEI]
Answer 3″

12 Describe one *approximate* and one *precise* method of establishing the azimuth of a survey line by gyro theodolite observations.

The following 'transit' observations were recorded with a gyro theodolite attachment on a laboratory base line bearing 128° 17.52′.

Observations east of true north
Horizontal circle reading during transit oscillations = 15° 30.00′
Horizontal circle reading to reference object   =143° 32.45′
Times of transit: 0 min 0 s, 03 min 57.7 s, 07 min 20.5 s, 11 min 18.5 s, 14 min 41.1 s
Amplitudes: −10.8, +8.3, −10.7, +8.2.

Observations west of true north
Horizontal circle reading during transit oscillations = 15° 00.00′

Horizontal circle reading to reference object   = 143° 32.45'
Times of Transit: 0 min 0 s, 04 min 05.7 s, 07 min 20.4 s, 11 min 26.0 s, 14 min 41.2 s
Amplitudes: +7.9, −5.6, +7.9, −5.5.

Determine the additive constant and the proportionality factor for this particular attachment stating carefully the units of both.   [CEI]
Answer   +0.58'; 0.044'/s

13   Describe carefully the 'amplitude' method of observation using a gyro theodolite.

Using the Schuler mean calculate the bearing of reference object B from the following observations taken at station A using a gyro theodolite.

Horizontal circle readings to B: face left   42° 26' 15"
                                  face right   222° 26' 25"

Angular readings of successive gyro turning points were as follows:

left   276° 20.1'   right   280° 32.4'
       276° 21.6'   280° 30.8'
       276° 23.3'   280° 29.5'

The calibration constant of the instrument is +2.6'.   [Bradford]
Answer   124° 02.3'

14   Given that the co-ordinates of two survey stations A and B are ea,na and eb,nb on a local arbitrary grid and EA,NA and EB,NB on the National Grid, derive expressions for the co-ordinates of the original of the local grid relative to National Grid origin and for the sine and cosine of the angle between the two sets of co-ordinate axes.

If the co-ordinates of A and B relative to the two grids are as tabulated below, derive the National Grid co-ordinates of point P from the values of the local co-ordinates given.

Point A   ea   374.62   na   615.88
          EA 49 624.31   NA 21 315.74

Point B   eb   47.23   nb   566.37
          EB 46 676.24   NB 20 988.72

Point P   ep   573.19   np   127.38

[Bradford]
Answer   50 111.34 mE, 21 517.92 mN

15   An oil company wishes to locate accurately the position of a drilling rig. Unfortunately, no Ordnance Survey control is visible from the rig, but from a point (S) nearby, two second-order triangulation stations, Castle Hill and Beckwith Mount, may be seen. The co-ordinates of these stations are as in Table 4.14.

The approximate co-ordinates of the rig are scaled from the Ordnance Survey sheet as 3662 mE, 4400 mN. The distance from station S to the rig is carefully measured as 18.835 m and the following angles observed.

Table 4.14

	Easting (m)	Northing (m)
Castle Hill	2619.43	2803.03
Beckwith Mount	4142.49	2005.50

Beckwith Mount → Castle Hill → S   85° 03′ 11″
Castle Hill → Beckwith Mount → S   50° 36′ 32″
Rig → S → Castle Hill   108° 30′ 43″
Rig → S → Beckwith Mount   64° 10′ 26″

Calculate the co-ordinates of the rig.   [Leeds]
Answer 3661.56 mE, 4400.32 mN

16 Explain the significance of the local scale factor, indicating how and when it should be used.

If you had measured a distance between two Ordnance Survey points whose mean height is 325 m and the horizontal distance is equal to 6500 m, what length would you expect to calculate from the co-ordinates of the two points? (Curvature correction $D^3/33R^2$ if $R = 6380$ km and a LSF = 1.000 14.)

Explain why the true bearings from the two ends of the line are not 180° different.   [Leeds]
Answer 6500.579 m

17 The National Grid co-ordinates of stations A and B are as in Table 4.15.

Table 4.15

	E (m)	N (m)
A	449 674.24	321 315.74
B	449 624.31	320 988.72

Given that the local scale factor is 0.999 63, determine the spheroid distance between A and B.
Answer 331.24 m

18 A straight line AB, 18.5 km long, was set out at 90° to the meridian at A in latitude 52° 22′ N. Determine the bearing of AB at B if 1″ of longitude has a length of 18.9222 m.
Answer 90° 12′ 54″

19 The slope distances and mean reciprocally observed vertical angles of the three sides of a triangle ABC, forming part of a surface network, are given below.

     AB   1424.954 m + 06° 48′ 29″

AC 2011.865 m + 00° 17′ 14″
CB 1398.453 m + 06° 31′ 20″

Assuming that the level of station A is 91.069 m above mean sea level, the local mean radius of the earth to be 6383.393 km and the local scale factor to be 0.999 672, calculate the grid lengths of the three sides and the internal horizontal angles of the triangle. [Eng. Council]
Answer AB = 1414.403 m, Â = 43° 39′ 16″

20 Your firm has won a contract to construct some 30 km of road running approximately in an east–west direction. A control traverse is to be run, establishing stations approximately 5 km apart, between two Ordnance Survey triangulation points lying near the two ends of the road.

What corrections might have to be made to the traverse observations to achieve a satisfactory closure and calculation of National Grid co-ordinates of the control points if they are to be calculated to the nearest centimetre?

The longest line is measured as 6593.455 m between two points whose heights are 40.172 m and 168.866 m. What length should be used in calculating the co-ordinates if the curvature correction is $D^3/43R^2$ and the local SF is 1.0036?

Discuss whether any of these corrections will be required for local traverses connecting adjacent control points. [Leeds]
Answer 6615.823 m

21 Two stations have National Grid co-ordinates as in Table 4.16:

Table 4.16

	E (m)	N (m)
A	527 398.249	304 854.427
B	551 477.378	315 465.283

Determine the $(t - T)$ values for the two points.
Answer −3.64″; +3.85″

22 (*a*) Compare the essential features of the three main methods of observing azimuth with a gyro theodolite.

(*b*) Describe the procedures of measurement, and outline the data analysis required when using a gyro theodolite to determine the azimuth of:

(*i*) a line of 150 m length in a mine gallery with one end of the line beneath a vertical shaft from the surface;

(*ii*) a line above ground, approximately 30 km in length, and extending in a generally north-easterly direction. [RICS]

23 A, B and C are the corners of a triangulation survey, and O is a station within the triangle. The co-ordinates of O are 1000 E, 1000 N, of A are 260 E, 75 N and of B are 1930 E, 56 N. If the whole circle bearing of OC is 285° 10′ 20″ and CA is 187° 10′ 25″, what is

(a) the whole circle bearing of OA
(b) the whole circle bearing of BO
(c) the co-ordinates of C?

[Salford]

Answer (a) 218° 39′ 35″; (b) 315° 25′ 40″; (c) 397.00 mE, 1163.52 mN

24 In a geodetic survey the following mean angles were observed in a triangle

P 67° 46′ 16.9″
Q 57° 34′ 38.3″
R 54° 39′ 07.4″

The spherical excess of the triangle was estimated to be 1.4″. Determine the correction to be applied to each angle if they had all been observed with equal precision.

Answer Correction = −0.4″

25 Measurements were made in spherical triangle ABC as follows:

A   61° 36′ 12.4″
B   52° 22′ 24.2″
C   66° 02′ 27.2″
AB 45 986.248 m
AC 34 257.189 m.

If the radius of the earth is 6367 km calculate the correct values of the angles.

Answer Spherical excess = 3.5′; A = 61° 35′ 12.3″

26 Two radio towers are to be erected, one at a point X, latitude 56° 15′ 23″ N, longitude 4° 04′ 20″ W, and the other at a point Y, latitude 56° 06′ 39″ N, longitude 4° 07′ 41″ W. A microwave aerial is to be fitted to the top of each tower and so aligned that each aerial is precisely directed axially at the other.

Determine to the nearest second the direction in which each aerial should be oriented relative to the meridian in order that this condition may be achieved. Assume that the earth is spherical.

Answer 192° 04′ 35″ at X, 12° 01′ 48″ at Y.

5

Areas and volumes

The measurement of areas of sites, areas of cross-sections and volumes plays an important part in civil engineering construction; various examples are illustrated in this chapter.

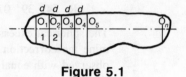

Figure 5.1

Two fundamental rules exist for the determination of areas of irregular figures typified in Fig. 5.1. Offsets, o_1, etc., have been taken at equal intercepts, d, along the length of a survey line. In the case of a narrow strip of land, this line can be run down the centre, measurements being taken either side or continuously between boundaries.

Trapezoidal rule

$$\text{Area} = d \left(\frac{o_1 + o_n}{2} + o_2 + o_3 + \dots + o_{n-1} \right).$$

In the trapezoidal rule the area is divided into trapezoids, boundaries being assumed to be straight between pairs of offsets. The areas of individual trapezoids, i.e.

$$\left(\frac{o_1 + o_2}{2} \right) d, \; \left(\frac{o_2 + o_3}{2} \right) d$$

are added together to derive the whole area.

Simpson's rule

$$\text{Area} = \frac{d}{3} (o_1 + 4o_2 + 2o_3 + \dots + 2o_{n-1} + o_n).$$

In Simpson's rule it is assumed that the irregular boundary is made up of parabolic arcs, so that the area of 1 and 2 combined is

$$\frac{d}{3} (o_1 + 4o_2 + o_3).$$

The areas of successive pairs are added together when formulating the rule. Since pairs of intercepts are taken, it will be evident that an even number of intercepts is required when using the rule to determine an area. If an odd number of intercepts is present then the first or last intercept is treated as a trapezoid. Simpson's rule gives more accurate results for a given d than does the trapezoidal rule.

Areas of straight-sided figures

In so far as straight-sided figures are concerned there are standard expressions available, for example:

Triangle   area $= \frac{1}{2}ab \sin C$,

in which C is the angle included between sides a and b.

Trapezium   area $= \dfrac{a+b}{2}p$,

in which a and b are the parallel sides, separated at perpendicular distance p.

Looped traverse   area $= \Sigma y_n (x_{n+1} - x_{n-1})$

in which x_n and y_n are the easting and northing co-ordinates of a station.

Planimeter

The planimeter, an integrating device, is available for the direct measurement of all shapes, irregular and regular, and high accuracy can be attained.

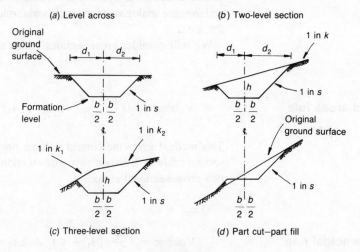

(a) Level across   (b) Two-level section

(c) Three-level section   (d) Part cut–part fill

Figure 5.2

Volumes

Volumes can be determined via cross-sections, contours or spot heights. Providing that the original ground surface is reasonably uniform in respect of the cross-fall, or gradient transverse to the longitudinal centre line, it is convenient to work from 'standard-type' cross-sections. Expressions for side-widths (d) and area (A) are easily derived, as indicated in various examples later. As an illustration, we will however consider the 'two-level' section shown in Fig. 5.2. Here

$$d_1 = \left(\frac{b}{2} + sh \right) \frac{k}{k+s}$$

$$d_2 = \left(\frac{b}{2} + sh \right) \frac{k}{k-s}$$

$$A = \frac{1}{2s} \left[\left(\frac{b}{2} + sh \right) (d_1 + d_2) - \frac{b^2}{2} \right].$$

If cross-fall k is very large compared to s (the side-slope of the excavation), i.e. the ground surface is tending towards the horizontal, we have

$$d_1 = \left(\frac{b}{2} + sh \right) = d_2$$

$$A = h(b + sh)$$

which is the 'level-across' case in Fig. 5.2.

Having computed the cross-sections at given intervals of chainage along the centre line by standard expressions, as above, or by planimeter, etc., volumes of cut in the case of excavation or volumes of fill in the case of embankments can be determined. The end-areas rule or the prismoidal rule are applicable, and they are analogous to the trapezoidal rule and Simpson's rule, respectively, for areas.

We will consider cross-sections which are D apart.

End-areas rule

$$\text{Volume} = D \left[\frac{A_1 + A_n}{2} + A_2 + A_3 + \ldots + A_{n-1} \right].$$

This method gives the correct volume providing that the area of the cross-section midway between two cross-sections D apart equals the mean of the two cross-sectional areas.

Prismoidal rule

$$\text{Volume} = \frac{2D}{6} [A_1 + 4A_2 + 2A_3 + \ldots + 2A_{n-1} + A_n].$$

The prismoidal rule is based on the assumption that the earth forms a prismoid between the two cross-sections $2D$ apart. For this to apply the linear dimensions of the section midway between them have to be the mean of the corresponding dimensions at the outer sections. It is accepted that the prismoidal rule is the more accurate, but in practice 'end-areas' is frequently adopted for estimation purposes, since some irregularities are likely to occur between various cross-sections. In addition, there is the problem of bulking and shrinking when soil is removed from its virgin state and later replaced or placed elsewhere. Quantities for the construction of mass-haul curves are normally established using 'end-areas'.

Contours and spot heights

Contours are used in a manner similar to cross-sections when calculating volumes, the contour intercept value for the scheme being adopted for D. Spot heights have a role to play in the computation of volumes of excavations for basements, etc. A network or grid of levels is taken over the site, the net or grid being such that the ground surfaces contained within the grids can be taken to be plane surfaces. Square or rectangular grids can be divided into triangles to facilitate this, and a series of triangular prisms results. The volume of such a prism, whose ends must be plane but not necessarily parallel, equals $\frac{1}{3}(h_1 + h_2 + h_3)A$ in which h_1, h_2, and h_3 are the lengths of the parallel edges and A is the area of the cross-section normal to those edges. In the case of volumes based on spot heights the triangular prisms are vertical.

5.1 Scaling from plans with an allowance for shrinkage

(a) On a certain Ordnance Survey sheet an area of 8.965 hectares on the ground covers 143.44 cm^2. What is the scale?

(b) The plan of an old chain survey, plotted to 1/500 scale on linen cloth, was found to have shrunk such that a line originally 100 mm long was now 98 mm. In addition, a footnote on the plan stated the 20 m chain used during the survey was later found to be 20.02 m long after completion of the plot. If a certain area on the plan is measured by planimeter to be 2143 mm^2 estimate the correct area on the ground. Assume uniform shrinkage.

Introduction. If the scale of a plan is 1:s then 1 mm on the plan is equivalent to s mm on the ground. Thus in so far as area is concerned 1 mm^2 on the plan is equivalent to s^2 mm^2 on the ground.

Solution. (a) Let the scale of the Ordnance Sheet be 1:s.

Then $\qquad$ 143.44 cm^2 $\equiv$ 143.44 $\times$ s^2 cm^2 on the ground.

Therefore $\quad$ 143.44 $\times$ s^2 = 89 650 $\times$ 100^2

since $\qquad$ 8.965 hectare = 89 650 m^2 and 1 m^2 = 100^2 cm^2.

Whence $\qquad\qquad\qquad s$ = 2500 (i.e. scale of sheet = **1 in 2500**).

(b) Adjust the scale. Since a line 100 mm long has 'shrunk' to 98 mm, lengths measured on the plan must be 'scaled up' by a factor $\frac{100}{98}$ to obtain the original plotted value. Hence 1 mm on the plan = 500 $\times$ $\frac{100}{98}$ mm on the ground. But the chain was found to be 20 mm too long when standardized after the survey. Thus each 'booked' length of 20 m was actually 20.02 m on the ground and a correction factor $\dfrac{20.02}{20.00}$ needs to be supplied to all linear measurements.

Therefore the ground area = $2143 \times 500^2 \times \left(\dfrac{100}{98}\right)^2 \times \left(\dfrac{20.02}{20.00}\right)^2$

$$= 558\ 956\ 700 \text{ mm}^2$$
$$= \textbf{559 m}^2.$$

5.2 Using Simpson's rule and the trapezoidal rule

State Simpson's rule and the trapezoidal rule for the determination of areas. Figure 5.3 indicates a field with two straight boundaries AB and BC and an irregular third boundary AC. The lengths of the straight lines between the stations and the offsets from AC at defined chainages from A are as follows:

Line	AB	BC	CA
Length (m)	470.0	550.0	770.0

Chainage from A (m)	0	110	220	330	440	550	660	770
Offset (m)	0	12.5	15.0	10.7	19.6	8.7	5.0	0

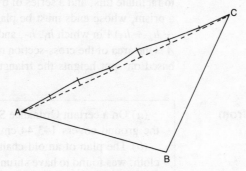

Figure 5.3

Determine the enclosed area by calculating the area of triangle ABC and adding on the area contained within the irregular boundary and straight line AC. This latter area should be calculated using both Simpson's rule and the trapezoidal rule, and a comparison made of the resulting areas.　　　　　　　　　　　　　　　　　　　　　　　　　　　　　[Salford]

Introduction. The example states that triangle ABC is to be treated separately. Such a figure is the simplest shape used in surveying. Any area enclosed by straight lines can be broken down into a number of triangles, whose areas can be aggregated to give the total area.

Solution. (*a*) First calculate the area of triangle ABC. A standard expression for the area of a triangle whose sides are known is

$$\text{area} = \sqrt{S(S-a)\ (S-b)\ (S-c)}$$

in which a, b, c, are the lengths of sides and $S = \dfrac{a+b+c}{2}$.

In ABC,　　$2S = 470.0 + 550.0 + 770.0$
　　　　　　　　$= 1790.0$ m
Therefore　　$S = 895.0$ m,
whence　　$S-a = 425.0$
　　　　　　$S-b = 345.0$
　　　　　　$S-c = 125.0$

Therefore area $= \sqrt{(895.0 \times 425.0 \times 345.0 \times 125.0)}$
$$= 128\ 076.82\ \text{m}^2$$
$$= 12.8077\ \text{hectares}.$$

(b) Next calculate the area of the irregular figure. Simpson's rule requires an even number of increments in its application, whereas the trapezoidal rule can be used with either an even number or an odd number. In this example there are seven points at which offsets have been measured. Simpson's rule is applied over the first six increments to 660.0 with the final increment from 660 m to 770 m being treated by the trapezoidal rule.

(a) Use Simpson's rule.

Table 5.1

Offset	Offset length	Simpson multiplier	Product
O_1	0	1	0.0
O_2	12.5	4	50.0
O_3	15.0	2	30.0
O_4	10.7	4	42.8
O_5	19.6	2	39.2
O_6	8.7	4	34.8
O_7	5.0	1	5.0
			201.8

Therefore Area (0 m to 660.0 m) $= \frac{110}{3} \times 201.8 \doteq 7399.3\ \text{m}^2$.

$$\text{Area (660.0 to 770.0 m)} = 110\left(\frac{5.0+0}{2}\right) = 275.0\ \text{m}^2.$$

Total area by Simpson's rule $= 7674.3\ \text{m}^2$.

(b) Use the trapezoidal rule.

$$\text{Area} = 110\left[\left(\frac{0+0}{2}\right) + 12.5 + 15.0 + 10.7 + 19.6 + 8.7 + 5.0\right]$$
$$= 110 \times 71.5$$
$$= 7865\ \text{m}^2.$$

Total area enclosed is:

135 751.1 m², i.e. **13.575 hectares** by Simpson's rule

or 135 941.8 m², i.e. **13.594 hectares** by the trapezoidal rule.

Show that the area, abcd, enclosed by a traverse having four stations $a(x_1\ y_1)$, $b(x_2\ y_2)$, $c(x_3\ y_3)$ and $d(x_4\ y_4)$ is given by the equation

$$\tfrac{1}{2}\left[y_1(x_2-x_4)+y_2(x_3-x_1)+y_3(x_4-x_2)+y_4(x_1-x_3)\right].$$

The co-ordinates of a closed traverse survey are given in Table 5.2.

Table 5.2

Station	E (m)	N (m)
A	0	0
B	+ 150	+ 300
C	+ 450	+ 70
D	+ 320	− 130
E	− 60	− 240

Calculate the area of the figure enclosed by the traverse.　　　[CEI]

Introduction. In the previous example it was pointed out that a straight-sided figure could be broken down into triangles whose individual areas could be determined, thereby leading to a total. In the specific case of a closed traverse with stations of known co-ordinates, the area can be determined directly from the co-ordinates.

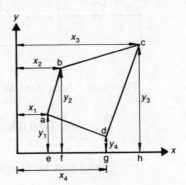

Figure 5.4

Solution. Establish the expression. Figure 5.4 shows the four stations abcd, lettered in a clockwise manner. By inspection: Area abcda = Area (abfea + bchfb − dchgd − adgea). Now

$$\text{Area abfea} = \left(\frac{y_1+y_2}{2}\right)(x_2-x_1)$$

$$\text{Area bchfb} = \left(\frac{y_2+y_3}{2}\right)(x_3-x_2)$$

$$\text{Area dchgd} = \left(\frac{y_3 + y_4}{2}\right)(x_3 - x_4)$$

$$\text{Area adgea} = \left(\frac{y_4 + y_1}{2}\right)(x_4 - x_1).$$

Therefore area abcda $= \frac{1}{2}[y_1x_2 - y_1x_1 + y_2x_2 - y_2x_1 + y_2x_3 - y_2x_2 + y_3x_3$
$- y_3x_2 - y_3x_3 + y_3x_4 - y_4x_3 + y_4x_4 - y_4x_4 + y_4x_1$
$- y_1x_4 + y_1x_1]$

$= \frac{1}{2}[y_1(x_2 - x_4) + y_2(x_3 - x_1) + y_3(x_4 - x_2)$
$+ y_4(x_1 - x_3)]$

which can also be written as

$= \frac{1}{2}[(y_1x_2 + y_2x_3 + y_3x_4 + y_4x_1)$
$- (y_1x_4 + y_2x_1 + y_3x_2 + y_4x_3)].$

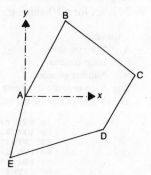

Figure 5.5

Next determine the area ABCDEA. In this example the basic expression derived above must be extended to include the fifth station, i.e.,

$$\text{Area} = \frac{1}{2}[y_1(x_2 - x_5) + y_2(x_3 - x_1) + y_3(x_4 - x_2) + y_4(x_5 - x_3) + y_5(x_1 - x_4)]$$

in which x and y refer to easting and northing co-ordinates, respectively. Figure 5.5 shows the five stations together with the co-ordinate axes x and y. It is convenient to tabulate the data with respect to the above expression, as in Table 5.3.

Therefore area $= \frac{1}{2}\Sigma\, y_n\,(x_{n+1} - x_{n-1})$

$= \frac{1}{2} \times 290\,000$

$= 145\,000\ \text{m}^2$

$= \textbf{14.50 hectares.}$

In the table n stands for the particlar station under consideration, $(n+1)$ relating to the next station and $(n-1)$ relating to the preceding station as lettered within

Table 5.3

Station (n)	y_n (m)	x_n (m)	x_{n+1} (m)	x_{n-1} (m)	$y_n(x_{n+1}-x_{n-1})$ (m²)
A	0	0	+150	−60	0 (+150 + 60) = 0
B	+300	+150	+450	0	300 (+450 − 0) = +135 000
C	+70	+450	+320	+150	70 (+320 − 150) = + 11 900
D	−130	+320	−60	+450	−130 (− 60 − 450) = + 66 300
E	−240	−60	0	+320	−240 (0 − 320) = + 76 800
					+ 290 000

the traverse. A further point of interest is that if the traverse is lettered in an anti-clockwise direction a negative total arises in the product column. Naturally the negative sign is ignored.

The following computer program will calculate the area within a closed-loop traverse from the co-ordinates of the stations. The DIM statement in line 20 is set for a 10-sided traverse, the minimum requirements are X(N+1), Y(N).

Variables
A   Traverse area
I   Loop counter
N   Number of sides
X(i)   X co-ordinate of station i
Y(i)   Y co-ordinate of station i

```
10 REM AREA INSIDE A TRAVERSE
20 DIM X(11),Y(10)
30 INPUT"NUMBER OF STATIONS ",N
40 FOR I=1 TO N
50 PRINT"INPUT CO-ORDINATES OF STATION ;"I
60 INPUT X(I),Y(I)
70 NEXT I
80 X(0)=X(N)
90 X(N+1)=X(1)
100 FOR I=1 TO N
110 A=A+(Y(I)*(X(I+1)-X(I-1))/2)
120 NEXT I
130 A=ABS(A)
140 PRINT"AREA =";A;"M2"
150 END
```

5.4 Splitting a traverse into two areas

A site is defined by the lines of traverse ABCDEA. It is to be crossed by a straight boundary QP on a bearing of 17° 39′. Q lies on DE and P on BC. Calculate the co-ordinates of P and Q which will give an area of 30 000 m² for PCDQP. Co-ordinates are given in Table 5.4.

Table 5.4

Station	A	B	C	D	E
E	350	500	900	650	470
N	400	650	650	150	270

[Bradford]

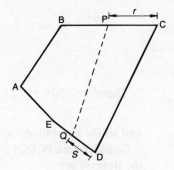

Figure 5.6

Introduction. It is unnecessary to calculate the whole area ABCDEA in this instance since the area of PCDQP has been specified. However, in certain problems when, say, PQ has to divide the area into equal parts it can form the first step of the solution. The computer program listed with Example 5.3 could be used to perform the task.

In this example there are two unknown sets of co-ordinates which depend upon lengths r and s in Fig. 5.6. The solution is based, first, on the relationship between r and s and then, second, using that relationship to evaluate the co-ordinates of either P or Q, knowing the area PCDQP.

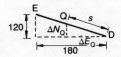

Figure 5.7

Solution. First establish the relationship between distances r and s. Consider Q which lies on DE at a distance s from D (see Fig. 5.6). Then, by similar triangles, Fig. 5.7,

$$\frac{\Delta E_Q}{180} = \frac{s}{DE} \quad \text{and} \quad \frac{\Delta N_Q}{120} = \frac{s}{DE}$$

in which

$$(DE)^2 = (650-470)^2 + (270-150)^2.$$

Therefore   DE = 216.33 m.

Whence $\Delta E_Q = 0.832 \times s$ and $\Delta N_Q = 0.555 \times s$ and so the co-ordinates of Q are $(650-0.832 \times s)$ E, $(150+0.555 \times s)$ N.

Also, since P lies at a distance r from C, its co-ordinates are $(900-r)$ E, 650 N.

Now, the bearing of QP is 17° 39′, and so

$$\tan 17° 39′ = 0.318\ 18$$

$$= \frac{E_P - E_Q}{N_P - N_Q}$$

$$= \frac{(900-r) - (650-0.832 \times s)}{650 - (150+0.555 \times s)}$$

Therefore $250 - r + (0.832 \times s) = 0.318\ 18\ [500 - (0.555 \times s)]$

$$r = 90.91 + (1.009 \times s)$$

and so the co-ordinates of P are $(809.09 - 1.009s)$ E, 650 N.

Consider area PCDQP. The co-ordinates of the stations of this section of the traverse are

Table 5.5

	Station			
	P	C	D	Q
E(m)	$(809.09 - 1.009s)$	900	650	$(650 - 0.832s)$
N(m)	650	650	150	$(150 + 0.555s)$

The area enclosed by the traverse lines is given by the expression $\frac{1}{2}\Sigma y_n (x_{n+1} - x_{n-1})$ and, so on, tabulating we have the values in Table 5.6.

Table 5.6

Stn (n)	y_n	x_n	$x_{n+1}-x_{n-1}$	$y_n (x_{n+1}-x_{n-1})$
P	650	$(809.09 - 1.009s)$	$900 - (650 - 0.832s)$	$162\ 500.0 + 540.80s$
C	650	900	$650 - (809.09 - 1.009s)$	$-103\ 408.5 + 655.85s$
D	150	650	$(650 - 0.832s) - 900$	$-37\ 500.0 - 124.80s$
Q	$(150 + 0.555s)$	$(650.0 - 0.832s)$	$(809.09 - 1.009s) - 650$	$23\ 863.5 - 63.06s\ -0.56s^2$
				$45\ 455.0 + 1008.79s\ -0.56s^2$

Now, area

$$\text{PCDQP} = \tfrac{1}{2}\Sigma y_n (x_{n+1} - x_{n-1})$$
$$= 30\ 000\ \text{m}^2.$$

Therefore $45\ 455.0 + 1008.79s - 0.56s^2 = 60\ 000$,

i.e. $s = 14.54$ m

and $r = 90.91 + 1.009s$
$$= 105.58\ \text{m}.$$

From this:

co-ordinates of P $= (900 - r)$ E, 650 N
$$= 794.42\ \text{E},\ 650\ \text{N};$$

$$\text{co-ordinates of } Q = (650 - 0.832s)\text{E}, \ (150 + 0.555s)\text{N}$$
$$= 637.91 \text{ E}, \ 158.07 \text{ N}$$

These could be rounded off as

P **794 mE, 650 mN**
Q **638 mE, 158 mN,**

since the traverse co-ordinates were given to 10 m units. The reader can check that the 'rounded-off' co-ordinates result in an area of 30 075 m^2 for area PCDQP.

5.5 Cross-sections and volumes from cutting and embankment details

Prove that on uniformly sloping ground the cross-section area of a cutting or an embankment is given by the following formula:

$$\frac{1}{s}\left(d_1\, d_2 - \frac{b^2}{4}\right),$$

where d_1 and d_2 are the horizontal distances from the centre line to the limits of the side slopes, b is the formation width of the cutting or embankment and 1 vertical in s horizontal is the gradient of the side slopes.

An embankment for a new road is to be 20 m wide at the top, with side slopes of 1 vertical to 3 horizontal. The heights of fill at the centre line of three successive cross-sections, 50 m apart, are 2.4 m, 3.2 m and 4.0 m respectively and the existing ground has a uniform cross fall of 1 in 12. Calculate the volume of fill required.   [Salford]

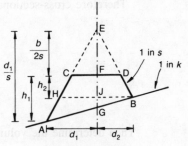

Figure 5.8

Introduction. In Fig. 5.8 the side slopes are shown meeting the original ground surface at A and B so that the side widths are d_2 and d_1, respectively, and h_2 and h_1 denote the differences in height between these points and the formation level. Points E, F and G, lie on the centre line of the embankment.

Now
$$d_2 = \frac{b}{2} + sh_2$$

and
$$h_2 = h - \frac{d_2}{k} \text{ where } h = \text{FG}.$$

Therefore $d_2 = \dfrac{b}{2} + s\left(h - \dfrac{d_2}{k}\right)$

$$= \left(\dfrac{b}{2} + sh\right)\left(\dfrac{k}{k+s}\right).$$

Similarly $d_1 = \dfrac{b}{2} + sh_1$

and $\qquad h_1 = h + \dfrac{d_1}{k}.$

Therefore $d_1 = \left(\dfrac{b}{2} + sh\right)\left(\dfrac{k}{k-s}\right).$

ACDBA is known as a 'two-level' section because two observations of height will define cross fall 1 in k.

Solution. Derivation of the general expression for the cross-sectional area.
Area of cross-section ABDCA = area of triangle AEBA − area of triangle CEDC

$$= \left(\dfrac{BH}{2} \times \dfrac{d_1}{s}\right) - \left(\dfrac{2b}{2} \times \dfrac{EF}{2}\right).$$

Now $\qquad\qquad$ BH = HJ + JB = $2d_2$

and $\qquad\qquad$ EF = $\dfrac{b}{2s}.$

Therefore cross-sectional area $= \left(\dfrac{2d_2}{2} \times \dfrac{d_1}{s}\right) - \left(\dfrac{2b}{2} \times \dfrac{b}{2s} \times \dfrac{1}{2}\right)$

$$= \dfrac{d_1 d_2}{s} - \dfrac{b^2}{4s}$$

$$= \dfrac{1}{s}\left(d_1 d_2 - \dfrac{b^2}{4}\right).$$

Then determine the volume between the cross-sections.

In the expressions $\qquad d_1 = \left(\dfrac{b}{2} + sh\right)\left(\dfrac{k}{k-s}\right)$

$$d_2 = \left(\dfrac{b}{2} + sh\right)\left(\dfrac{k}{k+s}\right)$$

and $\qquad\qquad A = \dfrac{1}{s}\left(d_1 d_2 - \dfrac{b^2}{4}\right)$

the following are given: $s = 2$, $k = 12$ and $b = 20$ m; h has values of 2.4 m, 3.2 m and 4.0 m, respectively, at 50 m intervals.

It is convenient to tabulate the data as follows in Table 5.7.

Table 5.7

Cross-section	h (m)	sh (m)	$\dfrac{b}{2} + sh$ (m)	d_1 (m)	d_2 (m)	$d_1 d_2$ (m²)	$\dfrac{b^2}{4}$ (m²)	$A = \dfrac{1}{s}\left(d_1 d_2 - \dfrac{b^2}{4}\right)$ (m²)
1	2.4	7.2	17.2	22.93	13.76	315.52	100.00	71.84
2	3.2	9.6	19.6	26.13	15.68	409.72	100.00	103.24
3	4.0	12.0	22.0	29.33	17.60	516.21	100.00	138.74

By the 'end-areas' rule

$$V = D\left(\frac{A_1 + A_3}{2} + A_2\right)$$

$$= 50\left(\frac{71.84 + 138.74}{2} + 103.24\right)$$

$$= 10\ 426.50 \text{ m}^3, \text{ say } \mathbf{10\ 426\ m^3}.$$

By the prismoidal rule

$$V = \frac{2D}{6}(A_1 + 4A_2 + A_3)$$

$$= \frac{100}{6}[71.84 + (4 \times 103.24) + 138.74]$$

$$= 10\ 392.33 \text{ m}^3, \text{ say } \mathbf{10\ 392\ m^3}.$$

By means of a correction known as the prismoidal correction (PC), the volume (V_P) determined by the prismoidal rule can be deduced directly from the volume (V_{EA}) determined by the 'end-areas' rule, since $V_{EA} - PC = V_P$. For the 'two-level' section

$$PC = \frac{D}{6}\left(\frac{k^2}{k^2 - s^2}\right)s(h_1 - h_2)^2.$$

In this example

$$\frac{D}{6}\left(\frac{k^2}{k^2 - s^2}\right) = 26.67$$

$$(h_1 - h_2) = -0.8 \text{ m and } (h_2 - h_3) = -0.8 \text{ m}.$$

Therefore $PC = 2 \times 26.67 \times (-0.8)^2$

$$= 34.14 \text{ m}^3.$$

Therefore $V_p = 10\ 426.50 - 34.14$

$$= 10\ 392.36 \text{ m}^3, \text{ say } \mathbf{10\ 392\ m^3}$$

Prismoidal corrections can only be derived for regular sections, as illustrated in Fig. 5.2.

5.6 Calculating the area of a 'three-level' section

Calculate the area of cut in section 1 in Fig. 5.9 when BE = 1.0 m.

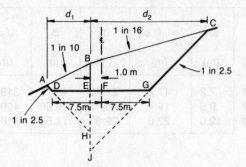

Figure 5.9

The areas of further sections at successive 20 m intervals along the line of a straight road have been calculated. These are:

Section	2	3	4	5
Area of cut (m²)	20.3	21.7	28.8	30.2

Calculate the volume of cut between sections 1 and 5 by the end-areas and the prismoidal methods. [Bradford]

Introduction. Section 1 shown in Fig. 5.9 is an example of the 'three-level' section. The original ground surface has two crossfalls, requiring three reduced levels for their determination. Had the change occurred on the centre line, the standard expressions for side-widths and area could have been used directly, substituting the k values as applicable. In this example the change occurs off the centre line, but line BE can be used as a substitute centre line for two cross-sections whose half-widths at formation level are:

$$DE = (7.5 - 1.0) = 6.5 \text{ m}$$
$$EG = (7.5 + 1.0) = 8.5 \text{ m}.$$

Solution. First determine the side widths. The two relevant expressions for side widths are

$$d_1 = \left(\frac{b_1}{2} + sh\right)\left(\frac{k_1}{k_1 + s}\right)$$

and

$$d_2 = \left(\frac{b_2}{2} + sh\right)\left(\frac{k_2}{k_2 - s}\right)$$

in which $\frac{b_1}{2} = 6.5$ m, $k_1 = 10$, $s = 2.5$ and $h = 1.0$ m and $\dfrac{b_2}{2} = 8.5$ m, $k_2 = 16$.

Therefore $d_1 = (6.5 + 2.5 \times 1.0) \left(\dfrac{10}{10 + 2.5} \right) = 7.20$ m

$$d_2 = (8.5 + 2.5 \times 1.0) \left(\dfrac{16}{16 - 2.5} \right) = 13.04 \text{ m.}$$

Next determine the area of section 1.

$$\text{Area ABCGDA} = \text{area ABHA} - \text{area DEHD} + \text{area BJCB} - \text{area EJGE}$$

$$= (\tfrac{1}{2} d_1 \times \text{BH}) - (\tfrac{1}{2}\text{DE} \times \text{EH})$$
$$+ (\tfrac{1}{2} d_2 \times \text{BJ}) - (\tfrac{1}{2}\text{EG} \times \text{EJ}).$$

Now
$$\text{BH} = \text{BE} + \text{EH} = \text{BE} + \dfrac{\text{DE}}{s} = 1.0 + \dfrac{6.5}{2.5}$$
$$= 3.6 \text{ m}$$
$$\text{EH} = 2.6 \text{ m}$$

$$\text{BJ} = \text{BE} + \text{EJ} = \text{BE} + \dfrac{\text{EG}}{s} = 1.0 + \dfrac{8.5}{2.5}$$
$$= 4.4 \text{ m}$$
$$\text{EJ} = 3.4 \text{ m.}$$

Therefore area ABCGDA $= (\tfrac{1}{2} \times 7.20 \times 3.6) - (\tfrac{1}{2} \times 6.5 \times 2.6)$
$$+ (\tfrac{1}{2} \times 13.04 \times 4.4) - (\tfrac{1}{2} \times 8.5 \times 3.4)$$
$$= 12.95 - 8.45 + 28.69 - 14.45$$
$$= 18.75 \text{ m}^2, \text{ say } \mathbf{18.8 \text{ m}^2.}$$

Now calculate the volume of cut. There are five cross-sections and so the prismoidal rule can be applied directly.

Now
$$V_{\text{P}} = \dfrac{2D}{6} (A_1 + 4A_2 + 2A_3 + 4A_4 + A_5).$$

Therefore $V_{\text{P}} = \dfrac{2 \times 20}{6} [18.8 + (4 \times 20.3) + (2 \times 21.7)$
$$+ (4 \times 28.8) + 30.2]$$
$$V_{\text{P}} = 1925.3 \text{ m}^3, \text{ say } \mathbf{1925 \text{ m}^3}$$

$$V_{\text{EA}} = \text{D} \left(\dfrac{A_1 + A_5}{2} + A_2 + A_3 + A_4 \right)$$

$$= 20 \left(\dfrac{(18.8 + 30.2)}{2} + 20.3 + 21.7 + 28.8 \right)$$

$$= \mathbf{1906 \text{ m}^3.}$$

A road having a formation width of 15 m is to be constructed in ground having a cross fall of 1 in 10. The depths of fill at the centre lines of cross-sections X and Y are 0.25 m and 0.38 m, respectively. Given that the side slopes are 1 vertical to 1 horizontal in cut and 1 vertical to 2 horizontal in fill, determine the volume of cut and fill between X and Y which are 20 m apart, allowing for prismoidal excess.

Introduction. This example refers to a cross-section which is part in cut and part in fill, and the respective volumes will be calculated initially using the end-areas methods. Prismoidal excess (or the prismoidal correction) will be applied to give the volumes which would be derived by the prismoidal rule, assuming that the dimensions of the cross-section midway between X and Y are the mean of the corresponding dimensions at X and Y. The first stage will be to establish general expressions for the areas of cut and fill from Fig. 5.10.

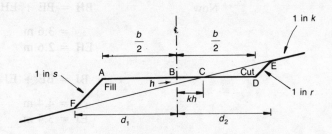

Figure 5.10

Solution. First calculate the sidewidths and areas. Consider the difference in level between formation ABCD and points E and F which define the limits of the sidewidths.

$$\text{Rise from C to E} = h_2 = \frac{d_2 - kh}{k}$$

$$= \text{Rise from D to E}$$

$$= \frac{d_2 - b/2}{r}.$$

Therefore $\dfrac{d_2 - kh}{k} = \dfrac{d_2 - b/2}{r}$

$$d_2 = \left(\frac{b}{2} - rh\right)\left(\frac{k}{k - r}\right).$$

In a similar manner, for F,

$$h_1 = \frac{d_1 + kh}{k} = \frac{d_1 - b/2}{s}.$$

Therefore $d_1 = \left(\dfrac{b}{2} + sh\right)\left(\dfrac{k}{k-s}\right)$.

The cross-sectional areas are as follows:

Area in cut $= \frac{1}{2} \times \text{CD} \times h_2$

$\qquad = \frac{1}{2}\left(\dfrac{b}{2} - kh\right)\left(\dfrac{d_2 - b/2}{r}\right)$

$\qquad = \frac{1}{2}\left(\dfrac{b}{2} - kh\right)^2\left(\dfrac{1}{k-r}\right)$.

Area in fill $= \frac{1}{2} \times \text{AC} \times h_1$

$\qquad = \frac{1}{2}\left(\dfrac{b}{2} + kh\right)\left(\dfrac{d_1 - b/2}{s}\right)$

$\qquad = \frac{1}{2}\left(\dfrac{b}{2} + kh\right)^2\left(\dfrac{1}{k-s}\right)$.

Note that if h had been the depth of cut at the centre line rather than fill, the insertion of $-h$ for h in the above expressions would give the respective formulae for side widths and areas.

Next, calculate the volumes of cut and fill. As mentioned previously the relevant volumes will be determined initially by end-areas.

Table 5.8

Section	h	k	kh	$\left(\dfrac{b}{2} - kh\right)$	$\left(\dfrac{b}{2} + kh\right)$	$\dfrac{1}{k-r}$	$\dfrac{1}{k-s}$	Area (m²)	
	(m)		(m)	(m)	(m)			cut	fill
X	0.25	10	2.50	5.00	10.00	$\frac{1}{9}$	$\frac{1}{8}$	1.39	6.25
Y	0.38	10	3.80	3.70	11.30	$\frac{1}{9}$	$\frac{1}{8}$	0.76	7.98

In Table 5.8, $b = 15$ m, $r = 1$ and $s = 2$. By the end-area method

volume of cut between X and Y $= D\left(\dfrac{A_1 + A_2}{2}\right)$

$\qquad\qquad = 20\left(\dfrac{1.39 + 0.76}{2}\right)$

$\qquad\qquad = 21.5 \text{ m}^3$,

volume of fill between X and Y $= 20\left(\dfrac{6.25 + 7.98}{2}\right)$

$\qquad\qquad = 142.3 \text{ m}^3$.

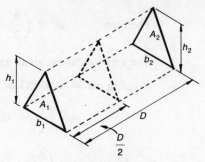

Figure 5.11

Establish the prismoidal correction. Consider two parallel cross-sections, triangular in shape, D apart as in Fig. 5.11, their areas being A_1 and A_2. Also consider the cross-section midway between them, formed by the straight connecting them.

Then $A_1 = \frac{1}{2} b_1 h_1 \qquad A_2 = \frac{1}{2} b_2 h_2$.

A_m = area of mid-section

$$= \frac{1}{2}\left(\frac{b_1+b_2}{2}\right)\left(\frac{h_1+h_2}{2}\right).$$

The volumes contained between the outer cross-sections are: by the end-area method

$$V_{EA} = \frac{D}{2}\left(\frac{b_1h_1}{2} + \frac{b_2h_2}{2}\right).$$

By the prismoidal rule

$$V_P = \frac{D}{6}\left[\frac{b_1h_1}{2} + \frac{4}{2}\left(\frac{b_1+b_2}{2}\right)\left(\frac{h_1+h_2}{2}\right) + \frac{b_2h_2}{2}\right].$$

Now, the prismoidal correction or prismoidal excess is $V_{EA} - V_P$

$$= \frac{D}{2}\left(\frac{b_1h_1}{2} + \frac{b_2h_2}{2}\right) -$$

$$\frac{D}{6}\left[\frac{b_1h_1}{2} + 2\left(\frac{b_1+b_2}{2}\right)\times\left(\frac{h_1+h_2}{2}\right) + \frac{b_2h_2}{2}\right]$$

$$= \frac{D}{12}(b_1-b_2)(h_1-h_2).$$

This is the standard expression for triangular cross-sections. Expressions for other cross-sectional shapes can be derived by breaking them down into triangles, and using the above.

Apply the prismoidal correction. The prismoidal correction has been shown to be

$$\frac{D}{12}\left(\begin{array}{c}\text{difference in}\\ \text{base lengths}\end{array}\right)\left(\begin{array}{c}\text{difference in}\\ \text{heights}\end{array}\right)$$

for the triangular section. Using the expressions previously derived for the cut section

$$\text{base length} = CD = \frac{b}{2} - kh$$

$$\text{and height} = h_2 = \frac{d_2 - b/2}{r}$$

in which $d_2 = \left(\frac{b}{2} - rh\right)\left(\frac{k}{k-r}\right)$.

Now, $b/2 = 7.50$ m, $k = 10$ and $r = 1$. Whence, on tabulating, we get the values in Table 5.9.

Table 5.9

Cross-section	h (m)	$\frac{b}{2} - kh$ (m)	$\frac{b}{2} - rh$ (m)	d_2 (m)	h_2 (m)
X	0.25	5.00	7.25	8.06	0.56
Y	0.38	3.70	7.12	7.91	0.41

$$\text{Therefore prismoidal correction} = \frac{20}{12}(5.00 - 3.70)(0.56 - 0.41)$$

$$= 0.33 \text{ m}^3.$$

Therefore volume when treating the earth between the two sections as a prismoid

$$= 21.5 - 0.3$$
$$= \mathbf{21.2 \text{ m}^3}.$$

Similarly for the fill

$$\text{base length} = AC = \frac{b}{2} + kh$$

$$\text{height} = h_1 = \frac{d_1 - b/2}{s}$$

in which $d_1 = \left(\frac{b}{2} + sh\right)\left(\frac{k}{k-s}\right)$.

With $s = 2$ we get the results in Table 5.10.

$$\text{Therefore prismoidal correction} = \frac{20}{12}(10.00 - 11.30)(1.25 - 1.41)$$

$$= 0.35 \text{ m}^3.$$

Table 5.10

Cross-section	h (m)	$\left(\dfrac{b}{2}+kh\right)$ (m)	$\left(\dfrac{b}{2}+sh\right)$ (m)	d_1 (m)	h_1 (m)
X	0.25	10.00	8.00	10.00	1.25
Y	0.38	11.30	8.26	10.32	1.41

Therefore volume of fill $= 142.3 - 0.3$
$\qquad\qquad\qquad\qquad\qquad\quad = \mathbf{142.0\ m^3}.$

5.8 Calculating volumes along a curved section of road

In a cutting the width of the formation is 8 m and the side slopes are 1 in 2. The depths at the centre line, which lies on a circular curve of radius 150 m, at three cross-sections 20 m apart are 2.50 m, 3.10 m and 4.30 m, respectively. The cutting is to be widened by increasing the formation width to 11 m, the excavation being on the outside of the curve and retaining the original side slopes.

Calculate the volume of excavation between the cross-sections using the end-areas method. Assume that the transverse slope of the ground is 1 in 5 at each cross-section.

Introduction. In previous examples it has been assumed that the cross-sections were parallel and at right angles to a straight centre line. When the centre line lies in a curve, as in this example, the cross-sections are set out radially at stations established at chainages measured around the curve.

Use is made of Pappus's theorem when calculating volumes. This states that the volume swept out by an area revolving about an axis is given by the product of the length of travel of the centroid of the area and of the area, provided that the area is in the plane of the axis and to one side thereof. Therefore it is necessary to locate the centroid positions when estimating the volumes of earthworks.

Solution. First amend the end-areas and prismoidal formulae. If the cross-sections are unsymmetrical, e.g. have transverse ground slopes, but have the same shape and dimensions, the path of the centroids will be parallel to the centre line. When there is some variation, as in this example, the path is not parallel to the centre line. Some close approximations are made when using either the end-areas of prismoidal formulae. For example, consider the three successive cross-sections, D apart on the centre line, in Fig. 5.12, the centroids of the 'widening' areas being e_1, e_2 and e_3, respectively, from the centre line. The mean distances of the centroids of the two pairs of sections are $(e_1+e_2)/2$ and $(e_2+e_3)/2$ from that centre line. These values are taken to represent all respective centroid positions within the relevant cross-sections and therefore define the path.

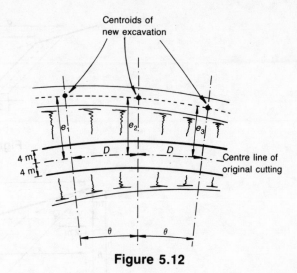

Figure 5.12

If θ is the angle subtended at the centre of the curve by the pairs of equidistant cross-sections

$$\theta = \frac{D}{R}.$$

Therefore the distance between centroids $= \theta \left(R + \dfrac{e_1 + e_2}{2} \right)$ and

$$\theta \left(R + \frac{e_2 + e_3}{2} \right)$$

$$= D \left(1 + \frac{e_1 + e_2}{2R} \right) \text{ and}$$

$$D \left(1 + \frac{e_2 + e_3}{2R} \right).$$

Had the widening been in the inside of the curve the above would change to

$$D \left(1 - \frac{e_1 + e_2}{2R} \right), \text{ and } D \left(1 - \frac{e_2 + e_3}{2R} \right).$$

An alternative approach is to treat each area separately, amending it by the factor $1 \pm (e_1/R)$, etc. Thus in this particular example if the areas of widening are A_1, A_2 and A_3 then the areas would be taken to be $A_1[1 + (e_1/R)]$, $A_2[1 + (e_2/R)]$ and $A_3[1 + (e_3/R)]$ spaced at intervals of D when using the end-areas or prismoidal formulae.

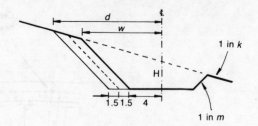

Figure 5.13

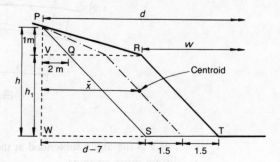

Figure 5.14

Next locate the centroids. In Figs 5.13 and 5.14

$$h = \frac{d}{k} + H = \frac{d-7}{m}$$

$$\frac{d+5H}{5} = \frac{d-7}{2}$$

Therefore $\qquad d = \dfrac{10H+35}{3}.$

Similarly, $\dfrac{w}{5} + H = \dfrac{w-4}{2}.$

Therefore $\qquad w = \dfrac{10H+20}{3}.$

Also $\qquad 2h = d-7.$

Therefore $\qquad h = \dfrac{10H+14}{6}.$

Also QRTS $= 3h_1$ m², and so we get the values (in metres) given in Table 5.11. It will be noted that $(d-w)$ is constant, and so triangle PQR has fixed dimensions, its height PV being $5.0/k = 1.0$ m, with QR = 3 m.

Therefore area PQR $= \frac{1}{2} \times 1.0 \times 3.0 = 1.5$ m².

Its centroid is $\frac{2}{3} \times 3.5$ m from PW, since VQ = $(5-3)$ m.

Table 5.11

H	w	d	d−w	d−7	h	h_1	$3h_1$
2.5	15.0	20.0	5.0	13.0	6.5	5.5	16.5
3.1	17.0	22.0	5.0	15.0	7.5	6.5	19.5
4.3	21.0	26.0	5.0	19.0	9.5	8.5	25.5

Area QRTS $= 3h_1$ m^2.

Its centroid is

$$d - \left(4.0 + 1.5 + \frac{2h_1}{2} \right) = (d - h_1 - 5.5) \text{ m}$$

from PW.

Therefore $\bar{x} = \dfrac{(1.5 \times \frac{2}{3} \times 3.5) + 3h_1 (d - h_1 - 5.5)}{1.5 + 3h_1}$.

From this we get the values in Table 5.12. We can now calculate the volume of excavation.

Table 5.12

H (m)	d (m)	h_1 (m)	$(d - h_1 - 5.5)$ (m)	$A = (1.5 + 3h_1)$ (m^2)	$\bar{x}$ (m)	$e = d - \bar{x}$ (m)
2.5	20.0	5.5	9.0	18.0	8.44	11.56
3.1	22.0	6.5	10.0	21.0	9.45	12.55
4.3	26.0	8.5	12.0	27.0	11.46	14.54

Using the end-areas method

$$V = \left(\frac{A_1 + A_2}{2} \right) D \left(\frac{1 + e_1 + e_2}{2R} \right) + \left(\frac{A_2 + A_3}{2} \right) D \left(\frac{1 + e_2 + e_3}{2R} \right)$$

$$= \left(\frac{18.0 + 21.0}{2} \right) 20 \left(1 + \frac{11.56 + 12.55}{2 \times 150} \right) +$$

$$\left(\frac{21.0 + 27.0}{2} \right) \times 20 \left(1 + \frac{12.55 + 14.54}{2 \times 150} \right)$$

$$= 944.7 \text{ m}^3, \text{ say } \textbf{945 m}^3.$$

Alternatively, the areas can be amended as in Table 5.13 overleaf.

$$V = \frac{20}{2} [19.39 + (2 \times 22.76) + 29.62]$$

$$= \textbf{945 m}^3.$$

Table 5.13

h (m)	A_2 (m^2)	e (m)	$A(1 + e/R)$ (m^2)
2.5	18.0	11.56	19.39
3.1	21.0	12.55	22.76
4.3	27.0	14.54	29.62

Had the influence of curvature been neglected the calculated volume would be 870 m^3.

5.9 Volumes from a rectangular grid of spot heights

Figure 5.15 shows the distribution of 12 spot heights with a regular 10 m spacing covering a rectangular area which is to be graded to form a horizontal plane. The spot height data (in metres) are:

1	17.06;	2	17.48;	3	17.63
4	17.37;	5	17.70;	6	17.96
7	17.58;	8	18.01;	9	18.25
10	17.83;	11	18.19;	12	18.42

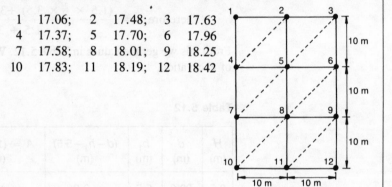

Figure 5.15

Neglecting any effects of side slopes determine a suitable design level such that cut and fill are balanced.   [London]

Introduction. The solution to this example will be based on the fact that the volume of a triangular prism is equal to the area of its normal section multiplied by the mean length of side. Dividing the square grid of Fig. 5.15 into triangles, as shown, produces a set of vertical triangular prisms, one plane end being at formation level and the other being the ground plane contained between the corners of each triangle. Squares could be used rather than triangles, but the ground plane should be better represented by the latter unless the square grid size is small.

The volume of excavation in triangle 124 will be $\dfrac{A}{3}$ $(d_1+d_2+d_4)$ m in which A is the plan area of the triangle and d_1, d_2 and d_4 are the depths from ground level to formation level at the corners. For triangle 245 the volume of excavation is $\frac{A}{3}(d_2+d_4+d_5)$, and so on.

It will be noted that d_4 appears in each of the volumes as it does in the volume of the prism related to triangle 457: d_5 appears in six triangles. Thus, rather than evaluating the volume of each triangular prism and then computing the grand total, it is more convenient to establish an expression based on the number of times particular depths are used.

$$\text{Total volume} = \frac{A}{3}(\Sigma D_1 + 2\Sigma D_2 + 3\Sigma D_3 + \ldots + \ldots n\Sigma D_n),$$

in which A is the plan area of a triangle, ΣD_1 is the sum of depths used once, ΣD_2 is the sum of depths used twice, ΣD_3 is the sum of depths used three times, etc.

Had squares (or rectangles) been used four depths would arise in each prism and the volume would be given by $\frac{a}{4}(\Sigma D_1 + 2\Sigma D_2 + 3\Sigma D_3 + 4\Sigma D_4)$ in which a is the plan area with D, etc., as before.

Solution. Determination of volume of excavation. In the first instance a general expression for the volume of excavation over this particular grid will be derived, assuming that the formation level is x m.

Table 5.14

Station	Depth of cut D (m)	Number of triangles in which station occurs n	Product $D \times n$
1	$17.06-x$	1	$17.06-x$
2	$17.48-x$	3	$52.44-3x$
3	$17.63-x$	2	$35.26-2x$
4	$17.37-x$	3	$52.11-3x$
5	$17.70-x$	6	$106.20-6x$
6	$17.96-x$	3	$53.88-3x$
7	$17.58-x$	3	$52.74-3x$
8	$18.01-x$	6	$108.06-6x$
9	$18.25-x$	3	$54.75-3x$
10	$17.83-x$	2	$35.66-2x$
11	$18.19-x$	3	$54.57-3x$
12	$18.42-x$	1	$18.42-x$
			$\Sigma D \times n = 641.15-36x$

It will be noted that in Table 5.14 the multiplying factors 2, 3 and 6 have been used directly, rather than adding the depths used twice, three times and six times, respectively, and then multiplying by those factors.

$$\text{Area } A \text{ of each triangle} = \frac{10 \times 10}{2}\ \text{m}^2.$$

Therefore volume of excavation $= \frac{1}{3} \times \dfrac{10 \times 10}{2} \Sigma Dn.$

Thus if the formation level was to be 10 m we could write $x = 10$ m and

$$V = \tfrac{50}{3}(641.15 - 36 \times 10)$$
$$= 4685.83 \text{ m}^3.$$

In practice, of course, it is simpler to deduct x from the original ground levels so as to enter specific depths of cut in the table for each station, i.e. 7.06 m for station 1, etc., when computing volumes.

However, in respect of the actual question, and assuming no bulking or shrinking of the excavation material, there will be no residual volume of excavation after grading has been carried out since it is required that cut balances fill. Hence

$$V = \tfrac{50}{3}(641.15 - 36x)$$
$$= 0$$

Therefore $x = 17.81$ m.

The required design level is therefore **17.81 m.**

The volumes for this type of problem can be calculated by the following computer program. If the formation level is input as 10.0 m then the value 4686 m^3 is output as above. This particular problem would need a trial and error solution, altering the formation level until the volume was zero. Lines 120 and 150 make use of the logic functions AND and OR; on computers that do not support these functions line 120 must be replaced by four separate IF ... THEN GOTO 150 statements and line 150 by

```
150 IF I = 1 THEN 153
151 IF I = X THEN 153
152 GOTO 160
153 IF J = 1 THEN 180
154 IF J = Y THEN 180
```

Variables

H	Height from formation to ground level at grid point	V	Final volume
		V1	Formation level
I, J	Counters	V2	Level of a point on the grid
L1	Grid interval X direction	X	Number points X direction
L2	Grid interval Y direction	Y	Number points Y direction
S	Sum of heights with appropriate multiplier		

```
10 REM VOLUMES FROM A GRID OF LEVELS
20 INPUT"GRID INTERVAL IN X DIRECTION ",L1
30 INPUT"GRID INTERVAL IN Y DIRECTION ",L2
40 INPUT"NUMBER OF POINTS IN X DIRECTION ",X
50 INPUT"NUMBER OF POINTS IN Y DIRECTION ",Y
60 INPUT"FORMATION LEVEL ",V1
70 FOR I=1 TO X
80 FOR J=1 TO Y
90 PRINT"INPUT LEVEL AT GRID POINT";I;J
100 INPUT V2
110 H=V2-V1
```

```
120 IF I=1 OR I=X OR J=1 OR J=Y THEN GOTO 150
130 S=S+(4*H)
140 GOTO 190
150 IF (I=1 OR I=X) AND (J=1 OR J=Y) THEN GOTO 180
160 S=S+(2*H)
170 GOTO 190
180 S=S+H
190 NEXT J
200 NEXT I
210 V=INT((L1*L2*S/4)+0.5)
220 PRINT" VOLUME =";V
230 END
```

5.10 Volume of a truncated cone

Excavation is required for the construction of a tank, the existing ground sloping at 1 in 5. The tank is to have a diameter of 12 m at formation level, its centre being 3.50 m below ground level. If the side slopes are at 2 vertical to 1 horizontal, determine the volume of excavation to the nearest m^3.

Introduction. The trace of the perimeter of the excavation at ground level (see Fig. 5.16) will be elliptical, the major axis DF being in the direction of maximum slope (1 vertical to 5 horizontal) of the ground. The centre of the ellipse is at G.

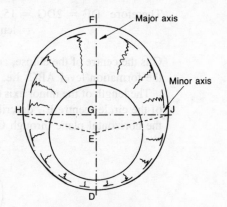

Figure 5.16

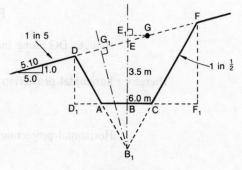

Figure 5.17

Figure 5.17 indicates an oblique cone of height B_1G_1 whose base is the ellipse. The required volume of excavation is the difference between the volume of that cone and the volume of the cone of height BB_1 on a circular base 12 m in diameter. To determine the volume of the oblique cone the lengths of the major axis and minor axis of the ellipse have to be found.

Solution. First calculate the lengths of major and minor axes. From Fig. 5.17, in which the ground slope is 1 vertical in 5 horizontal,

$$\frac{FF_1 - 3.50}{CF_1 + 6.00} = \frac{1}{5} = \frac{3.50 - DD_1}{6.00 + AD_1}.$$

But $CF_1 = FF_1/2$ and $AD_1 = DD_1/2$ since the side slopes are 2 vertical to 1 horizontal.

Whence $\qquad FF_1 = 5.22$ m and $DD_1 = 2.09$ m.

Also $\qquad \dfrac{DF}{D_1F_1} = \dfrac{5.10}{5.00}$

and $\qquad\qquad D_1F_1 = 12.00 + \dfrac{5.22}{2} + \dfrac{2.09}{2}$

$$= 15.66 \text{ m}$$

Therefore $\quad DF = 2DG = 15.97$ m

$$= \text{length of major axis.}$$

G is the centre of the ellipse, and thus it is at a height of $(FF_1 + DD_1)/2$ above the formation level ABC, i.e. 3.66 m.

The length of the minor axis of the ellipse is equal to the length of the chord of the circle, centre E_1, described by the intersection of the side slopes with the horizontal plane through G (Fig. 5.18).

Figure 5.18

Length DG along the slope $= \dfrac{15.97}{2}$ m.

Therefore horizontal projection of DG $= \dfrac{15.97}{2} \times \dfrac{5.00}{5.10}$

$$= 7.83 \text{ m.}$$

Horizontal projection of DE $= BD_1$

$$= 6.00 + \frac{2.09}{2}$$

$$= 7.05.$$

Therefore $\qquad GE_1 = 7.83 - 7.05 = 0.78$ m.

Since the side slopes of the excavation are 2 vertical to 1 horizontal and GE_1 is 3.66 m above formation level, the radius of the circle is

$$\left(6.00 + \frac{3.66}{2}\right) = 7.83 \text{ m.}$$

Therefore minor axis $= 2\sqrt{(7.83^2 - 0.78^2)}$

$$= 15.58 \text{ m.}$$

Next calculate the volume of excavation. In Fig. 5.17

$$\frac{BB_1}{AB} = \frac{2}{1}.$$

Therefore $\qquad BB_1 = 12$ m

and $\qquad EB_1 = EB + BB_1 = 3.50 + 12 = 15.50$ m.

B_1G_1 is perpendicular to DF, being the height of the oblique cone.

Therefore $\qquad B_1G_1 = 15.50 \times \dfrac{5.00}{5.10}$

$$= 15.20 \text{ m.}$$

Volume of excavation $= \dfrac{\pi}{3} \text{ GF} \times \text{GH} \times B_1G_1 - \dfrac{\pi}{3} \times AB^2 \times BB_1$

$$= \frac{\pi}{3} \times \frac{15.97}{2} \times \frac{15.58}{2} \times 15.20 -$$

$$\frac{\pi}{3} \times 6^2 \times 12$$

$$= 537.72 \text{ m}^3, \text{ say } \mathbf{538 \text{ m}^3}.$$

5.11 Triangular ground surface model

Two stations were established on the top of a spoil heap. The following readings were recorded using a total station instrument sighting on to points at the base of the heap. The height of instrument and height of prism were equal for both sets of readings.

From A the values in Table 5.15 were obtained.

Table 5.15

Point	Horizontal circle	Horizontal distance (m)	Difference in level (m)
1	220° 10′ 11″	50.23	−5.24
2	271° 21′ 5″	41.69	−4.93
3	312° 42′ 19″	37.28	−4.72
4	14° 36′ 43″	46.17	−5.03
5	94° 12′ 33″	48.33	−5.35
6	173° 31′ 57″	20.13	+0.24

From B the values in Table 5.16 were obtained.

Table 5.16

Point	Horizontal circle	Horizontal distance (m)	Difference in level (m)
1	118° 18′ 12″	42.89	−4.86
6	181° 8′ 47″	28.72	+0.62
5	236° 38′ 22″	58.96	−4.97
7	313° 41′ 19″	45.41	−5.25
8	52° 25′ 38″	34.35	−5.31

What is the volume above a horizontal plane 5.8 m below station A?

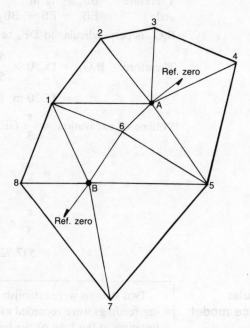

Figure 5.19

Introduction. Figure 5.19 shows the spoil heap; the reader should note that the top of the heap would be truncated if point 6 did not lie on the ridge joining stations A and B. The volume can be calculated by establishing a ground surface model consisting of a network of triangles. The volume under each triangle is the plan area of the triangle multiplied by the mean height of the corners above datum.

Solution. To calculate the plan area of the triangles we have the horizontal length of two sides and the included angle, and

$$\text{Area} = \tfrac{1}{2} \, ab \sin C.$$

For triangle A12:

$$\text{angle} = 271° \ 21' \ 5'' - 220° \ 10' \ 11''$$
$$= 51° \ 10' \ 54''$$
$$\text{area} = \tfrac{1}{2} \times 50.23 \times 41.69 \times \sin 51° \ 10' \ 54''$$
$$= 815.7914 \ \text{m}^2.$$

$$\text{Height of 1} = 5.8 - 5.24 = 0.56$$
$$\text{Height of 2} = 5.8 - 4.93 = 0.87$$
$$\text{Height of A} = 5.8 \qquad\qquad = \underline{5.8}$$
$$7.23$$
$$\div 3 = 2.41.$$
$$\text{Volume} = 815.7914 \times 2.41 = 1966.06 \ \text{m}^2.$$

The calculation is best computed as a table, noting that B is 0.38 m lower than A when the level differences for Point 1 are considered.

Table 5.17

Triangle	Angle	L_1	L_2	Area	H_1	H_2	H_3	H	Volume
A12	51° 10′ 54″	50.23	41.69	815.7914	0.56	0.87	5.8	2.41	1966.06
A23	41° 21′ 14″	41.69	37.28	513.4372	0.87	1.08	5.8	2.58	1324.67
A34	61° 54′ 24″	37.28	46.17	759.2133	1.08	0.77	5.8	2.55	1935.99
A45	79° 35′ 50″	46.17	48.33	1097.3590	0.77	0.45	5.8	2.34	2567.82
A56	79° 19′ 24″	48.33	20.13	478.0203	0.45	6.04	5.8	4.10	1959.88
A61	46° 38′ 14″	20.13	50.23	367.5563	6.04	0.56	5.8	4.13	1518.01
B16	62° 50′ 35″	42.89	28.72	548.0033	0.56	6.04	5.42	4.01	2197.49
B65	55° 29′ 35″	28.72	58.96	697.7012	6.04	0.45	5.42	3.97	2769.87
B57	77° 2′ 57″	58.96	45.41	1304.6343	0.45	0.17	5.42	2.01	2622.31
B78	98° 44′ 19″	45.41	34.35	770.8633	0.17	0.11	5.42	1.90	1464.64
B81	65° 52′ 34″	34.35	42.89	672.3010	0.11	0.56	5.42	2.03	1364.77
									$\Sigma = 21\ 692 \ \text{m}^3$

The following computer program can be used to solve this problem. The DIM statement in line 20 is set for 20 readings at any one station. For computers with limited memory this can be reduced, as appropriate; the value 6 is required for this problem. The program is designed to work with ground and formation levels. To solve this example input the formation level as zero, and the heights as positive values above formation, i.e. level of station A = 5.8, B = 5.42, point 1 = 0.56, etc.

Important: for any station, input should start with the point with the lowest horizontal circle reading and proceed in increasing values of horizontal circle, i.e. for A Point 4, 5, 6, ... 3; for B Point 8, 1, 6, ... 7. This allows the instrument to be referenced on to an object or bearing that is not one of the points, since line 270 calculates the included angle either side of the zero reference line.

Variables

A(i) Horizontal circle reading to point i

A Included angle in triangle

D Input angle, degrees

H(i) Level of point i

H1 Height of instrument above formation

I, J Loop counters

L(i) Horizontal length to point i

M Input angle, minutes

N1 Number of stations

N2 Number of readings at station under consideration

P Plan area of triangle

S Input angle, seconds

V Volume

V1 Formation level

V2 Level of station under consideration

V3 Level of point under consideration

```
10 REM VOLUMES FROM TRIANGLES
20 DIM A(20),L(20),H(20)
30 INPUT"FORMATION LEVEL ",V1
40 INPUT"HOW MANY INSTRUMENT STATIONS ",N1
50 FOR I=1 TO N1
60 PRINT"HOW MANY READINGS AT STATION";I;
70 INPUT N2
80 PRINT"INPUT LEVEL OF STATION";I;
90 INPUT V2
100 H1=V2-V1
110 FOR J=1 TO N2
120 PRINT"HORIZ. CIRCLE READING TO POINT";J;"IN DEG,MIN,SEC ";
130 INPUT D,M,S
140 A(J)=((D*3600)+(M*60)+S)/206264.8
150 PRINT"INPUT DISTANCE FROM STATION";I;"TO POINT";J;
160 INPUT L(J)
170 PRINT"INPUT LEVEL OF POINT";J;
180 INPUT V3
190 H(J)=V3-V1
200 NEXT J
210 FOR J=1 TO (N2-1)
220 A=A(J+1)-A(J)
230 P=L(J+1)*L(J)*0.5*SIN(A)
240 H=(H1+H(J)+H(J+1))/3
250 V=V+(H*P)
260 NEXT J
270 A=2*3.1415962 -A(N2)+A(1)
280 P=L(N2)*L(1)*0.5*SIN(A)
290 H=(H1+H(N2)+H(1))/3
300 V=V+(H*P)
310 NEXT I
320 PRINT"VOLUME =";V
330 END
```

5.12 Calculating heights

A digital terrain model is being established, the characteristic points taking the form of a square grid of side L, shown in Fig. 5.20. Show that the level, Z_Q, of Q may be expressed by

$$Z_Q = \frac{Z_A(L-x)(L-y)}{L^2} + \frac{Z_B x(L-y)}{L^2} + \frac{Z_C xy}{L^2}$$
$$+ \frac{Z_D y(L-x)}{L^2}.$$

Spot heights at the nodes of squares of 20 m side were obtained during the photogrammetric survey for a motorway, and Fig. 5.21 shows heights established in the vicinity of the site PQR of a proposed permanent subsidiary works unit. Determine the heights of P, Q and R and hence make a first approximation of the volume of fill required to raise the site level to 57.3 m above datum. The co-ordinates within the squares are given in Table 5.18.

Table 5.18

Point	X (m)	Y (m)
P	10.0	16.0
Q	6.0	7.0
R	13.0	17.0

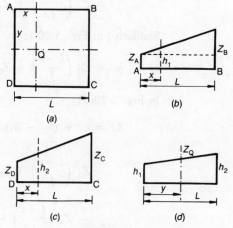

Figure 5.20

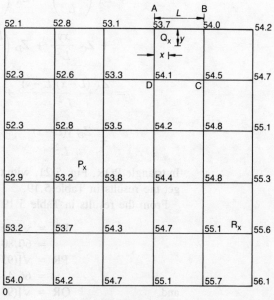

Figure 5.21

Introduction. A digital terrain model (DTM) is a statistical sampling of the x, y and z co-ordinates of the ground. By interpolation, z co-ordinates can be estimated when the corresponding x and y co-ordinates are known. The position of the nodal points of the model have to be such that the terrain is well represented.

Solution. Interpolating linearly between A and B, in Fig. 5.20 (*b*)

$$h_1 = Z_A + (Z_B - Z_A) \frac{x}{L}$$

$$= Z_A \left(\frac{L-x}{L} \right) + Z_B \frac{x}{L}.$$

Similarly, in Fig. 5.20 (*c*)

$$h_2 = Z_D \left(\frac{L-x}{L} \right) + Z_C \frac{x}{L}$$

In Fig. 5.20(*d*)

$$Z_Q = h_1 + (h_2 - h_1) \frac{y}{L}$$

$$= Z_A \left(\frac{L-x}{L} \right) + Z_B \frac{x}{L} + \frac{y}{L} \left[Z_D \left(\frac{L-x}{L} \right) + \frac{Z_C x}{L} \right.$$

$$\left. - Z_A \left(\frac{L-x}{L} \right) - Z_B \frac{x}{L} \right]$$

$$= Z_A \left(\frac{L-x}{L} \right) - Z_A \left(\frac{L-x}{L} \right) \frac{y}{L} + Z_B \frac{x}{L} - Z_B \frac{xy}{L^2}$$

$$+ Z_C \frac{xy}{L^2} + Z_D \left(\frac{L-x}{L} \right) \frac{y}{L}$$

$$= \frac{Z_A (L-x)(L-y)}{L^2} + \frac{Z_B x(L-y)}{L^2} + \frac{Z_C xy}{L^2}$$

$$+ \frac{Z_D y(L-x)}{L^2}.$$

In triangle PQR, Fig. 5.21, with the origin for the grid co-ordinates at 0 we get the results in Table 5.19.

From the results in Table 5.19 we get

$$PQ = \sqrt{[(66.0 - 30.0)^2 + (93.0 - 44.0)^2]}$$
$$= 60.80 \text{ m},$$
$$PR = \sqrt{[(93.0 - 30.0)^2 + (23.0 - 44.0)^2]}$$
$$= 66.41 \text{ m},$$

and
$$QR = \sqrt{[(93.0 - 66.0)^2 + (23.0 - 93.0)^2]}$$
$$= 75.03 \text{ m}.$$

Table 5.19

Point	x co-ordinate (m)	y co-ordinate (m)
P	30.0	44.0
Q	66.0	93.0
R	93.0	23.0

Now $PQ + PR + QR = 2S = 202.24$ m
and so
$$S = 101.12 \text{ m}$$
$$S - PQ = 40.32 \text{ m}$$
$$S - PR = 34.71 \text{ m}$$
$$S - QR = 26.09 \text{ m.}$$

Therefore area of triangle $PQR = \sqrt{(101.12 \times 40.32 \times 34.71 \times 26.09)}$
$$= 1921.5 \text{ m}^2.$$

For the calculation of Z_P, Z_Q and Z_R we get the values in Table 5.20.

Table 5.20

Point	x (m)	y (m)	L − x (m)	L − y (m)	(L − x) (L − y) (m²)	x(L − y) (m²)	xy (m²)	y(L − x) (m²)
P	10.0	16.0	10.0	4.0	40.0	40.0	160.0	160.0
Q	6.0	7.0	14.0	13.0	182.0	78.0	42.0	98.0
R	13.0	17.0	7.0	3.0	21.0	39.0	221.0	119.0

From the values in Table 5.20 we get

$$Z_P = \frac{(52.8 \times 40) + (53.5 \times 40) + (53.8 \times 160) + (53.2 \times 160)}{400}$$
$$= 53.4 \text{ m}$$

$$Z_Q = \frac{(53.7 \times 182) + (54.0 \times 78) + (54.5 \times 42) + (54.1 \times 98)}{400}$$
$$= 53.9 \text{ m}$$

$$Z_R = \frac{(54.8 \times 21) + (55.3 \times 39) + (55.6 \times 221) + (55.1 \times 119)}{400}$$
$$= 55.4 \text{ m}$$

Let h_P, h_Q and h_R be the depths of fill required at P, Q and R, respectively.

$$h_P = 57.3 - 53.4 = 3.9 \text{ m}$$
$$h_Q = 57.3 - 53.9 = 3.4 \text{ m}$$
$$h_R = 57.3 - 55.4 = 1.9 \text{ m.}$$

$$\text{Therefore volume of fill} = \frac{A}{3}\Sigma h$$

$$= \frac{1921.5}{3} \times (3.9+3.4+1.9)$$

$$= 5892.6 \text{ m}^3$$
$$= \textbf{5893 m}^3\textbf{, say.}$$

5.13 Mass-haul diagram

The volumes in cubic metres between successive cross-sections 50 m apart along a proposed road are given below, positive volumes denoting cut and negative volumes denoting fill.

Chainage (m)	2000	2050	2100	2150	2200	2250	2300
Volume (m³)	−2400	−2700	−1700	+2100	+2500	+2400	

Chainage (m)	2300	2350	2400	2450	2500	2550	2600
Volume (m³)	+2000	+1500	+400	−700	−2600	−2800	

Plot a mass-haul diagram for this length of road, assuming that the earthworks were balanced at 2000 m. Indicate the positions of balancing lines such that for:

Scheme 1 there is balance at chainage 1000 with borrow at chainage 2600,

Scheme 2 there is equal borrow at chainages 2000 and 2600.

Determine the costs arising in the two schemes using the following rates

Excavate, cart and fill within a free-haul distance of 200 m	£1.20/m³
Excavate, cart and fill for overhaul	£1.70/m³
Borrow and fill at chainage 2000	£2.20/m³
Borrow and fill at chainage 2600	£2.40/m³

What change occurs in the cost of Scheme 2, should a rate of £0.30/m³ × 50 m be proposed for overhaul?

Introduction. A mass-haul diagram or curve can be drawn subsequent to the calculation of earthwork volumes, its ordinates showing cumulative volumes at specific points along the centre line. In Fig. 5.22 the ordinates have been plotted from base line abc such that ordinate gg_1 represents the volume of fill between a and g. Volumes of cut and fill are treated as positive and negative, respectively; compensation can be made as necessary, for shinkage or bulking of the excavated material when placed finally in an embankment.

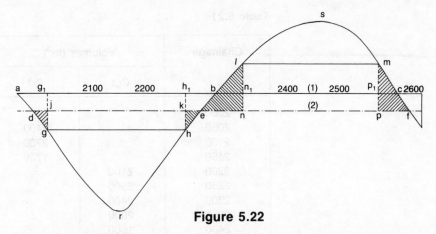

Figure 5.22

Referring to Fig. 5.22, which has been drawn using the data of the example, it will be noted that:

- (*a*) decreasing aggregate volumes, i.e. a to g, s to m, imply the formation of an embankment;
- (*b*) a minimum point occurs in the curve at the end of an embankment, e.g. r;
- (*c*) when the curve rises cut is involved, i.e. from r towards b;
- (*d*) a maximum ordinate will occur on the mass-haul curve at the end of a cut, i.e. at s.

If a horizontal line is drawn, i.e. gh, the ordinates of the mass-haul curve are equal at g and h, and so the volumes of cut and fill balance over that length. When the curve lies below that trace line earth is moved to the left, i.e. h–r–g, and, similarly, when the curve lies above, earth is moved to the right, i.e. b–s–c. The length of such a line, termed a balancing line, indicates the maximum distance that earth will be transferred within the particular loop of the diagram formed by that line. It will be seen that the base line gives continuous balancing lines ab and bc, but continuity is not essential, the balancing lines being arranged to ensure the most economical solution.

Haul is defined as the total of the products of increments of volumes of cut and their distances of travel to the embankment, i.e. area grh gives the haul in length gh. Two other terms are of importance: namely free haul and overhaul. Their definitions are given later in the solution.

Solution. First determine the accumulated volumes. From the data, accumulated volumes can be tabulated as in Table 5.21 and then plotted as in Fig. 5.22.

We now plot the balancing lines for the schemes, Scheme 1 first. For no borrow (or, for that matter, no surplus) at chainage 2000 the balancing must commence at the origin (*a*) of the mass-haul curve, the earthworks having balanced to that chainage. Accordingly, being horizontal, the balancing lines ab, bc will pass through b and c to give a borrow requirement of 2000 m^3 at chainage 2600.

For Scheme 2, the accumulated volume at chainage 2600 is -2000 m^3, and therefore to give equal borrow at chainage 2000 and 2600 the balancing

Table 5.21

Chainage	Volumes (m³)		Accumulated volume (m³)
	Cut	Fill	
2000			0
2050		2400	−2400
2100		2700	−5100
2150		1700	−6800
2200	2100		−4700
2250	2500		−2200
2300	2400		+ 200
2350	2000		+2200
2400	1500		+3700
2450	400		+4100
2500		700	+3400
2550		2600	+ 800
2600		2800	−2000

line must bisect that particular ordinate to give 1000 m³ borrow at the two chainages. This is satisfied by balancing lines passing through d, e and f, as shown.

Finally we estimate the costs incurred. The rate for excavation generally allows for transport up to a specified distance, known as the free-haul distance or free haul. For transport beyond this distance we have overhaul and a different unit rate applies. Free-haul distances can be plotted on the mass-haul diagram as shown in Fig. 5.22, balancing lines gh and lm being 200 m long in this case. The total volumes of excavation involved are given by the intercepts from gh to r and from lm to s. Since cut balances fill over the length of a balancing line earth would be carted a maximum distance of 200 m from h to g and l to m, respectively.

The volumes remaining, gg_1 and ln_1 in Scheme 1, and gj with ln in Scheme 2, i.e. the ordinates from the respective balancing lines, are the volumes to which the unit cost of overhaul will refer.

In Scheme 1, balancing line abc:

$$\text{free-haul volumes (based on gh and lm)} = (6800 - 2100) + (4100 - 1700)$$
$$= 7100 \text{ m}^3,$$
$$\text{overhaul volumes (intercepts } gg_1 \text{ and } ln_1) = 2100 + 1700$$
$$= 3800 \text{ m}^3,$$
$$\text{borrow at chainage } 2600 = 2000 \text{ m}^3.$$
$$\text{Cost} = (7100 \times 1.20) + (3800 \times 1.70) + (2000 \times 2.40)$$
$$= £19\ 780.$$

In Scheme 2, balancing line def:

$$\text{free-haul volumes (based on gh and lm)} = (6800 - 2100) + (4100 - 1700)$$
$$= 7100 \text{ m}^3,$$
$$\text{overhaul volumes (intercepts gj and ln)} = (2100 - 1000) + (1700 + 1000)$$
$$= 3800 \text{ m}^3,$$

borrow at chainage 2000 = 1000 m^3,
borrow at chainage 2600 = 1000 m^3.

Cost = (7100 × 1.20)+(3800 × 1.70)+(1000 × 2.20)+(1000 × 2.40)
= **£19 580.**

A further method of costing earthworks involves the use of a rate for overhaul based on unit volume multiplied by distance, referred to as a station distance and generally meaning the interval at which the ordinates are plotted on the mass-haul diagram. This rate is 'extra over' the rate for free-haul movement.

It will be remembered that haul is defined by volume times distance and that haul within a loop is given by the area of that loop on the mass-haul diagram. In Scheme 2 the areas for free haul are gjkhrg and nlsmpn, and so the corresponding areas for the calculation of overhaul are (dehrgd − gjkhrg) and (elsmfe − nlsmpn). The cost of overhaul is obtained by multiplying these areas by the rate for overhaul and adding the result to the costs of 'free-haul' volumes.

In Fig. 5.22 the overhaul may be taken as being

$$= \left(\frac{jg}{2} \times dj\right) + \left(\frac{kh}{2} \times ke\right) + \left(\frac{ln}{2} \times en\right) + \left(\frac{np}{2} \times pf\right)$$

$$= \left(\frac{1100}{2} \times 25\right) + \left(\frac{1100}{2} \times 20\right) + \left(\frac{2700}{2} \times 65\right)$$

$$+ \left(\frac{2700}{2} \times 45\right) m^4$$

$$= 173\ 250\ m^4.$$

Hence:

freehaul volume (based on intercepts from de to r and ef to s, respectively)
= (6800 − 1000) + (4100 + 1000)
= 10 900 m^3;
borrow at chainage 2000
= 1000 m^3;
borrow at chainage 2600
= 1000 m^3;
cost = (10 900 × 1.20) + (1000 × 2.20) + (1000 × 2.40)

$$+ \left(173\ 250 \times \frac{0.30}{50}\right)$$

= **£18 720.**

Problems

1 A road of formation width 12 m is to be constructed with side slopes of 1 vertical to 2 horizontal in cut and 1 vertical to 3 horizontal in fill. The existing ground surface has a cross-fall of 1 vertical to 8 horizontal and it will intersect the formation 1.5 m to the left of the centre line and 1.0 m to the right of the centre line at two cross-sections 20 m apart.

Calculate the areas related to the cut and fill at the two sections. Hence calculate the outstanding volume of cut or fill remaining after the establishment of the formation between the two sections.

[Bradford]

Answer Fill 2.025 m², cut 4.688 m²; fill 4.900 m², cut 2.083 m²; fill 1.54 m³

2 A straight road is being constructed, its formation width being 16 m and its side slopes being 1 vertical in 2.5 horizontal. Two sections, 20 m apart, are in cutting, the details in Table 5.22 applying. Calculate the

Table 5.22

Section	Transverse cross — fall	Depth to formation at centre line
A	1 vertical in 10 horizontal	2.2 m
B	1 vertical in 8 horizontal	4.6 m

volume of cut between the two sections by both the end-area and prismoidal methods.

[Bradford]

Answer 1951.6 m³ (end areas); 1893.9 m³ (prismoidal)

3 An embankment on a line due north and laid on level ground has a uniform height of 16 m. The width at the top is 30 m and at the base 80 m. A road, formed at ground level is to be cut through the embankment with WCB of 60°. The width of the cutting at ground level is to be 25 m, and the sides are to slope at a gradient of 1 vertical to $1\frac{1}{2}$ horizontal. Calculate the volume of material in m³ to be removed from the embankment in forming the cutting for the road. Check your calculations by means of an accurately scaled plan and cross-sectional drawings, and draw an isometric view of the cutting.   [Eng. Council]

Answer 46 096 m³

4 The values in Table 5.23 relate to a traverse survey. Find, in hectares, the area of the figure enclosed by the traverse.   [CEI]

Table 5.23

Station	E (m)	N (m)
A	0	0
B	E 168.83	N 146.88
C	E 306.09	S 14.10
D	E 266.71	S 38.26
E	E 177.74	S 105.56

Answer 3.832 hectares

5 Data from a looped traverse ABCDEA is given below in Table 5.24. The area is to be divided into two equal parts by a line through D which will intersect AB at X.

Table 5.24

Line	AB	BC	CD	DE	EA
ΔE	30	−140	50	−20	80
ΔN	−90	−20	60	70	−20

Calculate the distance BX.　　　　　　　　　　　　[Bradford]
Answer 32.34 m

6 The co-ordinates of 3 points are given in Table 5.25. A line PQ is to be set out parallel to AC where P lies on AB and Q lies on CB and the area APQC is 1200 m^2. Calculate the lengths AP and CQ.

Table 5.25

Point	E	N
A	50.000	100.000
B	200.000	200.000
C	150.00	100.000

[Bradford]

Answer AP = 23.11 m; CQ = 14.33 m

7 (*a*) Give reasons why the estimation of areas and volumes is important in most engineering schemes.

(*b*) Describe the planimeter and explain briefly how it is used to measure areas from plans.

(*c*) State Simpson's rule for the determination of areas and mention the assumption which underlies the rule.

(*d*) In a survey of a field enclosed by a fence, offsets were taken to the fence from a chain line as follows:

Chainage (m)	0	10	20	30	40	50	60	70	80	90
Offset (m)		4.41	6.61	9.08	11.14	11.20	9.16	7.08	4.82	2.56	0

Use Simpson's rule to determine the area between the fence and the chain line.　　　　　　　　　　　　[Salford]
Answer 641.5 m^2

8 A certain cutting ABCDEA has a formation width of 14 m and is to be made in ground which has transverse slopes BA and BC falling away from B respectively at 1 vertical in 8 horizontal and 1 vertical in 14 horizontal. B is 2.6 m above formation ED and is 5.8 m horizontally from E. If the side slopes from D and E are at 1 vertical to 2 horizontal determine the cross-sectional area of the cutting.

If the area of the next section 20 m along the centre line is 45 m^2 calculate the volume of cut between the sections by the end-areas method.　　　　　　　　　　　　[Bradford]
Answer 38.24 m^2; 832.4 m^3

9 Two successive cross-sections of a 10 m wide road formation have been constructed part in cut and part in fill, the original ground having a traverse slope of 1 vertical in 5 horizontal. At one section there is 0.40 m cut at the centre line and at the other 0.26 m of fill.

Given that the respective side slopes of cut and fill are 1 vertical in 1.5 horizontal and 1 vertical in 2 horizontal determine the net volume of earthworks contained between the two sections, which are 20 m apart. The centre line of the road has a radius of 160 m in plan.

Answer 2.23 m^3 (cut)

10 A circular tank of diameter 10 m is to be constructed on a plane ground surface which has a maximum slope of 1 in 8. The centre point of the tank is to be 2.75 m above existing ground level, being placed on fill having a side slope of 1 vertical to 2 horizontal.

Determine the volume of fill required (to the nearest m^3).

Answer 602 m^3

11 The centre line of a highway cutting lies in a circular curve in plan. This cutting is to be widened by increasing the formation width of 20 m to 26 m, the excavation being on the inside of the curve and retaining the side slopes of 2 horizontal to 1 vertical. The ground surface and the formation are each horizontal and the depth to formation over a length of 400 m increases uniformly from 3 m to 5 m at the centre line.

Determine the radius of the centre line, if the volume of excavation is overestimated by 5% when the influence of curvature is neglected.

[Salford]

Answer 342 m

12 A highway of 30 m formation width and with side slopes of 1 in 1.5 on embankments and 1 in 1.75 in cuttings, falls with a gradient of 1 in 75 longitudinally. The original ground slopes at 1 in 150 in the opposite direction, but is level in a direction at right angles to the centre line of the highway.

At a certain point the highway is on an embankment 2.7 m in height measured at the centre line. Calculate the quantities of earthworks for a distance of 250 m from this point in the direction of the falling highway and plot the results in the form of a mass-haul diagram. [Salford]

Answer +955.6 m^3 at 250 m

13 A tank, whose horizontal base ABCD is 50 m long by 40 m wide, is to be constructed in ground having a slope of 1 in 14 in the direction of the larger dimensions BA and CD. The depths of excavation at A and D are to be 5.7 m, the side slopes being 1 vertical to 2 horizontal. Calculate the volume of excavation required.

Answer 27 007 m^3

14 The provisional centre line for a new road has been pegged out but no further survey operations have yet taken place.

(a) Describe how you would construct a mass-haul diagram, including any field measurements and calculations that may be necessary.

(b) Give one application of a mass-haul diagram:
   (i) at the design stage of a highway project,
   (ii) during the construction stage of a highway project.

[RICS]

15 Plot a mass-haul diagram for a proposed road of length 1500 m given the cross-sectional areas at intervals of 100 m indicated in Table 5.26.

Table 5.26

Chainage (m)	Cross-sectional area (m^2)
0	0
100	+ 20.2
200	+ 36.2
300	+ 55.6
400	+ 61.7
500	+ 40.8
600	+ 15.6
700	+ 0
800	− 30.5
900	− 58.7
1000	− 72.3
1100	− 45.6
1200	− 30.0
1300	− 22.5
1400	− 13.0
1500	0

The ground between the given values of cross-sectional area can be assumed to be uniform. The volume between the sections should be calculated using the 'mean area method'. Positive values of area denote cut and negative values denote areas of fill. It is intended to re-use the excavated material for filling and to allow for consolidation a shrinkage factor of 0.9 should be adopted.

Determine:

(i) the accumulated volume at the end of the proposed road;

(ii) the accumulated volume coinciding with the end of the cutting operation;

(iii) the haul distance over which 10 000 m^3 of excavated material has to be moved to an area of fill so that the quantities of cut and fill balance. [Salford]

Answer (i) − 6551 m^3; (ii) 20 709 m^3; (iii) 610 m

16 Estimate the cost arising in the previous question when the following rates apply:

Excavate, cart and fill within a free haul distance of 500 m
£1.40/m^3

Overhaul £0.70/m^3 × 100 m

Borrow and fill £2.50/m^3

Answer £70 280

17 Figure 5.23 shows a 20 m square grid of levels set out on an open-cast site from a temporary bench mark. The enclosed area is to be excavated to the level of the top of a coal seam known to be 52.50 m AOD at point E, and to dip downwards in the direction EH at a gradient of 1 vertically to 16 horizontally.

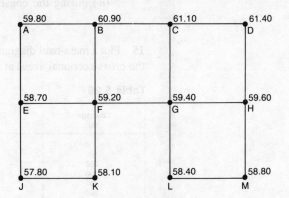

Figure 5.23

(*i*) If the excavation has vertical sides, calculate the volume of material to be removed. (Assume no bulking.)

(*ii*) It was subsequently discovered that the grid actually had been set out at 20.1 m centres, and an error in the adopted temporary bench mark had resulted in all the ground levels being shown 0.50 m too high. Calculate the corrected volume of excavation.   [Salford]

Answer   (*i*) 17 820 m^3; (*ii*) 16 989 m^3

18   A hillside, which may be considered to be a plane surface, has a slope of 1 vertical in 9 horizontal. A straight road is to be constructed thereon at a gradient of 1 vertical in 12 horizontal such that it is wholly in embankment. Determine the volume of fill required between two cross-sections 50 m apart if a side slope of 1 vertical to 2 horizontal obtains and the width of formation is 16 m.

Answer   551.6 m^3

19   Table 5.27 shows the equivalent areas enclosed by contour loops on the plan of a tip at a construction site. The tip may be assumed to have been levelled off at 262 m AOD.

160 000 m^3 of fill is required at another location on the site. If this fill is removed uniformly downwards from the top by motorized scrapers,

Table 5.27

Contour level (m AOD)	262	260	258	256	254	252	250
Area in hectares	0.468	0.712	1.031	1.509	2.603	4.066	6.741

estimate the surface level of the tip after completion. Check your answer using an alternative method. [Salford]

Answer 252.1 m AOD

20 A rectangular area 100 m by 80 m has to be excavated on a construction site. In order to assess the quantity of excavation a grid of levels has been taken over the area at the corners of squares of side 20 m with values as follows (all in m AOD):

152.36	152.43	152.25	152.11	151.92	151.76
152.18	152.21	152.07	151.84	151.61	151.41
152.02	152.08	151.85	151.56	151.24	151.04
151.89	151.86	151.62	151.28	150.93	150.67
151.78	151.75	151.45	151.12	150.72	150.26

The formation level for the excavation is 150.00 m AOD.

Calculate the volume of excavation assuming vertical sides. Check your answer using an alternative method. [Salford]

Answer 13 254.7 m^3

21 A dam is to be constructed across a valley to form a reservoir, and the areas given in Table 5.28 enclosed by contour loops were obtained from a plan of the area involved.

Table 5.28

Contour line (metres AOD)	Area enclosed (hectares)
640	5.2
645	9.4
650	16.3
655	22.4
660	40.7
665	61.5
670	112.2
675	198.1
680	272.4

(*a*) If the 640 m level represents the level floor of the reservoir, use the prismoidal formula to calculate the total volume of water impounded when the water level reaches 680 m.

(*b*) Determine the level of water at which one-third of the total capacity is stored in the reservoir.

(*c*) On checking the calculations, it was found that the original plan from which the areas of contour loops had been measured had shrunk evenly by approximately 1.2% of linear measurement. What was the corrected total volume of water? [Salford]

Answer (*a*) 29.693 × 10^6 m^3; (*b*) 669.8 m; (*c*) 30.406 × 10^6 m^3

22 The following spot height data (in m) apply in Fig. 5.15:

1	17.83;	2	18.19;	3	18.42
4	17.58;	5	18.01;	6	18.25

7	17.37;	8	17.70;	9	17.96
10	17.06;	11	17.48;	12	17.63

The area is to be graded with a uniform downgrade of 1 in 15 from line 1, 2, 3 towards line 10, 11, 12. Neglecting the effects of sideslopes and bulking calculate the design levels at those lines such that cut and fill are balanced. [London]

Answer 18.81 m; 16.81 m

23 A sewer is to be laid between two points having co-ordinates 12.0 m, 12.0 m and 85.0 m, 85.0 m respectively in Fig. 5.21, the depth of invert being 1.5 m at each point. Using 'end areas' estimate the volume of excavation for a trench, 1 m wide, taken to invert level between the points.

Answer 149 m^3

24 The plan of a very old chain survey, plotted to 1/500 scale on linen cloth, was found to have shrunk so that a line originally 200 mm long was only 197 mm. Furthermore, a note on the plan stated that the 30 m chain used for the survey had been found to be 20 mm too long after completion of the plot. If a certain area on the plan is measured by planimeter as 0.225 m^2, estimate the correct area on the ground in hectares. Uniform shrinkage of the plan may be assumed. [Salford]

Answer 5.805 hectares

6

Curve ranging

Circular curves

The circular curve, shown in Figs 6.1 and 6.2, connects two straights between tangent points T_1 and T_2 at a constant radius R. The straights meet at the intersection point I, crossing at angle θ which is known either as the total angle of deflection or the angle of deviation or the angle of intersection.

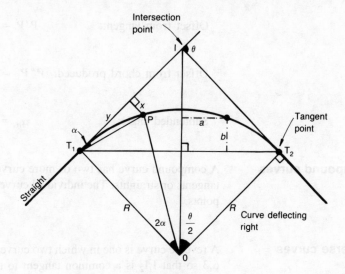

Figure 6.1

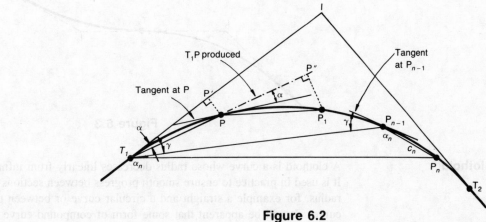

Figure 6.2

Formulae for the design and setting out of this curve include the following.

Tangent length: $\qquad$ $IT_1 = IT_2 = R \tan (\theta/2)$.
Curve length: $\qquad$ $T_1T_2 = R\theta$.
Chord length: $\qquad$ $T_1P = 2R \sin \alpha$.

Offset from tangent: $\qquad$ $y = \dfrac{x^2}{2R}$.

Offset from long chord: $\qquad$ $b = \sqrt{(R^2 - a^2)} - \sqrt{\left[R^2 - \left(\dfrac{T_1T_2}{2} \right)^2 \right]}$.

Tangential angle: $\qquad$ $\alpha = 1718.9 \dfrac{T_1P}{R}$ (minutes), where

$\qquad\qquad\qquad\qquad\qquad\qquad\qquad$ T_1P is less than $R/10$ and preferably less than $R/20$.

Offset from tangent: $\qquad$ $P'P = \dfrac{T_1P^2}{2R}$.

Offset from chord produced: $\quad P''P_1 = \dfrac{PP_1}{2R} (T_1P + PP_1)$.

Subtended angle: $\qquad\qquad\qquad$ $\alpha_n = 1718.9 \dfrac{c_n}{R}$ (minutes).

Compound curves

A compound curve has two or more curves contained between the two main tangents or straights. The individual curves meet tangentially at their junction points.

Reverse curves

A reverse curve is one in which two curves change direction as shown in Fig. 6.3 so that I_1I_2 is a common tangent to the curves.

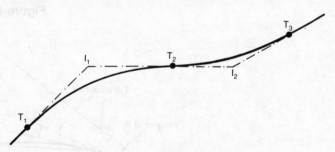

Figure 6.3

Clothoid

A clothoid is a curve whose radius decreases linearly from infinity to zero. It is used in practice to ensure smooth progress between sections of different radius, for example a straight and a circular curve or between two circular curves. It will be apparent that some form of compound curve arises.

Transition curve

A transition curve is the part of the clothoid whose maximum radius occurs at one junction point, and whose minimum radius, R, occurs at a second junction point. These junction points can be the tangent points with the straights or with a circular curve or another transition curve.

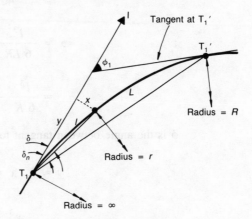

Figure 6.4

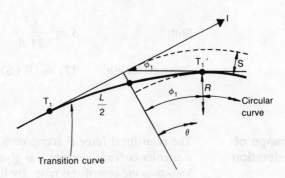

Figure 6.5

In Figs 6.4 and 6.5 for transition curve $T_1 T_1'$

$$l \times r = L \times R = K.$$

Deflection angle of curve: $\qquad \phi_1 = \dfrac{L}{2R}$ (radian).

Deflection angle for a specific chord: $\delta = \dfrac{1800\, l^2}{\pi\, R\, L}$ (minutes).

$$\delta_n = \frac{1800\, L}{\pi\, R} \text{ (minutes)}$$

$$= \frac{\phi_1}{3} \text{ (radians) when } \phi_1 \text{ is small.}$$

Offset from tangent: $y = l \left(1 - \dfrac{\phi^2}{10} + \dfrac{\phi^4}{216} + \ldots \right)$

$$= l - \frac{l^5}{40\,K^2} + \frac{l^9}{3456\,K^4} + \ldots$$

$$x = \frac{l^3}{6\,LR} \left(1 - \frac{\phi^2}{14} + \frac{\phi^4}{440} + \ldots\right)$$

$$= \frac{l^3}{6\,K} - \frac{l^7}{336\,K^3} + \frac{l^{11}}{42\,240\,K^5}.$$

ϕ is the angle between tangent to curve point and $T_1 I$ as in Fig. 6.17.

$$y \approx l, \quad x \approx \frac{l^3}{6\,LR} \text{ (cubic spiral).}$$

$$x \approx \frac{y^3}{6\,LR} \text{ (cubic parabola).}$$

Shift: $\qquad S = \dfrac{L^2}{24\,R}.$

Tangent length: $\qquad IT_1 = (R+S)\tan(\theta/2) + L/2$

Rate of change of radial acceleration

The centrifugal force P acting on a vehicle of weight W as it moves along a circular curve at velocity V is given by $P = (WV^2/gR)$ in which (P/W) is known as the centrifugal ratio. By lifting (super-elevating) the outer edge of the road or rail the resultant force can be made to act perpendicularly to the running surface, Fig. 6.6. In practice to avoid large super-elevations an allowance (fB) for friction is made. Radial acceleration is given by the expression (V^2/R).

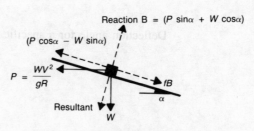

Figure 6.6

In the case of a transition curve it changes as the vehicle moves along the curve because the radius is variable. For constant velocity V the rate of change of radial acceleration (assumed uniform) is

$$a = \left(\frac{V^2/R}{L/V} \right) = \frac{V^3}{LR}.$$

Vertical curves

Vertical curves are introduced at the intersection of two gradients, either as summit curves or sag curves. Usually they are parabolic in form, and so in Fig. 6.7

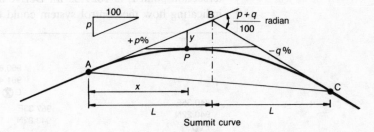

Figure 6.7

Vertical offset: $y = \dfrac{[p-(-q)]x^2}{400\,L}$

$$= \frac{(p+q)x^2}{400\,L}.$$

In many published texts the term $(p+q)$, or $(p-q)$ in the case of gradients of an equal sense, is replaced by one variable, G. For flat gradients it is normal to assume that the length along the gradients equals the length of the curve $(2L)$ and that these in turn equal the long chord AC and its horizontal projection. This assumption is not to be made in the design of horizontal curves.

Although the parabola is adopted for the vertical curve, it can be approximated by a circular curve in respect of radial (centrifugal) acceleration when estimating the length of curve at the design stage. For the curve connecting gradients of opposite sense, i.e. $+p\%$ and $-q\%$

Curve length: $2L = \dfrac{V^2(p+q)}{100\,f} = K\,(p+q).$

Tables of K values are published by the Department of Transport for design purposes, to satisfy different road conditions and velocities. Also for parabolic summit curves:

Change of gradient over $2L = \dfrac{(p+q)}{100}$

$$\text{Change over unit distance} = \frac{(p+q)}{200\,L}$$

$$\text{Slope of tangent at P} = \frac{p}{100} - \frac{(p+q)\,x}{200\,L}.$$

6.1 Setting out a circular curve from the tangent points and control points

The centre line of a road is to be set out as part of a new development. A circular curve of radius 350.00 m, deflecting right through 32° 40′, is to be incorporated within its length and Fig. 6.8 shows its relationship with control points established previously. The chainage of the intersection point I, is 1029.35 m. Derive data for setting out the curve indicating how the control system could be utilized.

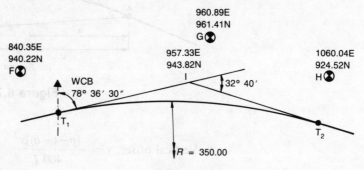

Figure 6.8

Solution.

$$\text{Curve length} = R\theta = 350.00 \times \frac{\pi}{180} \times 32.667$$
$$= 199.55 \text{ m.}$$

$$\text{Tangent length} = IT_1 = IT_2 = R \tan \theta/2$$
$$= 350 \tan 16° 20′$$
$$= 102.57 \text{ m.}$$

$$\text{Chainage of } T_1 = 1029.35 - 102.57$$
$$= 926.78 \text{ m.}$$

$$\text{Chainage of } T_2 = 926.78 + 199.55$$
$$= 1126.33 \text{ m.}$$

Now determine the co-ordinates of T_1 and T_2.

Easting of T_1 = 957.33 + 102.57 sin 258° 36′ 30″ = 856.78 m.
Northing of T_1 = 943.82 + 102.57 cos 258° 36′ 30″ = 923.56 m.
WCB of IT_2 = 78° 36′ 30″ + 32° 40′ 00″ = 111° 16′ 30″.
Easting of T_2 = 957.33 + 102.57 sin 111° 16′ 30″ = 1052.91 m.
Northing of T_2 = 943.82 + 102.57 cos 111° 16′ 30″ = 906.60 m.

Now we set out from the tangent point. The procedure is:

(*a*) locate T_1, measured back 102.57 m from I; and
(*b*) calculate the deflection (tangential) angles using the formula

$$\delta = 1718.9 \, \frac{C}{R} \text{ minutes},$$

where C is the length of chord selected. It should be in excess of $R/10$ and should preferably be of the order of $R/20$. Then the chord length is virtually the arc length, and the sum of the individual chords is essentially the length of the circular curve.

Table 6.1

Chainage (m)	Chord length (m)	Tangential angle (minutes)	Total angle (minutes)	Angle set on 20″ theodolite	Angle set on 1″ theodolite
926.78	0.0	0.0	0.0	0.0	0.0
940.00	13.22	64.925	64.925	1° 05′ 00″	1° 04′ 55″
960.00	20.00	98.223	163.148	2° 43′ 00″	2° 43′ 09″
980.00	20.00	98.223	261.371	4° 21′ 20″	4° 21′ 22″
1000.00	20.00	98.223	359.594	5° 59′ 40″	5° 59′ 36″
1020.00	20.00	98.223	457.817	7° 37′ 40″	7° 37′ 49″
1040.00	20.00	98.223	556.040	9° 16′ 00″	9° 16′ 02″
1060.00	20.00	98.223	654.263	10° 54′ 20″	10° 54′ 16″
1080.00	20.00	98.223	752.486	12° 32′ 20″	12° 32′ 29″
1100.00	20.00	98.223	850.709	14° 10′ 40″	14° 10′ 43″
1120.00	20.00	98.223	948.932	15° 49′ 00″	15° 48′ 56″
1126.33	6.33	31.087	980.019	16° 20′ 00″	16° 20′ 00″

Notice (from Table 6.1) that curve points have been located in 'running-chainage' form or 'through-chainage' form. This facilitates the levelling of the longitudinal and cross sections over the complete length of road.

The following computer program will calculate data to set out a circular curve from the tangent points. It is an extract from the complete program for a compound curve listed with Example 6.5; the variables are defined with the main program.

```
10 REM SET OUT A CIRCULAR CURVE FROM THE TANGENT POINTS
20 INPUT"DEFLECTION ANGLE    DEG,MIN,SEC     ",D,M,S
30 A1=((D*3600)+(M*60)+S)/206264.8
40 INPUT"RADIUS OF CURVE      M            ",R
50 INPUT"CHAINAGE OF INTERSECTION POINT   M ",C
60 INPUT"SETTING OUT CHAINAGE INTERVAL    M ",I
70 C1=C-R*TAN(A1/2)
80 C2=C1+(R*A1)
90 PRINT"CHAINAGE","POLAR D.","SETTING OUT ANGLE"
100 PRINT C1,F,F;F;F,"FIRST TANGENT POINT"
110 L3=(INT(C1/I)+1)*I-C1
120 C5=L3+C1
130 A4=A4+(L3/(R*2))
140 L4=INT(2*R*SIN(A4)*1000+0.5)/1000
150 A =INT(206264.8*A4+0.5)
160 D =INT(A/3600)
170 M =INT((A-(D*3600))/60)
180 S =A-(D*3600)-(M*60)
190 IF C5=C2 THEN 280
```

```
200 PRINT C5,L4,D;M;S
210 L3=I
220 C5=C5+I
230 IF C5>C2 THEN GOTO 250
240 GOTO 130
250 A4=A4+(C2-C5+I)/(2*R)
260 C5=C2
270 GOTO 150
280 PRINT C2,L4,D;M;S,"FINAL TANGENT POINT"
290 END
```

When setting out from the control points the polar method can be used to establish all points on the centre line of the road. The control information is:

FG: length 122.39 m, bearing 80° 01′ 47″
GH: length 105.79 m, bearing 110° 24′ 30″.

In respect of T_1 whose co-ordinates are 856.78 E, 923.56 N, we have

$$\text{bearing } FT_1 = \tan^{-1}\left(\frac{856.78 - 840.35}{923.56 - 940.22}\right)$$

$$= 135° 23′ 54″$$
$$\text{length } FT_1 = \sqrt{[(856.78 - 840.35)^2 + (923.56 - 940.22)^2]}$$
$$= 23.40 \text{ m.}$$

In respect of point P (chainage 1000.00) the bearing of T_1P is

$$78° 36′ 30″ + 05° 59′ 36″ = 84° 36′ 06″.$$

From Fig. 6.9

$$T_1P = 2R \sin 05° 59′ 36″$$
$$= 73.09 \text{ m.}$$

Note that the chord length is used, not the curve length. Co-ordinates of P are

easting 856.78 + 73.09 sin 84° 36′ 06″ = 929.55 mE
northing 923.56 + 73.09 cos 84° 36′ 06″ = 930.44 mN.

$$\text{Bearing } GP = \tan^{-1}\frac{929.55 - 960.89}{930.44 - 961.41}$$

$$= 225° 20′ 25″.$$

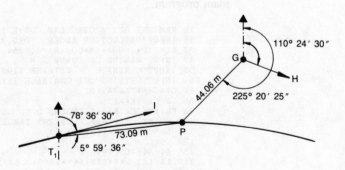

Figure 6.9

A clockwise angle of $(225° 20' 25'' - 110° 24' 30'') = 114° 55' 55''$ could be set off at G from GH to give the required direction. In practice, the surveyor would probably sight H and set the horizontal circle reading on the theodolite to $110° 24' 30''$, a circle reading of $225° 20' 25''$ then gives the pointing on P.

$$\text{Length GP} = \sqrt{[(929.55 - 960.89)^2 + (930.44 - 961.41)^2]}$$
$$= 44.06 \text{ m.}$$

6.2 Fitting a circular curve through a fixed point

Two straights intersect at B. Straight AB has a bearing of 115° and straight BC a bearing of 135°. The straights are to be connected by a circular curve which passes through a point D which lies 45.72 m from B on a bearing of 285°. Calculate the radius of the curve, the tangent length, the length of the curve and the deflection angle for a 20 m chord.

[Bradford]

Solution. Review the data with respect to triangle DBO in Fig. 6.10. The circular curve has to pass through point D which lies between tangent points A and C, 45.72 m from B. O is the centre of the circular curve.

$$\text{Bearing of BA} = 115° + 180° = 295°.$$
$$\text{Angle A}\hat{\text{B}}\text{D} = 295° - 285° = 10°.$$

The total deflection angle at B is

$$\theta = 135° - 115° = 20°.$$

$$\text{A}\hat{\text{B}}\text{O} = \frac{180 - 20}{2} = 80°,$$

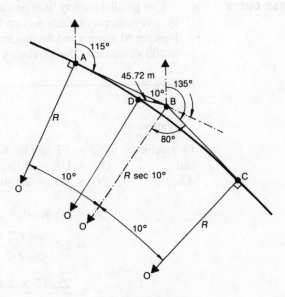

Figure 6.10

and so
$$\hat{DBO} = 80° - 10° = 70°$$
$$\hat{AOB} = 90° - 80° = 10° = \frac{\theta}{2}.$$

Next determine R and tangent length AB. In triangle DBO, OD = R, DB = 45.72 m, OB = R sec 10° and $\hat{DBO} = 70°$.

Therefore $R^2 = (R \sec 10)^2 + 45.72^2 - 2(R \sec 10) \times 45.72 \times \cos 70$

i.e. $0.0311R^2 - 31.7569R + 2090.3184 = 0$.

This solves for R to equal either **950.402 m** or 70.723 m. The higher value must be adopted in view of the position of D, since the respective tangent lengths will be either (950.402 tan 10) or (70.723 tan 10), i.e. **167.584 m** or 12.470 m. The latter value is obviously too low since BD is 45.72 m in length.

Length of curve $= R\theta$

$$= 950.402 \times \frac{\pi}{180} \times 20$$

$$= \textbf{331.753 m.}$$

Deflection angle for a 20 m chord $= 1718.9 \dfrac{1}{R}$ min

$$= \frac{1718.9 \times 20}{950.402} \text{ min}$$

$$= \textbf{36' 10''.}$$

6.3 Reverse curve

Two parallel railway lines on the surface at a mine are to be connected by a reverse curve, each section having the same radius. If the centre lines are 50 m apart and the maximum distance between the tangent points is 200 m calculate the maximum allowable radius that can be used.

[CEI]

Solution.

By symmetry PQ = ST in Fig. 6.11
and QU = US = $R \tan \theta/2$.
Also, PU = UT = $2R \sin \theta/2 = \sqrt{(25^2 + 100^2)}$
$$= 25\sqrt{17}.$$

$$50 = \text{QS} \sin \theta = 2R \tan \theta/2 \times \sin \theta$$

$$= 2R \frac{\sin \theta/2}{\cos \theta/2} \times \sin \theta$$

$$= \frac{25\sqrt{17} \times 2 \sin \theta/2 \times \cos \theta/2}{\cos \theta/2}.$$

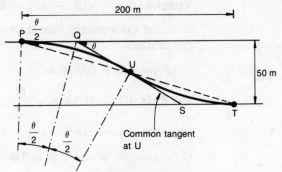

Figure 6.11

Therefore $\quad \sin\dfrac{\theta}{2} = \dfrac{1}{\sqrt{17}}$

and so $\quad \cos\dfrac{\theta}{2} = \dfrac{4}{\sqrt{17}}\;$ and $\tan\dfrac{\theta}{2} = \dfrac{1}{4}$.

But $2R\sin\theta/2 = 25\sqrt{17}$

Therefore $\quad R = \dfrac{25\sqrt{17}}{2 \times \dfrac{1}{\sqrt{17}}}$

$\qquad\qquad = \mathbf{212.5\ m.}$

6.4 Setting out a circular curve from the long chord

> Calculate the data needed to set out one of the curves of Example 6.3 using offsets from the long chord.

Solution. In Fig. 6.12
$$R^2 = (y + R\cos\theta/2)^2 + x^2.$$

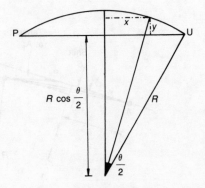

Figure 6.12

Therefore $y = \sqrt{(R^2 - x^2)} - R \cos \theta/2$.

Note that x is measured from the mid-point of PU which must be located. In this example

$R \cos \theta/2 = 206.16$ m

and $R \sin \theta/2 = 51.54$ m $= \dfrac{PU}{2}$.

Therefore $y = \sqrt{(R^2 - x^2)} - 206.16$ m.

Offsets can be tabulated as in Table 6.2.

Table 6.2

Distance x (m)	0.0	12.5	25.0	37.5	50.0
$\sqrt{(R^2 - x^2)}$	212.5	212.13	211.02	209.16	206.53
Offset y (m)	6.34	5.97	4.86	3.00	0.37

6.5 Compound curve — setting out from the tangent points

A compound curve, shown in Fig. 6.13, consisting of a circular curve with transition curves at each end, is to connect two straights having a total deflection angle of $32° 24' 00''$. The radius of the circular curve is 1000 m and the transition curves are to be designed for a rate of change of radial acceleration of 0.3 m/s^3 and a velocity of 110 km/h. Determine:

(a) the chainages of the tangent points on the straights given that the chainage of the intersection point is 1350.468 m;
(b) offsets required to locate the first transition curve; and
(c) setting-out data, using theodolite and tape, for the transition curve at 20 m intervals and the circular curve at 50 m intervals.

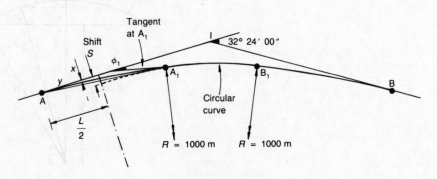

Figure 6.13

Solution.   First calculate the curve lengths and the tangent lengths. The rate of change of radial acceleration (α) is related to the design velocity (V) by

$$\alpha = \frac{V^3}{LR},$$

where L is the length of the transition curve and R is the radius of the circular curve (and the transition curve) at A_1.

$V = 110$ km/h $= 30.556$ m/s.

Therefore $\quad 0.3 = \dfrac{30.556^3}{L \times 1000}$

$$L = 95.097 \text{ m}.$$

Shift $\qquad S = \dfrac{L^2}{24R} = \dfrac{95.097^2}{24 \times 1000}$

$$= 0.377 \text{ m}.$$

Tangent length IA $= (R+S) \tan \theta/2 + L/2$

$$= (1000.377) \tan 16° \, 12' + \frac{95.097}{2}$$

$$= 338.185 \text{ m}.$$

Now $\qquad \phi_1 = \dfrac{L^2}{2LR}$ radian

$$= \frac{95.097^2}{2 \times 95.097 \times 1000}$$

$$= 0.047\,548\,5 \text{ radians}.$$

Therefore $\quad \phi_1 = 02° \, 43' \, 28''$

$2\phi_1 = 05° \, 26' \, 56''$.

Therefore angle subtended by circular curve $= 32°24'00'' - 05°26'56''$

$$= 26° \, 57' \, 04''$$

$$= 26.9511°.$$

Length of circular curve $= R \times \theta$

$$= 1000 \times \frac{\pi}{180} \times 26.9511$$

$$= 470.385 \text{ m}.$$

(*a*) We can now determine the chainages of the tangent points:

Chainage of intersection point $\quad = 1350.468$ m
Deduct tangent length $\qquad\qquad\quad = \underline{338.185 \text{ m}}$

Chainage of tangent point A $\quad = \textbf{1012.283 m}$
Add transition length $\qquad\qquad\quad\; = \underline{95.097 \text{ m}}$

Chainage of junction point $A_1 \;\; = 1107.380$ m
Add circular curve length $\qquad\quad = \underline{470.385 \text{ m}}$

Chainage of junction point B_1 = 1577.765 m
Add transition length = 95.097 m

Chainage of tangent point B = **1672.862 m**

(b) Determine data for setting-out the transition curve by offsets from the straight AI. The expression for offsets to locate a transition curve of cubic spiral form is

$$x = \frac{l^3}{6LR}$$

in which x is the offset from the main straight, and l is the distance along the transition curve approximated by the sum of the chord lengths.

$$6\,LR = 6 \times 95.097 \times 1000$$
$$= 570\,582$$

$$x = \frac{l^3}{570\,582}\ \text{m.}$$

The chord lengths used for setting out the transition curve may be taken as one-half to one-third of the corresponding chord to set out the circular curve (say $R/20$ m). Since $R = 1000$ m a chord length of 20 m is acceptable for the transition curves. The required offsets can be tabulated as in Table 6.3.

Table 6.3

Point	Chainage (m)	Chord length (m)	l (m)	x (m)
A	1012.283	0.0	0.0	0.0
	1020.00	7.717	7.717	0.001
	1040.00	20.000	27.717	0.037
	1060.00	20.000	47.717	0.190
	1080.00	20.000	67.717	0.544
	1100.00	20.000	87.717	1.183
	1107.38	7.380	95.097	1.507

(c) Determine data to set out the curve with theodolite and tape. For the transition curve the deflection angle is

$$\delta = \frac{1800\,l^2}{\pi\,LR} = 0.006\,025\,l^2\ \text{min.}$$

For the transition curve from A the data can be tabulated as in Table 6.4.
For the circular curve,

$$\delta = 1718.9\,\frac{l}{R} = 1.7189\,l\ \text{min,}$$

where l is the chord length. Although 50 m chords satisfy the requirements of $R/20$, 20 m chord lengths would normally be adopted. This example is

Table 6.4

Chainage (m)	Chord length (m)	l (m)	δ
1012.283	0.0	0.0	0.0
1020.000	7.717	7.717	00′ 22″
1040.000	20.000	27.717	04′ 38″
1060.000	20.000	47.717	13′ 43″
1080.000	20.000	67.717	27′ 38″
1100.000	20.000	87.717	46′ 21″
1107.380	7.380	95.097	54′ 29″

illustrative and the use of 50 m chords ensures that the table of setting-out data (Table 6.5) is not too long.

Table 6.5

Chainage (m)	Chord length (m)	δ	Σδ
1107.380	0.0	0.0	0.0
1150.000	42.620	1° 13′ 15.5″	1° 13′ 16″
1200.000	50.000	1° 25′ 56.7″	2° 39′ 12″
1250.000	50.000	1° 25′ 56.7″	4° 05′ 09″
1300.000	50.000	1° 25′ 56.7″	5° 31′ 06″
1350.000	50.000	1° 25′ 56.7″	6° 57′ 02″
1400.000	50.000	1° 25′ 56.7″	8° 22′ 59″
1450.000	50.000	1° 25′ 56.7″	9° 48′ 56″
1500.000	50.000	1° 25′ 56.7″	11° 14′ 52″
1550.000	50.000	1° 25′ 56.7″	12° 40′ 49″
1577.765	27.765	47′ 43.5″	13° 28′ 33″

The final transition must be set out from tangent point B. The data can be tabulated as follows in Table 6.6, δ values being set off from BI anti-clockwise.

Table 6.6

Chainage (m)	Chord length (m)	l (m)	δ
1580.000	20.000	92.862	51′ 57″
1600.000	20.000	72.862	31′ 59″
1620.000	20.000	52.862	16′ 50″
1640.000	20.000	32.862	6′ 30″
1660.000	12.862	12.862	1′ 00″
1672.862	0.0	0.0	0.0

This problem can be solved by the following computer program. The output data is in the form of a table and may need reformatting for computers with a limited screen display. The polar distance from the tangent point to the point in question is also output so that setting out can alternatively be carried out with an EDM.

Variables

A	Angle for DMS subroutine	F	Variable with value zero for
A1	Total angle of deflection of the		output
	curve, θ	I	Setting out chainage interval
A2	Deflection of the transition	L1	Length of the transition curve
	curve, ϕ_1	L2	Length of the circular curve
A3	Deflection of the circular	L3	Running curve length on tran-
	curve, $\theta - 2\phi_1$		sition: chord length on circular
A4	Setting out angle, δ		curve
C	Chainage of the intersection	L4	Setting out polar length
	point	M	Input/output angle, minutes
C1	Chainage of the first tangent	R	Radius of the circular curve
	point	S	Input/output angle, seconds
C2	Chainage of the second tangent	S1	Shift
	point	T	Tangent length (IT)
C3	Chainage of the third tangent	V	Design speed
	point	X	Offset of the transition from
C4	Chainage of the final tangent		the tangent
	point	X1	Offset at the tangent point
C5	Running chainage	Y	Offset of the transition along
D	Input/output angle, degrees		the tangent
		Y1	Offset at the tangent point

```
10 REM SET OUT A COMPOUND CURVE FROM THE TANGENT POINTS
20 INPUT"DESIGN SPEED  KM/H                 ",V
30 V=V/3.6
40 INPUT"DEFLECTION ANGLE    DEG,MIN,SEC    ",D,M,S
50 A1=((D*3600)+(M*60)+S)/206264.8
60 INPUT"RADIUS OF CIRCULAR CURVE     M     ",R
70 INPUT"CHAINAGE OF INTERSECTION POINT   M ",C
80 INPUT"SETTING OUT CHAINAGE INTERVAL    M ",I
90  L1=V^3/(0.3*R)
100 S1=L1^2/(24*R)
110 T =(L1/2)+(R+S1)*TAN(A1/2)
120 K=L1*R
130 X1=(L1^3/(6*K))-(L1^7/(336*K^3))
140 Y1=L1-(L1^5/(40*K*K))
150 A2=3*ATN(X1/Y1)
160 A3=A1-(A2*2)
170 L2=R*A3
180 C1=C-T
190 C2=C1+L1
200 C3=C2+L2
210 C4=C3+L1
220 PRINT"CHAINAGE","POLAR D.","SETTING OUT ANGLE"
230 PRINT C1,F,F;F;F,"FIRST TANGENT POINT"
240 L3=(INT(C1/I)+1)*I-C1
250 C5=C1+L3
260 GOSUB 550
270 L3=L3+I
280 C5=C5+I
290 IF C5>C2 THEN GOTO 310
300 GOTO 260
310 A=INT((206264.8*A2/3)+0.5)
320 GOSUB 660
330 L4=INT(SQR((X1*X1)+(Y1*Y1))*1000+0.5)/1000
340 PRINT C2,L4,D;M;S,"SECOND TANGENT POINT"
350 L3=C5-C2
360 A4=0
370 GOSUB 600
380 L3=I
390 C5=C5+I
```

```
400 IF C5>C3 THEN GOTO 420
410 GOTO 370
420 A4=A4+(I-C5+C3)/(2*R)
430 L4=INT(2*R*SIN(A4)*1000+0.5)/1000
440 A=INT(206264.8*A4+0.5)
450 GOSUB 660
460 PRINT C3,L4,D;M;S,"THIRD TANGENT POINT"
470 L3=L1-(C5-C3)
480 GOSUB 550
490 L3=L3-I
500 C5=C5+I
510 IF C5>C4 THEN 530
520 GOTO 480
530 PRINT C4,F,F;F;F,"FINAL TANGENT POINT"◄
540 END
550 X =(L3^3/(6*K))-(L3^7/(336*K^3))
560 Y =L3-(L3^5/(40*K*K))
570 L4=INT(SQR((X*X)+(Y*Y))*1000+0.5)/1000
580 A4=(ATN(X/Y))
590 GOTO 620
600 A4=A4+(L3/(R*2))
610 L4=INT(2*R*SIN(A4)*1000+0.5)/1000
620 A =INT(206264.8*A4+0.5)
630 GOSUB 660
640 PRINT C5,L4,D;M;S
650 RETURN
660 D=INT(A/3600)
670 M=INT((A-(D*3600))/60)
680 S=A-(D*3600)-(M*60)
690 RETURN
```

6.6 Compound curve — setting out the transition curve with offsets from the tangent

A circular curve of radius 540 m leaves a straight at through-chainage 740.40 m and meets a second circular curve of radius 450 m at chainage 1192.95 m and terminates on a second straight at chainage 1365.90 m. The compound curve is to be replaced by one of 660 m radius with transition curves 120 m long at each end.

Calculate the chainages of the two new tangent points and the quarter-point offsets for the transition curves. [London]

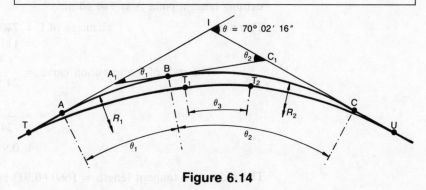

Figure 6.14

Solution. Determine the existing tangent lengths. For the first circular curve

$$R_1\,\theta_1 = 1192.95 - 740.40 \text{ m}$$
$$= 452.55 \text{ m}$$

Since $R_1 = 540$ m

$$\theta_1 = 0.838\,055\,6 \text{ rad}$$
$$= 48° 01' 01''$$

Therefore tangent length $AA_1 = R_1 \tan \theta_1/2 = 240.52$ m.

For the second circular curve

$$R_2 \, \theta_2 = 1365.90 - 1192.95$$
$$= 172.95 \text{ m.}$$

Since $R_2 = 450$ m

$$\theta_2 = 0.384\ 333\ 3 \text{ rad}$$
$$= 22°\ 01'\ 15''.$$

Therefore tangent length $CC_1 = R_2 \tan \theta_2/2 = 87.56$ m.

Next determine the distance from the first tangent point to the intersection point. The length of the common tangent A_1C_1 at the junction point of the two existing circular curves

$$= 240.52 + 87.56$$
$$= 328.08 \text{ m.}$$

Also $\qquad \theta = \theta_1 + \theta_2 = 70°\ 02'\ 16''.$

By the sine rule

$$\frac{IA_1}{\sin \theta_2} = \frac{A_1C_1}{\sin 109°\ 57'\ 44''}.$$

Therefore $\quad IA_1 = 328.08 \, \dfrac{\sin 22°\ 01'\ 15''}{\sin 109°\ 57'\ 44''}$

$$= 130.87 \text{ m.}$$
$$IA = 130.87 \text{ m} + 240.52 \text{ m}$$
$$= 371.39 \text{ m}$$

Next determine the chainages of the new tangent points. The chainage of the existing tangent point A is 740.40 m.

Therefore $\qquad$ chainage of I $= 740.40$ m $+ 371.39$ m
$$= 1111.79 \text{ m.}$$

The shift of the transition curve $= \dfrac{L^2}{24 R}$

$$= \dfrac{120^2}{24 \times 660}$$
$$= 0.91 \text{ m.}$$

Therefore new tangent length $= (660 + 0.91) \tan \left(\dfrac{70°02'16''}{2} \right) + \dfrac{120}{2}$

$$= 523.10 \text{ m.}$$

Therefore chainage of new tangent point $= 1111.79 - 523.10$
$$= \mathbf{588.69 \text{ m.}}$$

Now $\quad \phi_1 = \dfrac{L^2}{2LR} = \dfrac{120^2}{2 \times 120 \times 660} = 0.090\ 909\ 1 \text{ rad}$

$$= 05°\ 12'\ 31''.$$

Therefore the angle subtended by new circular curve at its centre is

$$70° \ 02' \ 16'' - (2 \times 05° \ 12' \ 31'') = 59° \ 37' \ 14''$$
$$= 59.6206°.$$

Therefore length of new circular curve $= 660 \times \dfrac{\pi}{180} \times 59.6206$

$$= 686.78 \text{ m.}$$

Chainage of second tangent point $= 588.69 + 120.00 + 686.78$
$$+ \ 120.00$$
$$= \textbf{1515.47 m.}$$

Now calculate offsets to locate the quarter points on the transition curve. The equation of the offsets to the transition curve is

$$x = \frac{l^3}{6LR}$$

$$= \frac{l^3}{475 \ 200}.$$

Table 6.7

l (m)	x (m)
30	0.057
60	0.455
90	1.534

6.7 Locating a transition curve from a traverse

A curve, wholly transitional, is to be set out in a built-up area. At the preliminary survey and design stage it is found that the intersection point I is inaccessible. A and B, E and F lie on the two straights and they are linked with a traverse survey as shown on Fig. 6.15. The data in Table 6.8 were obtained.

Table 6.8

Line	Horizontal distance (m)		Horizontal clockwise angle
BC	50.34	AB̂C	196° 11′ 40″
CD	69.27	BĈD	189° 45′ 20″
DE	64.76	CD̂E	204° 23′ 20″
		DÊF	193° 19′ 40″

Assuming the design value for the rate of change of radial acceleration to be 1/3 m/s³ for a velocity of 50 km/h determine the positions of the tangent points.

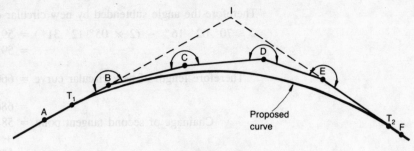

Proposed curve

Figure 6.15

Solution. From the traverse survey, let the bearing of ABI be $00° 00' 00''$.

$$
\begin{array}{ll}
\text{Angle A}\hat{\text{B}}\text{C} & 196° \ 11' \ 40'' \\
\hline
& 196° \ 11' \ 40'' \\
\text{Deduct} & 180° \ 00' \ 00'' \\
\hline
\text{Bearing BC} & 16° \ 11' \ 40'' \\
\text{Angle B}\hat{\text{C}}\text{D} & 189° \ 45' \ 20'' \\
\hline
& 205° \ 57' \ 00'' \\
\text{Deduct} & 180° \ 00' \ 00'' \\
\hline
\text{Bearing CD} & 25° \ 57' \ 00'' \\
\text{Angle C}\hat{\text{D}}\text{E} & 204° \ 23' \ 20'' \\
\hline
& 230° \ 20' \ 20'' \\
\text{Deduct} & 180° \ 00' \ 00'' \\
\hline
\text{Bearing DE} & 50° \ 20' \ 20'' \\
\text{D}\hat{\text{E}}\text{F} & 193° \ 19' \ 40'' \\
\hline
& 243° \ 40' \ 00'' \\
\text{Deduct} & 180° \ 00' \ 00'' \\
\hline
\text{Bearing I}\hat{\text{E}}\text{F} & 63° \ 40' \ 00''
\end{array}
$$

From this we obtain the values in Table 6.9. Thus in Fig. 6.16 $EI_1 = 94.21$ m, $BI_1 = 151.96$ m and $I_1\hat{I}E = 63° \ 40' \ 00''$.

Table 6.9

Line	Length l	Bearing θ	Easting Difference $l \sin \theta$	Northing Difference $l \cos \theta$
BC	50.34	16° 11′ 40″	14.04	48.34
CD	69.27	25° 57′ 00″	30.31	62.29
DE	64.76	50° 20′ 20″	49.86	41.33
			94.21 m	151.96 m

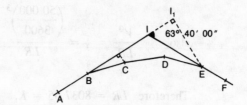

Figure 6.16

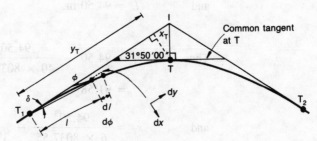

Figure 6.17

For the transition curves. The curve is to be wholly transitional and therefore, by symmetry, for each of the two transition curves

$$\phi_1 = \frac{63° \ 40' \ 00''}{2} = 31° \ 50' \ 00'' = 0.555 \ 596 \ 5 \ \text{radians}.$$

When the curve turns through an angle of this magnitude the first-order equations used between ϕ_1 and $(63° \ 40' \ 00'')/2$ in previous examples are not valid. In Fig. 6.17 we can write $dx = dl \sin \phi$ and $dy = dl \cos \phi$.

$$dx = \left(\phi - \frac{\phi^3}{3!} + \frac{\phi^5}{5!} \right) dl \qquad dy = \left(1 - \frac{\phi^2}{2!} + \frac{\phi^4}{4!} \right) dl$$

$$= \left(\frac{l^2}{2K} - \frac{l^6}{48K^3} + \frac{l^{10}}{3840K^5} \right) dl \qquad = \left(1 - \frac{l^4}{8K^2} + \frac{l^8}{384K^4} \right) dl.$$

Integrate

$$x = \frac{l^3}{6K} - \frac{l^7}{336K^3} + \frac{l^{11}}{42 \ 240K^5}$$

and

$$y = l - \frac{l^5}{40K^2} + \frac{l^9}{3456K^4}.$$

There are no constants of integration since $\phi = 0$ when $l = 0$. If L is the length of each transition curve and the minimum radius at junction point T is R, then

$$\phi_1 = 0.555 \ 596 \ 5 = \frac{L}{2R}$$

also,
$$\frac{V^3}{LR} = \tfrac{1}{3} = \frac{\left(\dfrac{50\,000}{3600}\right)^3}{LR}.$$

Therefore $LR = 8037.55 = K$.

Since $L = 1.111\,193R$

            $R = 85.05$ m

and $L = 94.50$ m.

Thus $y_T = 94.50 - \dfrac{94.50^5}{40 \times 8037.55^2}$

       $= 91.58$ m

and $x_T = \dfrac{94.50^3}{6 \times 8037.55} - \dfrac{94.50^7}{336 \times 8037.55^3}$

       $= 17.11$ m.

Tangent length $T_1I = T_2I = y_T + x_T \tan 31°\ 50'\ 00''$

$= 91.58 + 17.11 \tan 31°\ 50'\ 00''$

$= 102.20$ m.

From Table 6.9 $EI_1 = 94.21$ m.

From Fig. 6.16 $EI = \dfrac{94.21}{\sin 63°\ 40'\ 00''} = 105.12$ m

and $II_1 = 105.12 \cos 63°\ 40'\ 00'' = 46.63$ m.

Therefore IB $= 151.96 - 46.63 = 105.33$ m

From this we can locate the tangent points since

$BT_1 = 105.33 - 102.20 = \mathbf{3.13\ m}$

and $ET_2 = 105.12 - 102.20 = \mathbf{2.92\ m.}$

6.8 Transition between two circular curves

A circular curve of radius 950 m is to be connected to a further circular curve of radius 550 m by a transition curve such that the rate of change of radial acceleration is 0.333 m/s³ when the velocity is 100 km/h. Determine the length of the transition curve and derive setting-out data for that curve.

Introduction. Figure 6.18 shows the transition curve ST connecting the two circular curves. The tangent at S_1, common to the circular curve of radius R_1 and curve ST, is also shown making an angle of ϕ with the tangent to transition curve ST produced back to its origin at P. This particular tangent is needed for design purposes only and not for the actual setting out.

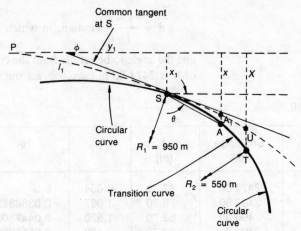

Figure 6.18

Curve ST can be set out either from the common tangent at S or by offsets from the circular curve, radius R_1, produced to U for final offset UT. Note that effectively just part of a whole transition curve is being laid down.

Solution. First establish the length of transition curve ST and the angle contained between the tangent at P and the common tangent at S. From the properties of the cubic spiral we have

$$lr = LR = K = l_1R_1,$$

where L is the total length PT, and l_1 is the length PS, the radii there being 550 m and 950 m, respectively.

The rate of change of radial acceleration

$$\alpha = \frac{V^3}{K}.$$

Now 100 km/h ≡ 27.78 m/s,

therefore $\frac{1}{3} = \dfrac{27.78^3}{l_1 \times 950} = \dfrac{27.78^3}{L \times 550}.$

Therefore $l_1 = 67.70$ m and $L = 116.94$ m.
Therefore ST $= 116.94 - 67.70 = $ **49.24 m.**
Also $LR = K = 64\,315.84.$

Therefore $\phi = \dfrac{l_1^2}{2l_1R_1} = 0.035\,631\,6$ rad

and $x_1 = \dfrac{l_1^3}{6K} = \dfrac{67.70^3}{6 \times 64\,315.84} = \dfrac{310\,288.73}{385\,895.04}$

$= 0.804$ m.

Now calculate the data for setting out from the common tangent at S. Inspection of Fig. 6.18 will show that

$$\theta = \frac{x - x_1}{l - l_1} \text{ radian, in which PA} = l,$$

and the angle to be set off from the common tangent $= \theta - \phi$. If the chainage of S is 2474.00 m, say, we get the values in Table 6.10.

Table 6.10

Station	Chainage (m)	l (m)	x	θ	$\theta - \phi$ (rad)	$\theta - \phi$
S	2474.00	67.70	0.804	0.0	0.0	0
	2480.00	73.70	1.037	0.0388333	0.0032017	00° 11′ 00″
	2490.00	83.70	1.520	0.0447500	0.0091184	00° 31′ 21″
	2500.00	93.70	2.132	0.0510769	0.0154453	00° 53′ 06″
	2510.00	103.70	2.890	0.0579444	0.0223128	01° 16′ 42″
	2520.00	113.70	3.809	0.0653261	0.0296945	01° 42′ 05″
T	2523.24	116.94	4.144	0.0678310	0.0321994	01° 50′ 42″

The data for setting out from the circular curve produced is derived as follows. Offset A_1A = offset from common tangent to transition curve minus offset from common tangent to circular curve

$$= (\theta - \phi)(l - l_1) - \frac{(l - l_1)^2}{2R_1}$$

$$= \left(\frac{x - x_1}{l - l_1} - \frac{l^2}{2K}\right)(l - l_1) - \frac{(l - l_1)^2}{2R_1}$$

$$= (x - x_1) - \frac{l^2(l - l_1)}{2K} - \frac{3l_1(l - l_1)^2}{6K} \left(\text{since } R_1 = \frac{K}{6l_1}\right)$$

Table 6.11

Station	Chainage (m)	$l - l_1$	$\dfrac{(l - l_1)^3}{6K}$ (m)
S	2474.00	0	—
	2480.00	6.00	—
	2490.00	16.00	0.011
	2500.00	26.00	0.046
	2510.00	36.00	0.121
	2520.00	46.00	0.252
T	2523.24	49.24	0.309

$$= \frac{l^3}{6K} - \frac{l_1{}^3}{6K} - \frac{3l^2(l-l_1)}{6K} - \frac{3l_1(l-l_1)^2}{6K}$$

$$= \frac{(l-l_1)^3}{6K}$$

Whence we get the values in Table 6.11.

6.9 Replacing a circular curve with a compound curve

A circular curve of radius 1600 m connects two straights having a total deflection angle of 65° 30′. It is to be shifted to allow a cubic spiral transition curve of length 125 m to be inserted at each end, the total route length remaining unchanged.

Calculate the distance between the new and the previous tangent points. Give data:

(a) for setting out the transition curve using 20 m chords;

(b) for locating the centre point of the new curve from the intersection point (I) of the straights whose chainage is 5264.50 m.

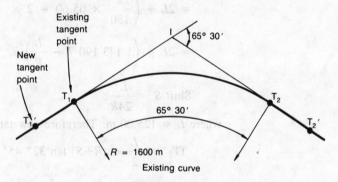

Figure 6.19

Solution. First determine the existing curve parameters.

$$\text{Curve length } T_1T_2 = R_1\theta = 1600 \times \frac{\pi}{180} \times 65.50$$

$$= 1829.11 \text{ m}.$$

$$IT_1 = R_1 \tan\left(\frac{65° 30′}{2}\right) = 1600 \times 0.643\,22$$

$$= 1029.15 \text{ m}.$$

For the new curve, Fig. 6.20, from the properties of the cubic spiral transition

$$\phi_1 = \frac{L}{2R} \text{ radian,}$$

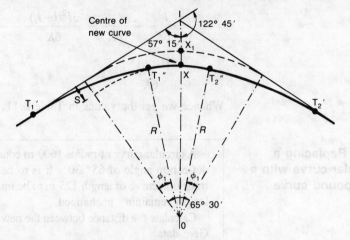

where R is the radius of the new circular curve.

New curve length

$$= 2L + \left(\frac{\pi}{180} \times 65.60 - 2 \times \frac{L}{2R} \right) R$$

$$= 2L + \left(1.143\ 190\ 7 - \frac{L}{R} \right) R.$$

Shift $S = \dfrac{L^2}{24R}$,

where $L = 125.00$ m. Therefore new tangent length

$$\mathrm{IT_1'} = \frac{L}{2} + (R+S) \tan 32°\ 45'$$

$$= \frac{L}{2} + \left(R + \frac{L^2}{24R} \right) \tan 32°\ 45'.$$

The total curve length $\mathrm{T_1'T_2'}$ has to equal the length of the existing circular curve plus $2\mathrm{T_1'T_1}$, since the total route length must remain unchanged. Thus

$$1829.11 + 2\mathrm{T_1T_1'} = 2L + \left(1.143\ 190\ 7 - \frac{L}{R} \right) R.$$

But $\mathrm{T_1T_1'} = \dfrac{L}{2} + \left(R + \dfrac{L^2}{24R} \right) \tan 32°\ 45' - 1029.15$

and so $1829.11 + 125.00 + 2 \left(R + \dfrac{125.00^2}{24R} \right) \times 0.643\ 221\ 6$

$$- 2058.30 = 250.00 + \left(1.143\ 190\ 7 - \frac{125.00}{R} \right) R.$$

Therefore $R^2 - 1599.90R + 5846.53 = 0$ and $R = 1596.24$ m.

Now shift $S = \dfrac{L^2}{24R} = \dfrac{125.00^2}{24 \times 1596.24 \text{ m}} = 0.41$ m.

Therefore $(R+S) = 1596.24 + 0.41 = 1596.65$ m.

$$IT_1' = \frac{125.00}{2} + 1596.65 \tan 32° 45'$$

$$= 1089.50 \text{ m}.$$

Therefore T_1T_1' = distance between the new and existing tangent points

$$= 1089.50 \text{ m} - 1029.15 \text{ m}$$

$$= \mathbf{60.35 \ m.}$$

Chainage of T_1' $= 5264.50 - 1089.50$

$$= 4175.00 \text{ m}.$$

(a) To set out transition curve $T_1'T_1''$ deflection angle (δ) can be calculated for chord lengths (l) using the expression

$$\delta = \frac{1800 \, l^2}{\pi \, RL} \text{ min}$$

$$= 0.002 \, 871 \, 5 \, l^2 \text{ min in this case.}$$

Table 6.12

Chainage (m)	l (m)	δ
4180.00	5.00	0.07′ = 0′ 04″
4200.00	25.00	1.79′ = 1′ 47″
4220.00	45.00	5.81′ = 5′ 49″
4240.00	65.00	12.13′ = 12′ 08″
4260.00	85.00	20.75′ = 20′ 45″
4280.00	105.00	31.69′ = 31′ 41″
4300.00	125.00	44.87′ = 44′ 52″

(b) Calculate data to locate X, the centre point of the curve (Fig. 6.20).

$$I0 = \frac{(R + S)}{\cos 32° 45'} = \frac{1596.24 + 0.41}{\cos 32° 45'} = 1898.43 \text{ m}$$

Therefore $IX_1 = 1898.43 - 1596.65 = 301.78$ m

and $IX = 301.78 + 0.41 = \mathbf{302.19 \ m.}$

Point X can be located from I by setting out an angle of $122° 45'$ from $T_1'I$ produced and measuring distance IX. Note that the centre point of the existing curve is 0.22 m further away from I.

A new road with a 100 km/h design speed has two straights joined by a curve consisting of two transitions and a 500 m radius circular curve. The rate of change of radial acceleration on the transitions is limited to 0.3 m/s^3. The co-ordinates in Table 6.13 have been fixed.

Table 6.13

	E (m)	N (m)
Remote control station A	1392.906	802.285
Remote control station B	1554.951	833.074
Intersection point of straights	1553.203	770.889
Tangent point with first straight (T)	1298.179	719.840
Tangent point with second straight (U)	1778.662	641.230

Calculate the setting out angles at A and B, related to line AB, for the mid-point of the curve.   [Salford]

Introduction. Most major roads are now set out from control points using either two theodolites and intersecting rays or, more usually, an EDM with a bearing and distance as indicated in Example 6.1. The computer program at the end of this problem allows suitable data to be calculated for the case of a compound curve.

Solution. First calculate the details for the transition curve.

$$\text{Design speed} = 100 \text{ km/h}$$

$$= \frac{100 \times 1000}{3600}$$

$$= 27.778 \text{ m/s}.$$

$$a = \frac{V^3}{LR},$$

so

$$L = \frac{27.78^3}{0.3 \times 500}$$

$$= 142.890 \text{ m}.$$

$$\phi_1 = \frac{L}{2R} = \frac{142.89}{2 \times 500}$$

$$= 0.142\,89 \text{ radians} = 8° \, 11' \, 13''.$$

This value of ϕ_1 is significantly large to warrant the inclusion of second-order terms in the formula for the offsets to the transition curve.

$$X = \frac{L^3}{6LR} - \frac{L^7}{336(LR)^3} \quad \text{(see Fig. 6.21)}$$

$$= \frac{142.89^2}{6 \times 500} - \frac{142.89^4}{336 \times 500^3}$$

$$= 6.796 \text{ m.}$$

$$Y = L - \frac{L^5}{40(LR)^2}$$

$$= 142.89 - \frac{142.89^3}{40 \times 500^2}$$

$$= 142.598 \text{ m.}$$

Calculate the co-ordinates of tangent point T_1. For the transition curve at T_1

$$\delta_1 = \tan^{-1} \frac{6.796}{142.598}$$

$$= 2° \ 43' \ 43''.$$

and   length $TT_1 = \sqrt{(6.796^2 + 142.598^2)}$

$$= 142.760.$$

From the co-ordinate data

$$\text{WCB of TI} = \tan^{-1} \frac{1553.203 - 1298.179}{770.889 - 719.840}$$

$$= 78° \ 40' \ 50''.$$

$$\text{WCB of TT}_1 = 78° \ 40' \ 50'' + 2° \ 43' \ 43''$$

$$= 81° \ 24' \ 33''.$$

E coordinate of T_1 = 1298.179 + 142.760 sin 81° 24′ 33″

$$= 1439.338 \text{ m.}$$

N coordinate of T_1 = 719.840 + 142.760 cos 81° 24′ 33″

$$= 741.165 \text{ m.}$$

Next find the deflection angle of the curve from the co-ordinate data.

$$\text{Deflection angle of curve} = \tan^{-1} \frac{770.889 - 719.840}{1553.203 - 1298.179}$$

$$+ \tan^{-1} \frac{770.889 - 641.230}{1778.662 - 1553.203}$$

$$= 11° \ 19' \ 10'' + 29° \ 54' \ 10''$$

$$= 41° \ 13' \ 20''.$$

Deflection angle of mid-point, $M = \dfrac{41° \ 13' \ 20''}{2} = 20° \ 36' \ 40''.$

Deflection angle of circular curve to M = 20° 36′ 40″ − 8° 11′ 13″

$$= 12° \ 25' \ 27''.$$

WCB of T_1M (Fig. 6.21) $= 81° \ 24' \ 33'' + \phi_1 - \delta_1 + \dfrac{12° \ 25' \ 27''}{2}$

$$= 81° \ 24' \ 33'' + 8° \ 11' \ 13'' - 2° \ 43' \ 43'' + 6° \ 12' \ 43''$$

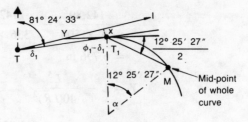

Figure 6.21

i.e. WCB of $T_1M = 93° 04' 46''$.

$$\text{Length of } T_1M = 2 R \sin \frac{\alpha}{2}$$

$$= 2 \times 500 \sin \frac{12° 25' 27''}{2}$$

$$= 108.209 \text{ m}$$

Co-ordinates of M:

$$\text{Easting} = 1439.338 + 108.209 \sin 93° 4' 46''$$
$$= 1547.391 \text{ m};$$
$$\text{Northing} = 741.165 + 108.209 \cos 93° 4' 46''$$
$$= 735.352 \text{ m}.$$

Now compute the setting-out data from the control points. In the triangle ABM:

$$AB = \sqrt{[(1392.906 - 1554.951)^2 + (802.285 - 833.074)^2]}$$
$$= 164.944 \text{ m};$$
$$AM = \sqrt{[(1392.906 - 1547.391)^2 + (802.285 - 735.352)^2]}$$
$$= 168.362 \text{ m};$$
$$BM = \sqrt{[(1554.951 - 1547.391)^2 + (833.074 - 735.352)^2]}$$
$$= 98.015 \text{ m}.$$

By cosine rule

$$\hat{BAM} = \cos^{-1}\left(\frac{AB^2 + AM^2 - BM^2}{2 \times AB \times AM}\right)$$

$$= \cos^{-1}\left(\frac{164.944^2 + 168.362^2 - 98.015^2}{2 \times 164.944 \times 168.362}\right)$$

$$= \mathbf{34° 11' 02''}$$

$$\hat{ABM} = \cos^{-1}\left(\frac{164.944^2 + 98.015^2 - 168.362^2}{2 \times 164.944 \times 98.015}\right)$$

$$= \mathbf{74° 49' 06''}$$

The following computer program will calculate the setting-out data for a compound curve located from two control stations. Co-ordinates of the control stations, the intersection point of the straights and the tangent points with the straights must be known. The data is output in a table and readers with programmable calculators will need to reformat the output. On computers that do not support the logic functions AND and OR lines 190 and 200 must be replaced by:

```
190 IF X4 = X3 THEN 210
191 IF X4 > X3 THEN 200
192 IF Y4 > Y3 THEN A4 = P + A4
193 GOTO 210
200 IF Y4 < Y3 THEN A4 = 2 * P + A4
201 IF Y4 > Y3 THEN A4 = P + A4
```

A similar routine is required to replace lines 620 and 630.

Variables

A1 Deflection of the whole curve, θ

A2 Deflection of the transition curves, ϕ_1

A3 Deflection of the circular curve, $\theta - 2\phi_1$

A4 WCB of first tangent/final tangent

A5 WCB of chord to transition curve

A6 Setting-out angle for the circular curve

A7 WCB of chord to the circular curve

A8, A9 Component of A1

C Chainage of the intersection point

C1 Chainage of the first tangent point

C2 Chainage of the second tangent point

C3 Chainage of the third tangent point

C4 Chainage of the final tangent point

C5, C6 Chainage of the current position

D Output angle, degrees

H ± counter

I Setting out chainage interval

K L1 * R for the transition

L1 Length of transition

L2 Length of circular curve

L3 Running curve length

L4 Length of a chord to the circular curve

L5 Distance between control points

L6, L7 Length from control points to points being calculated

L8 Length of a chord to the transition curve

M Output angle, minutes

P π

Q 'S' in the triangle formed by control points 1 and 2 and the point being calculated

Q1 Constant term in the expression for the angle

R Radius of the circular curve

S Output angle, seconds

T X offset of the transition from the tangent

U Y offset of the transition along the tangent

V Road design speed

W $\sin^{-1}$ of the angle at the control point

X X co-ordinate of current point

X1 X co-ordinate of control point 1

X2 X co-ordinate of control point 2

X3 X co-ordinate of intersection point

X4 X co-ordinate of tangent point with first straight

X5 X co-ordinate of tangent point with exit straight

Y Y co-ordinate of current point

Y1–Y5 Y co-ordinate as above

```
10 REM SET OUT A COMPOUND CURVE FROM REMOTE CONTROL POINTS
20 H=1
30 P=3.14159
40 INPUT"DESIGN SPEED   KM/H                    ",V
50 V=V/3.6
60 INPUT"RADIUS OF CIRCULAR CURVE     M          ",R
70 INPUT"CHAINAGE OF INTERSECTION POINT    M     ",C
80 INPUT"SETTING OUT CHAINAGE INTERVAL    M      ",I
90 INPUT"CO-ORDINATES OF CONTROL POINT 1        ",X1,Y1
100 INPUT"CO-ORDINATES OF CONTROL POINT 2        ",X2,Y2
110 INPUT"CO-ORDINATES OF INTERSECTION POINT     ",X3,Y3
120 INPUT"CO-ORDINATES OF TANGENT TO FIRST STRAIGHT ",X4,Y4
130 INPUT"CO-ORDINATES OF TANGENT TO EXIT  STRAIGHT ",X5,Y5
140 L5=SQR((X1-X2)^2+(Y1-Y2)^2)
150 L1=V^3/(0.3*R)
160 K =L1*R
170 A2=L1/(2*R)
180 A4=ATN((X3-X4)/(Y3-Y4))
190 IF X4>X3 AND Y4<Y3 THEN A4=2*P+A4
200 IF ( X4>X3 AND Y4>Y3 ) OR ( X4<X3 AND Y4>Y3 ) THEN A4=P+A4
210 A8=ABS(ATN((Y3-Y4)/(X3-X4)))
220 A9=ABS(ATN((Y3-Y5)/(X3-X5)))
230 A1=A8+A9
240 A3=A1-(A2*2)
250 L2=R*A3
260 C1=C-SQR((X3-X4)^2+(Y3-Y4)^2)
270 C2=C1+L1
280 C3=C2+L2
290 C4=C3+L1
300 C6=INT(C1*1000+0.5)/1000
310 L3=0
320 PRINT"CHAINAGE      CO-ORDINATES       DIST. 1     ANG. 1      DIST.2
    ANG.2"
330 PRINT" FIRST TANGENT POINT"
340 GOSUB 840
350 L3=(INT(C1/I)+1)*I-C1
360 C5=C1+L3
370 C6=INT(C5*1000+0.5)/1000
380 GOSUB 800
390 L3=L3+I
400 C5=C5+I
410 IF C5>C2 THEN GOTO 430
420 GOTO 370
430 L3=L1
440 C6=INT(C2*1000+0.5)/1000
450 PRINT" SECOND TANGENT POINT"
460 GOSUB 800
470 L3=C5-C2
480 X6=X
490 Y6=Y
500 C6=INT(C5*1000+0.5)/1000
510 GOSUB 870
520 L3=L3+I
530 C5=C5+I
540 IF C5>C3 THEN GOTO 560
550 GOTO 500
560 L3=L2
570 C6=INT(C3*1000+0.5)/1000
580 PRINT" THIRD TANGENT POINT"
590 GOSUB 870
600 H=-1
610 A4=ATN((X3-X5)/(Y3-Y5))
620 IF X5>X3 AND Y5<Y3 THEN A4=2*P+A4
630 IF ( X5>X3 AND Y4>Y3 ) OR ( X5<X3 AND Y5>Y3 ) THEN A4=P+A4
640 X4=X5
650 Y4=Y5
660 L3=L1-(C5-C3)
670 C6=INT(C5*1000+0.5)/1000
680 GOSUB 800
690 L3=L3-I
700 C5=C5+I
710 IF C5>C4 THEN 730
720 GOTO 670
730 L3=0
```

```
740 C6=INT(C4*1000+0.5)/1000
750 PRINT" FINAL TANGENT POINT"
760 X=X4
770 Y=Y4
780 GOSUB 920
790 END
800 T =(L3^3/(6*K))-(L3^7/(336*K^3))
810 U =L3-(L3^5/(40*K*K))
820 A5=A4+H*ATN(T/U)
830 L8=SQR((T*T)+(U*U))
840 X=INT((X4+L8*SIN(A5))*1000+0.5)/1000
850 Y=INT((Y4+L8*COS(A5))*1000+0.5)/1000
860 GOTO 920
870 A6=L3/R
880 A7=A4+A2+(A6/2)
890 L4=2*R*(SIN(A6/2))
900 X=INT((X6+L4*SIN(A7))*1000+0.5)/1000
910 Y=INT((Y6+L4*COS(A7))*1000+0.5)/1000
920 L6=INT(SQR((X1-X)^2+(Y1-Y)^2)*1000+0.5)/1000
930 L7=INT(SQR((X2-X)^2+(Y2-Y)^2)*1000+0.5)/1000
940 Q=(L5+L6+L7)/2
950 Q1=(Q-L5)/L5
960 PRINT C6;TAB(11);X;TAB(22);Y;
970 W=SQR((Q-L6)*Q1/L6)
980 GOSUB 1040
990 PRINT TAB(33);L6;TAB(44);D;M;S;
1000 W=SQR((Q-L7)*Q1/L7)
1010 GOSUB 1040
1020 PRINT TAB(60);L7;TAB(71);D;M;S
1030 RETURN
1040 A=INT(206264.8*2*ATN(W/SQR(1-W*W))+0.5)
1050 D=INT(A/3600)
1060 M=INT((A-(D*3600))/60)
1070 S=A-(D*3600)-(M*60)
1080 RETURN
```

6.11 Compound curve — checking a railway curve by the 'versine' method

A compound curve on an existing railway track is to be realigned to give two 100 m long transition curves and a circular curve of radius 900 m. A surveyor has measured versines from a 20 m chord and recorded the following data (in mm).

0, 0tp, 10, 15, 18, 20, 23, 29, 37, 39, 46, 58tp, 62, 60, 61, 62, 62, 63tp, 59, 46, 36, 32, 21, 17, 15, 10, 6, 1tp, −1, 0,

where tp indicates a tangent point on the curve. Calculate the necessary slew of the track using the moment method.

Introduction. In the UK there is very little new railway construction and the main job of the permanent way engineer is to maintain the alignment of the existing track. Since variations from the correct alignment are usually small the rail can be used as a datum. The standard procedure is to select a chord length appropriate to the design speed of the track, then set out regular intervals of half a chord length. Two offset devices, 'knives and forks', are attached to the track at the full chord points, and a wire is tensioned between them. The offset from the wire to the track at the mid-point of the chord (the versine of the chord) is measured. The wire is held 25 mm inwards from the correct alignment to ensure that zero and negative readings can be detected, and measurement is normally made with a Hallade rule that allows for this offset. Once the measurement has been made the apparatus is moved down the track by half a chord length and the measurement repeated (see Fig. 6.22).

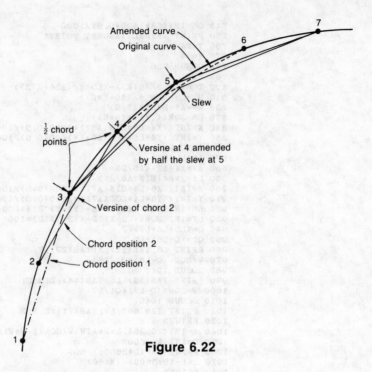

Figure 6.22

Solution. The versine of the circular curve can be approximated by

$$\frac{(\frac{1}{2}c)^2}{2R} = \frac{(\frac{1}{2} \times 20)^2}{2 \times 900} = 55.8 \text{ mm},$$

where c is the chord length.

For the transition curve a similar expression can be used.

$$\frac{(\frac{1}{2}c)^2 l}{2RL} = \frac{(\frac{1}{2} \times 20)^2 l}{2 \times 900 \times 100} = 0.000\ 56\ l$$

where L is the transition length and l is the running transition length. Note that the versine is a constant for the circular curve and for the cubic parabola transition curve it increases at a uniform rate.

Calculate the versines for the new curve. It is standard railway practice to ease the entry and exit from the transition by applying a versine at the tangent to the straight equal to 1/6 of the increase and by reducing the value at the tangent to the circular curve by the same amount.

Increase in versine for $\frac{1}{2}$ chord = $0.000\ 56 \times 10 = 5.6$ mm.

Thus first versine = $\dfrac{5.6}{6} = 0.9$ mm

Versine at 10 m = $5.6 + 0.9 = 6.5$ mm
Versine at 20 m = $6.5 + 5.6 = 12.1$ mm

Versine at 90 m $= 45.7 + 5.6 = 51.3$ mm
Versine at 100 m $= 55.8 - 0.9 = 54.9$ mm.

To correct the data by the method of moments a tabular format is used where:

Difference $=$ Required $-$ Actual versine;
Σ difference $=$ Sum of differences;
Moment $=$ Sum of Σ differences;
Slew $= -2 \times$ moment (sign convention outwards $+$ve, inwards $-$ve).

Note that from Fig. 6.22 a slew movement at point 5 amends the versine at points 4 and 6 by half of the slew, and the method of moments carries this correction cumulatively around the curve.

The 2 mm slew required for the exit straight would in practice be run out evenly at the end of the transition and the start of the straight.

Table 6.14

Point	Measured versine (mm)	l (m)	Required versine (mm)	Difference (mm)	Σ difference (mm)	Moment (mm)	Slew (mm)
1	0	—	0	0			
2	0	0tp	0.9	0.9	0	0	0
3	10	10	6.5	−3.5	0.9	0.9	−2
4	15	20	12.1	−2.9	−2.6	−1.7	+3
5	18	30	17.7	−0.3	−5.5	−7.2	+14
6	20	40	23.3	3.3	−5.8	−13.0	+26
7	23	50	28.9	5.9	−2.5	−15.5	+31
8	29	60	34.5	5.5	3.4	−12.1	+24
9	37	70	40.1	3.1	8.9	−3.2	+6
10	39	80	45.7	6.7	12.0	8.8	−18
11	46	90	51.3	5.3	18.7	27.5	−55
12	58	100tp	54.9	−3.1	24.0	51.5	−103
13	62	—	55.8	−6.2	20.9	72.4	−145
14	60	—	55.8	−4.2	14.7	87.1	−174
15	61	—	55.8	−5.2	10.5	97.6	−195
16	62	—	55.8	−6.2	5.3	102.9	−206
17	62	—	55.8	−6.2	−0.9	102.0	−204
18	63	100tp	54.9	−8.1	−7.1	94.9	−190
19	59	90	51.3	−7.7	−15.2	79.7	−159
20	46	80	45.7	−0.3	−22.9	56.8	−114
21	36	70	40.1	4.1	−23.2	33.6	−67
22	32	60	34.5	2.5	−19.1	14.5	−29
23	21	50	28.9	7.9	−16.6	−2.1	+4
24	17	40	23.3	6.3	−8.7	−10.8	+22
25	15	30	17.7	2.7	−2.4	−13.2	+26
26	10	20	12.1	2.1	0.3	−12.9	+26
27	6	10	6.5	0.5	2.4	−10.5	+21
28	1	0tp	0.9	−0.1	2.9	−7.6	+15
29	−1	—	0	1.0	2.8	−4.8	+10
30	0	—	0	0	3.8	−1.0	+2

6.12 Sag vertical curve

Two straights AB and BC falling to the right at 1 in 10 and 1 in 20, respectively, are to be connected by a parabolic vertical curve 200 m long. Given that the chainage and reduced level of B are 3627.00 m and 84.64 m, respectively, design the vertical curve.

What is the sighting distance for a car whose headlights are 0.70 m above road level and the beams are inclined upwards at an angle of 1°?

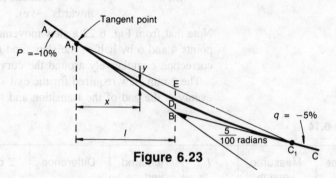

Figure 6.23

Solution.   First determine the design parameters. A sag vertical curve, Fig. 6.23, is required to connect the two gradients. Conventionally, a falling gradient is indicated by a negative sign and the gradient of 1 in 10 can be written in percentage terms as $\frac{100}{10}\% = 10\%$; similarly, that of 1 in 20 can be written as $\frac{100}{20}\% = 5\%$.

The general expression for the offsets (y) from a tangent to locate the curve is

$$y = \frac{(p-q)\,x^2}{400\,l},$$

where x is the distance from a tangent point; and l is the length of a tangent (half the length of the curve). Therefore for this vertical curve

$$y = \frac{(-10-(-5))}{400 \times 100}\, x^2$$

$$= \frac{5\,x^2}{400 \times 100},$$

when we ignore the negative sign. It is usual to take the curve length to be equal to the total length of the tangents and, for gradients of this order, to assume that they are equal to the horizontal lengths. When $x = 100$ m

$$\text{offset BD} = \frac{5}{400 \times 100} \times 100^2 = 1.25 \text{ m}.$$

Next establish the curve levels. The chainage of B has been given as 3627.00 m and, accordingly, the chainage of tangent point A_1 will be 3627.00 − 100.00 = 3527.00 m. The reduced level of A_1 will be that of B plus 100 × $\frac{1}{10}$ m, acknowledging that a length of 100 m is being covered at a gradient of 1 in 10. Alternatively, we can say that a gradient of 1 in 10 is equivalent to an angle of $\frac{10}{100}$ radian with the horizontal and this results in a change in level of $\dfrac{100 \times 10}{100}$ m over the distance of 100 m.

The reduced level of A_1 is therefore 84.64 m + 10.00 m = 94.64 m. Similarly, the reduced level of C_1 will be

$$84.64 - 100 \times \tfrac{1}{20} = 79.64 \text{ m}.$$

The offsets can be evaluated from both tangents or from either tangent, produced as necessary. The second alternative will be adopted in this example and tabulated in Table 6.15 using running chainages at 20 m intervals. Offsets must then be added to the corresponding grade (tangent) levels.

Table 6.15

Point	Chainage (m)	x (m)	Grade level $94.64 - (x/10)$ (m)	Offset $x^2/8000$ (m)	Curve level (m)
A_1	3527.00	0.00	94.64	0.00	94.64
	3540.00	13.00	93.34	0.02	93.36
	3560.00	33.00	91.34	0.14	91.48
	3580.00	53.00	89.34	0.35	89.69
	3600.00	73.00	87.34	0.67	88.01
	3620.00	93.00	85.34	1.08	86.42
B	3627.00	100.00	84.64	1.25	85.89
	3640.00	113.00	83.34	1.60	84.94
	3660.00	133.00	81.34	2.21	83.55
	3680.00	153.00	79.34	2.93	82.27
	3700.00	173.00	77.34	3.74	81.08
	3720.00	193.00	75.34	4.66	80.00
C_1	3727.00	200.00	74.64	5.00	79.64

With the car at tangent point A_1, the headlight beams will strike the road surface at a point where the offset is $(0.70 + x \tan 1°)$, x being the distance from A_1. Thus

$$0.70 + 0.0175x = \dfrac{5}{400 \times 100} x^2$$

and $\qquad x$ = sighting distance in this case
$\qquad\qquad\qquad$ = **172.47 m.**

6.13 Summit vertical curve

Two straights PB and BQ are to be connected at a summit by a parabolic vertical curve. P, reduced level 108.25 m, lies on the gradient rising to the right at 1 in 60 at chainage 1862.00. Q, reduced level 106.85 m, lies on the gradient falling to the right at 1 in 50 at chainage 2174.00. The vertical curve must pass through R, reduced level 109.68 m and chainage 1986.00.

Design the curve and determine the sighting distance between two points 1.05 m above road level.

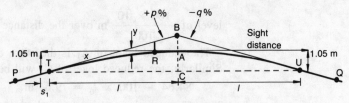

Figure 6.24

Solution. First determine the length of the curve. Let horizontal distance PB $= x_1$ in Fig. 6.24.

$$\text{Then level of B} = \text{level of P} + \frac{x_1}{60}$$

$$= \text{level of Q} + \frac{(2174 - 1862 - x_1)}{50}.$$

Therefore $108.25 + \dfrac{x_1}{60} = 106.85 + \dfrac{312 - x_1}{50}.$

and $\qquad\qquad x_1 = 132.00$ m.

Chainage of B is $1862.00 + 132.00 = 1994.00$ m. R lies at chainage 1986.00 and is thus 8.00 m from B and 124.00 m from P.

$$\text{Grade level at this chainage} = \frac{124.00}{60} + 108.25$$

$$= 110.32 \text{ m.}$$

$$\text{Curve level at R} = 109.68 \text{ m.}$$

$$\text{Therefore offset at R} = 0.64 \text{ m.}$$

If the tangent length is l

$$0.64 = \frac{(p+q)}{400l}(l-8)^2,$$

in which $p = \frac{100}{60}\%$ and $q = \frac{100}{50}\%$.

Therefore $l = 84.72$ m, say 85 m for design purposes and so the chainage of T is $1994.00 - 85.0 = 1909.00$ m.

The design data can be tabulated as in Table 6.16. All the values are based on horizontal distances and vertical offsets.

The level of a point on the curve above T is given by the expression

$$\frac{px}{100} - \frac{(p+q)}{400l}x^2,$$

where x is the horizontal distance from T. By differentiating we can determine the position of the highest point on the curve.

Thus $\dfrac{p}{100} - 2x\dfrac{(p+q)}{400l} = 0.$

Table 6.16

Chainage (m)	Grade level (m)	x (m)	Offset y (m)	Curve level (m)
1909.00	109.03	0.0	0.0	109.03
1920.00	109.21	11.00	0.01	109.20
1940.00	109.55	31.00	0.10	109.45
1960.00	109.88	51.00	0.28	109.60
1980.00	110.21	71.00	0.54	109.67
1994.00	110.45	85.00	0.78	109.67
2000.00	110.55	91.00	0.89	109.66
2020.00	110.88	111.00	1.33	109.55
2040.00	111.21	131.00	1.85	109.36
2060.00	111.55	151.00	2.46	109.09
2079.00	111.86	170.00	3.12	108.74

$$x_{max} = \frac{2p\,l}{(p+q)}$$

$$= \frac{2 \times 10/6 \times 85}{10/6 + 2} = 77.27 \text{ m.}$$

At this point the grade level is 110.32 m AOD

and the offset $= \dfrac{(10/6 + 2)}{400 \times 85} \times (77.27)^2 = 0.64$ m.

Therefore the highest point has a reduced level of 109.68 m.

Finally calculate the sight distance.

The offset AB at B = 0.78 m
$$= \text{AC.}$$

The sight line can be taken as the tangent to the vertical curve at A, (Fig. 6.24). It is parallel to TU whose slope is

$$\frac{\left(\dfrac{pl}{100} - \dfrac{ql}{100}\right)}{2l} \text{ rad.}$$

The angle contained between the tangent TB and chord TU at T is

$$\frac{p}{100} - \frac{(p-q)}{200} = \frac{p+q}{200} \text{ radian,}$$

whence $s_1 \dfrac{(p+q)}{200} = 1.05 - 0.78 = 0.27$

$$s_1 = 14.73 \text{ m.}$$

Therefore total sight distance = 2 (85.00 + 14.73)
$$= \textbf{200 m, say.}$$

Note, as mentioned previously, curve lengths can be established in practice using an expression of the type:

$$L = 2l = K (p+q) = KG,$$

in which K depends upon the value of centrifugal acceleration chosen for a particular design velocity and G is the relative change in gradient. In this particular example we have

$$-l = 85 \text{ m} \quad l = 170 \text{ m}$$

$$L = K \left(\frac{100}{60} + \frac{100}{50} \right).$$

Therefore $K = 46.4$.

Problems

1 A circular curve of radius 600 m is to connect two straights having a deviation angle of 11° 58′. The chainage of the intersection point is 2573.250. Tabulate the data needed to set out the curve by theodolite and chain using 20 m chords.

Explain carefully why 20 m chords are appropriate for setting out the curve.   [Bradford]

Answer Curve length 125.315 m; Tangent length 62.886 m

2 (*a*) Describe a method for setting out a circular curve of small radius, such as a minor road kerb line or a boundary wall, using only a chain and tape. You should derive any necessary formulae from first principles.

(*b*) Tabulate the data required to set out, by chain and 20″ theodolite, a circular curve of radius 500 m to connect two straights having an intersection angle 15° 20′ 00″. You may assume that the chainage of the intersection point is 2100.00 m and that the curve is to be defined in 20 m intervals.   [Salford]

Answer Chainages of tangent points 2032.69 m, 2166.50 m

3 Two straights AB and BC which have whole circle bearings of 45° and 7°, respectively, are to be connected by a circular curve passing through a point, 29 m from B, which lies on the bisector of angle ABC. Calculate the length of that curve.

If the chainage of the intersection point B is 4715 m determine (*i*) the chainages of the tangent points on AB and BC, and (*ii*) the deflection angles required to set out the first two points on the curve at through chainages of 20 m, using a theodolite reading to 20 seconds set up at the tangent point on AB.

Also find the length and bearing of a third straight which is tangential to the curve at chainage 4680 m.   [CEI]

Answer (*i*) 4541.70 m, 4875.50 m; (*ii*) 358° 57′ 20″, 357° 49′ 20″.
Length 168.58 m; 29° 15′ 17″

4 Straights UV and VW are to be connected by a circular curve. As V is inaccessible a point X was chosen on UV between U and V and a point Y on VW between V and W. The distance XY was 97.62 m. The following angles were measured: UXY = 164° 48′, XYW = 129° 12′. If X is used as a tangent point calculate the radius of the curve. Calculate the distance from Y to the tangent point on that straight.

[Bradford]

Answer Radius = 127.52 m; 54.79 m

5 Two straight sections of road intersect at chainage 2539.10 m. The straights are to be joined by a curve consisting of two transitions and a circular curve radius 500 m. If the transition curves are 56.32 m long and the tangent to the first curve is at 2417.20 m, what is the deflection angle of the whole curve and the chainage of the tangent point of the circular curve and second transition? [Salford]
Answer 21° 13′ 34″; 2602.43 m

6 (*a*) Name the purposes which a transition curve serves in road design and discuss briefly the factors which influence the length selected for it.
Establish the formula

$$v = \sqrt{\frac{gR(\tan \alpha + \mu)}{1 - \mu \tan \alpha}}$$

for the speed limit on a road curve, defining the terms on the right-hand side of the equation.

(*b*) One of the four loops of a 'cloverleaf' at the junction of two roads at different levels is to have a centre line formed by two symmetrical clothoids (intrinsic equaton $\phi = l^2/2RL$). The springing points of these clothoids have the same plan position and the horizontal angle between the approach and exit tangents is 110°. The maximum superelevation on the carriageway of the loop is specified as 1 in 10. The design speed is 40 km/h and the maximum coefficient of lateral friction is taken to be 0.15.

Calculate the total length in horizontal projection between the springing points along the centre line. *Aide mémoire:* $g = 9.81$ m/s^2

[London]

Answer 501.90 m

7 A wholly transitional curve with a constant rate of change of radial acceleration is to be inserted to connect two straight sections of highway AI and IB which intersect at I. The intersection point I is inaccessible and two points X and Y are established on AI and IB, respectively.

The co-ordinates of X and Y are 1693.41 m E, 1888.47 m N and 1904.07 m E, 1973.25 m N, respectively. The angle AXY is measured as 163° 15′ 20″ and XYB as 151° 12′ 00″. If the design speed of the road is 100 km/h and the minimum radius is to be 400 m, calculate:

(*i*) the total length of the curve;
(*ii*) the rate of change of radial acceleration; and
(*iii*) the co-ordinates of the two tangent points.

[Leeds]

Answer (*i*) 635.92 m; (*ii*) 0.17 m/s^3; (*iii*) 1555.08 mE, 1777.78 mN; 2141.12 mE, 1944.66 mN

8 The co-ordinates of two survey stations relative to a station A are

B E 500.00 N 550.00
C E 1300.00 N 700.00

Calculate the radius of the circular curve passing through the three stations and the co-ordinates of the centre of curvature.

Describe briefly how such a circular curve could be set out below ground. [CEI]

Answer 1223.67 m; 1112.67 mE, 509.25 mS

9 Two underground drivages AO and BO, on bearings of 117° 32′ 30″ and 57° 06′ 00″, respectively, are to be connected by a curve of radius 350 m by means of nine equal chords. Compute

(*i*) the common chord length,
(*ii*) the bearing of each chord, and
(*iii*) the tangent length.

If the co-ordinates of A are 56.9 mW, 1120.2 mN and the distance AO = 2230.5 m, compute the co-ordinates of the centre of the curve.

[CEI]

Answer (*i*) 81.0 m; (*ii*) 5th chord 177° 19′ 15″; (*iii*) 600.9 m; 1226.2 mE, 56.4 mN

10 A transition curve is required to join a straight to a circular curve of radius 450 m such that the maximum rate of change of radial acceleration = 0.3 m/s^3. The deflection angle of the curve at chainage 3634 m is 6′ 20″ and at 3664 m is 50′ 45″. What is the design speed of the curve in km/h?

Answer 69.6 km/h

11 Two straights AI and IB have bearings of 37° 15′ 00″ and 48° 19′ 40″, respectively, and are to be connected by means of a wholly transitional curve with a rate of change of radial acceleration of 0.3 m/s^3. If the design speed of the road is 100 km/h and the co-ordinates of point I are 2264.83 mE, 2742.88 mN calculate

(*a*) the minimum radius of the curve
(*b*) the total length of the curve and
(*c*) the co-ordinates of the two tangent points. [Leeds]

Answer (*a*) 332.96 m; (*b*) 128.75 m; (*c*) 2625.78 mE, 2691.53 mN; 2713.02 mE, 2785.77 mN

12 Two tangents intersecting at an angle of 35° at chainage 1450.361 are to be connected by means of a circular curve with a radius of 1000 m having similar transitions at each end. The transitions are to be cubic parabolas designed for a speed of 100 km/h and a rate of gain of radial acceleration of 0.25 m/s^3.

Calculate the chainage at the beginning of the first transition and the tangential angles for the first three points on the curve using 20 m increments. [Bradford]

Answer 1092.098 m; 02′ 40″; 10′ 42″; 24′ 03″

13 A curve connecting two straights which deflect through an angle of 8° is transitional throughout, and the junction of the two parts is 6.00 m from the intersecting point of the straights. Determine, from first principles, the total length of the curve, the tangent length and the rate of change of radial acceleration at a velocity of 110 km/h. [Salford]

Answer 514.58 m; 257.58 m; 0.6 m/s^3

14 A vertical parabolic summit curve is to be inserted to connect an upwards gradient of 2.4% to a downwards gradient of 1.0%. The height of the tangent point on the 2.4% gradient is to be 212.840 m above datum, and because of site restrictions, it is necessary that the height of the curve should be 215.070 m at a point 126.4 m from this tangent point. The *k*-value for the curve is 50 and it is to have equal tangent distances. Calculate:

(*i*) the length of the curve,

(*ii*) the height of the highest point on the curve and its distance from the tangent point on the 2.4% gradient,

(*iii*) the heights of points on the curve at 50 m intervals measured from the tangent point on the 2.4% gradient.

Show that the curve satisfies the required design standard.

[Leeds]

Answer (*i*) 337.82 m; (*ii*) 215.701 m

15 On a new road with an 80 km/h design speed, a summit joining an up grade of 1:50 to a down grade of 1:60, takes place on a horizontal curve consisting of two transitions and a 500 m radius circular curve. The road has been arranged so that the tangent points A and B, between the transitions and the straights are also the tangent points for the vertical curve. Tangent point A, at chainage 2356 m, lies on the 1:50 gradient and is at level 300.62 m. A sight distance between two points 1.05 m above the road surface is to be 300 m and the rate of change of radial acceleration on the transition is limited to 0.3 m/s^3.

(*a*) You have a theodolite set up on the tangent point between the first transition and the circular curve, on the 1:50 side of the summit. Calculate the data needed to reference the theodolite from tangent point A and then set out the horizontal position of the highest point on the curve. Pegs have already been placed at through 20 m chainage intervals.

(b) What is the road level at the tangent point between the circular curve and the transition at end B?

(Assume constant gradients.) [Salford]
Answer (a) 2570.29 m; (b) 302.25 m

16 A, B, C and D are four points marked on the centre-line of an existing straight road. From A to B and from C to D there are uniform rising gradients of 1 in 10 and 1 in 50, respectively. These gradients are connected smoothly by a vertical parabolic curve which extends from B to C. The through chainage of B, C and D measured from A are 240.0 m, 320.0 m, and 400.0 m respectively.

Visibility is to be improved over this length of road by connecting the existing gradients with a new parabolic curve 160.0 m long.

Compute the differences in level between the existing and new road surfaces at 20.0 m intervals of through chainage.

What will be the range of visibility on the new curve for a driver whose eyes are 1.25 m above the road surface? [CEI]
Answer Sight distance 141.42 m

17 A vertical summit curve is to be inserted to connect an upwards gradient of 2.5% to a downwards gradient of 2.0%. The height of the tangent point on the 2.5% gradient is to be 166.24 m above datum and, because of site restrictions, it is necessary that the height of a point 64.4 m along the curve, measured from the tangent point on the 2.5% gradient, must be 167.55 m above datum. Calculate

(i) the total length of the curve,
(ii) the position and height of the highest point on the curve, and
(iii) the height of the tangent point on the 2% gradient.

If the k value for the road is 50, show that the curve complies with the requirements in so far as sight distance is concerned. [Leeds]
Answer (i) 311.05 m; (ii) 172.80 m, 168.40 m; (iii) 167.02 m

18 An up gradient of 2% is to be joined to a down gradient of 1:75 by a vertical curve. A sight distance of 200 m at 1.05 m above ground level is required. If the highest point on the curve is at chainage 2532.00 m and level 100.23 m, what is the chainage and level of the two tangent points? [Salford]
Answer 2443.20 m; 2591.20 m; 99.34 m; 99.83 m

19 A vertical parabolic sag curve is to be designed to connect a down-gradient of 1 in 20 with an up-gradient of 1 in 15, the chainage and reduced level of the intersection point of the two gradients being 2659.00 m and 278.48 m respectively.

In order to allow for necessary headroom, the reduced level of the curve at chainage 2649.00 m on the down-gradient side of the intersection point is to be 280.00 m.

Calculate:

(a) the reduced levels and chainages of the tangent points and the lowest point on the curve,

(b) the reduced levels of the first two pegs on the curve, the pegs being set at the 20 m points of the through chainage. [ICE]

Answer (a) 2605.92 m, 281.13 m; 2651.39 m, 264.90 m; (b) 280.54 m, 280.07 m

20 Derive an expression for the versine of a chord on:

(a) a circular curve

(b) a transition curve.

21 The gradient of a straight length of rail track sloping at -1 in 150 is to be changed to a gradient of $+1$ in 100, the intervening curve being 400 m in length. Assume the curve to start at a point A on the negative gradient, calculate the reduced levels of pegs at 50 m intervals which you would set out in order to construct the curve, assuming the reduced level of A to be 100 m AOD. [CEI]

Answer 99.719; 99.541; 99.469; 99.500; 99.635; 99.875; 100.219; 100.667

22 The bearings of two straights AI and IB are 15° 26′ 12″ and 36° 19′ 20″ respectively. The co-ordinates of the intersection point I are 2440.43 mE, 2336.94 mN. The two straights are to be connected by a wholly transitional curve with a rate of change of radial acceleration of 0.15 m/s³. If the design speed of the road is 120 km/h, calculate the minimum radius of the curve, the co-ordinates of the two tangent points and the co-ordinates of the point on the curve 100 m from the tangent point on AI. [Leeds]

Answer $R = 823.021$ m; co-ordinates of T_1 2359.95 mE, 2045.48 mN; co-ordinates of T_2 2619.53 mE, 2580.55 mN

23 Two straights on a road with a 40 km/h design speed intersect at point I. Point A on the first straight is to be joined to point B on the second straight by a 100 m radius circular curve and two transition curves. The rate of change of radial accleration on the transitions is limited to 0.3 m/s³. Two control stations X and Y are located off the line of the road. The following co-ordinates are known:

X	878.229	928.503
Y	1051.275	937.212
I	1118.313	915.746
A	1061.579	892.347
B	1176.206	895.387

The intersection point I is at chainage 1827.921 m.

Calculate the co-ordinates of the tangent points on the curve and the intermediate points at a through chainage interval of 25 m.

You are in charge of the setting out of the curve and have two theodolites. Calculate the data needed to set out the curve from X and Y.

Answer T_1 1104.949, 906.495, 8° 25′ 31″, 147° 20′ 14″; T_2 1132.082, 907.214, 7° 40′ 29″, 156° 45′ 11″

24 A reverse circular curve is required to link two parallel straight pipelines 500 m apart.

The curve is to be formed from two circular arcs with a common tangent point, and a common radius of 1000 m. Calculate the following:

(a) the deviation angle
(b) the tangent lengths
(c) the total arc length of the double curve
(d) the tangential angle for setting out a 20 m chord.

[Salford]

Answer (a) 41° 24′ 34.6″; (b) 377.96 m; (c) 1445.50 m; (d) 34′ 23″

25 Two straights XB and BY are to be connected at a summit by a parabolic vertical curve. The reduced level at B is 68.18 m above mean sea level, and its chainage is 1078 m.

The gradient of XB and BY are +3% and −2%, respectively. If the vehicular design speed is 80 km/hour and the allowable vertical radial acceleration is 0.2 m/s^2 find

(a) the length of vertical curve required,
(b) the chainage and reduced levels of the tangent points A and C to the vertical curve,
(c) the chainage and reduced level of the highest point of the curve,
(d) the sight distance provided for a driver whose eye level is 1.05 m above the road surface.

If a level is set up at A and the staff is held with its base on the road surface at chainage 1130 m find the minimum height of the the telescope above the road surface which will enable the taking of a reading on the staff. What is the reading? [Salford]

Answer (a) 123.457 m; (b) A, 1016.272 m, 66.328 m, C, 1139.728 m, 66.945 m; (c) 1090.346 m, 67.439 m; (d) 144.017 m. Instrument height 1.111 m, Reading 319 mm

26 On a proposed road alignment, AT_1 and T_2B are straights that intersect at I, C_1C_2 is a circular curve and T_1C_1 and T_2C_2 transition curves of length L. Using the co-ordinates contained in Table 6.17, a maximum acceptable rate of change of radial acceleration of 0.75 m s^3, a design speed of 90 km/h and a minimum radius of curvature R = 300 m, compute

(i) a suitable value of L and

Table 6.17

Point	Easting	Northing
A	278 634.213	4 283 262.304
I	278 739.213	4 283 444.169
B	278 984.959	4 283 616.242

(*ii*) the length of the circular arc $C_1 C_2$ [London]

Answer   69.44 m; 61.45 m

7

Setting out and point location

Setting out

The procedure adopted to ensure that a specific design feature, i.e. a building, a road, a pipeline, etc., is correctly positioned at the construction stage is known as setting out. Control is provided by stations whose positions were fixed during the original survey or by subsidiary stations located by 'intersection', or by 'resection' from such stations. Setting out is thus the reverse of detail surveying in that the control stations are used to place points on the ground in their relative positions. If the setting out is referred to a control system based on National Grid co-ordinates the local scale factor should be considered.

Control points

If two control points A and B are known, a third point can be located in a number of ways (see Fig. 7.1).

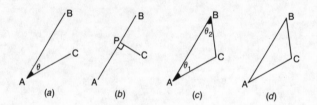

Figure 7.1

(a) Set off angle θ and distance AC.
(b) Set out distance AP and then perpendicular distance PC.
(c) Set off angles θ_1 and θ_2.
(d) Set out distances AC and BC.

Intersection

Figure 7.1(c) shows the method of intersection by which C can either be co-ordinated by observing θ_1 and θ_2 or be located if its co-ordinates are known, since θ_1 and θ_2 can be calculated.

Resection

By the method of resection a station or point can be located after pointings have been made on at least three known stations. This technique allows some flexibility when setting out, in addition to point location in hydrographic surveys.

Generally, plumb wires are used to transfer directions underground. Essentially, the plumb wires produce a vertical reference plane and on the surface the plane can be placed in the line of sight of a correctly oriented theodolite; below ground, the line of sight can be directed into that plane. This is known as 'co-planing' and the line of sight when established can be used to set up floor or roof stations within the tunnel.

Weisbach triangle

In an alternative approach the theodolite is positioned out of the vertical plane so that, in plan, each wire and the vertical axis of the theodolite form a triangle — the Weisbach triangle. The small angle subtended at the theodolite is measured and its value is included with other measurements to determine either the orientation of the vertical plane at ground level or the fixing of reference points below ground. Instrument positions can be changed (also when co-planing) to improve accuracy, and in deep shafts more than one pair of plumb wires can be put down.

7.1 Intersection

> A new control station C is to be located from two stations A and B, the horizontal clockwise angles recorded being $B\hat{A}C = 49°\ 27'\ 18''$ and $A\hat{B}C = 322°\ 45'\ 24''$. Determine the co-ordinates of C given that those of A and B are:
>
> A 950.00 mE 1200.00 mN
> B 983.50 mE 1340.00 mN

Introduction. Derive the standard expression. The co-ordinates of C can be derived readily by trigonometry as shown below. However, if many points are to be fixed it is useful to have recourse to general formulae.

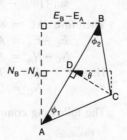

Figure 7.2

Solution. In Fig. 7.2 let the co-ordinates of C be E_C and N_C, etc.

$$N_C = N_D - CD \sin \theta$$

$$= \left[N_A + \frac{AD}{AB} (N_B - N_A) \right] - CD \frac{(E_B - E_A)}{AB}.$$

But $\qquad AB = AD + DB = CD \cot \phi_1 + CD \cot \phi_2.$

Therefore $N_C = \dfrac{N_A \cot \phi_2 + N_B \cot \phi_1 - (E_B - E_A)}{\cot \phi_1 + \cot \phi_2}$,

also $E_C = \dfrac{E_A \cot \phi_2 + E_B \cot \phi_1 + (N_B - N_A)}{\cot \phi_1 + \cot \phi_2}$.

Next determine the co-ordinates using the above expressions. From the data,

$$\phi_1 = 49° \ 27' \ 18''$$
$$\phi_2 = 360° - 322° \ 45' \ 24''$$
$$= 37° \ 14' \ 36''$$
$$E_B - E_A = 983.50 - 950.00 = 33.50 \text{ m}$$
$$N_B - N_A = 1340.00 - 1200.00 = 140.00 \text{ m}.$$

Whence $N_C =$

$$\dfrac{1200.00 \cot 37° 14' 36'' + 1340.00 \cot 49° 27' 18'' - 33.50}{\cot 37° 14' 36'' + \cot 49° 27' 18''}$$

$$= \mathbf{1239.74 \ m,}$$

and $E_C = \dfrac{950.00 \cot 37° 14' 36'' + 983.50 \cot 49° 27' 18'' + 140.00}{\cot 37° 14' 36'' + \cot 49° 27' 18''}$

$$= \mathbf{1027.69 \ m.}$$

Alternatively, the problem can be solved by trigonometry.

$$\text{Bearing } AB = \tan^{-1} \dfrac{983.50 - 950.00}{1340.00 - 1200.00} \qquad = 13° \ 27' \ 25''.$$

$$\text{Bearing } AC = 13° \ 27' \ 25'' + 49° \ 27' \ 18'' \qquad = 62° \ 54' \ 43''$$
$$AB = \sqrt{(33.50^2 + 140.00^2)} \qquad = 143.952 \text{ m.}$$

Thus $\dfrac{AB}{\sin (180° - 37° \ 14' \ 36'' - 49° \ 27' \ 18'')} = \dfrac{AC}{\sin 37° \ 14' \ 36''}$.

Therefore $AC = 87.264$ m.
$$E_C - E_A = AC \sin 62° \ 54' \ 43'' \qquad = 77.693 \text{ m}$$
$$N_C - N_A = AC \cos 62° \ 54' \ 43'' \qquad = 39.737 \text{ m.}$$
Hence $E_C = 950.00 + 77.693 \qquad = \mathbf{1027.69 \ m}$
$$N_C = 1200.00 + 39.737 \qquad = \mathbf{1239.74 \ m.}$$

The following computer program will solve this problem:

Variables

A1	Angle from station A	E1, N1	Co-ordinates of station A
A2	Angle from station B	E2, N2	Co-ordinates of station B
C1, C2	Cot of A1, Cot of A2	E3, N3	Co-ordinates of the point
D	Input angle, degrees	M	Input angle, minutes
		S	Input angle, seconds

```
10 REM INTERSECTION
20 INPUT"INPUT OBSERVATION ANGLE FROM STATION A IN DEG,MIN,SEC ",D,M,S
30 A1=((D*3600)+(M*60)+S)/206264.8
40 INPUT"INPUT OBSERVATION ANGLE FROM STATION B IN DEG,MIN,SEC ",D,M,S
50 A2=((D*3600)+(M*60)+S)/206264.8
60 INPUT"INPUT CO-ORDINATES OF A ",E1,N1
70 INPUT"INPUT CO-ORDINATES OF B ",E2,N2
```

```
80 C1=ABS(1/TAN(A1))
90 C2=ABS(1/TAN(A2))
100 E3=INT((((E1*C2)+(E2*C1)+N2-N1)/(C1+C2))*100+0.5)/100
110 N3=INT((((N1*C2)+(N2*C1)+E1-E2)/(C1+C2))*100+0.5)/100
120 PRINT"COORDINATES ARE ",E3,N3
130 END
```

7.2 Setting out by intersecting rays

The co-ordinates of two survey stations F and G are:

Station F   3812.07 mE   1631.32 mN
Station G   3669.35 mE   1746.89 mN

It is required to establish a subsidiary station H having co-ordinates 3700.00 mE, 1675.00 mN. Determine the angles to be set out at F and G to locate H, together with distances FH and GH for checking purposes.

Introduction. In the previous example an existing point was co-ordinated using the method of intersection. This example illustrates the reverse process in that a point has to be positioned on the ground, the setting out information being derived from its known co-ordinates. In the first instance the required angles at F and G will be evaluated directly from the given co-ordinates, together with the distances FG and GH. In addition reference will be made to the formulae used in the previous example to give a supplementary solution.

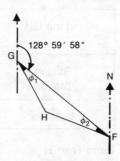

Figure 7.3

Solution. First calculate bearings of FG, FH and GH (Fig. 7.3).

Table 7.1

Line FG	Easting (m)	Northing (m)
Station F	3812.07	1631.32
Station G	3669.35	1746.89
Difference	− 142.72	115.57

From F to G

$$\text{Easting difference} = -142.72 \text{ m}$$
$$\text{Northing difference} = +115.57 \text{ m}$$

$$\tan \theta = \frac{-142.72}{115.57} = -1.234\ 922\ 6.$$

Therefore   bearing of FG = 308° 59′ 58″.

Line FH	Easting (m)	Northing (m)
Station F	3812.07	1631.32
Station H	3700.00	1675.00
Difference	−112.07	43.68

From F to H

$$\text{Easting difference} = -112.07 \text{ m}$$
$$\text{Northing difference} = +43.68 \text{ m}$$

Therefore   bearing of FH = $\tan^{-1} \dfrac{-112.07}{43.68}$ = 291° 17′ 37″.

Line GH	Easting (m)	Northing (m)
Station G	3669.35	1746.89
Station H	3700.00	1675.00
Difference	30.65	−71.89

From G to H

$$\text{Easting difference} = +30.65 \text{ m}$$
$$\text{Northing difference} = -71.89 \text{ m}$$

Therefore   bearing of GH = $\tan^{-1} \dfrac{30.65}{-71.89}$ = 156° 54′ 33″.

Next determine the angles to be set out at F and G. At F

$$\text{Bearing of FG} = 308° 59′ 58″$$
$$\underline{\text{Bearing of FH} = 291° 17′ 37″}$$

$$\text{Angle H}\hat{\text{F}}\text{G} = \textbf{17° 42′ 21″}$$

The angle will be set out with respect to line FG and so the clockwise angle

$$\text{G}\hat{\text{F}}\text{H} = 360° - 17° 42′ 21″ = \phi_2$$
$$= 342° 17′ 39″.$$

At G

$$\text{Bearing of GH} = 156° \, 54' \, 33''$$
$$\text{Bearing of GF} = 128° \, 59' \, 58''$$

$$\text{Angle } F\hat{G}H = \mathbf{27° \, 54' \, 35''} = \phi_1.$$

As an alternative to calculating and then setting off the clockwise angles at F and G the calculated bearings can be used directly. For instance, at G the horizontal circle can be set at $128° \, 59' \, 58''$ when pointing on F, then when the circle reading is $156° \, 54' \, 33''$ the line of sight of the theodolite will be directed on to H. This method is particularly advantageous when many pointings are being set out from one station.

It is good practice to check that the calculations are correct. At H

$$\text{Bearing of FH} = 291° \, 17' \, 37''$$
$$\text{Bearing of GH} = 156° \, 54' \, 33''$$

$$G\hat{H}F = 134° \, 23' \, 04''$$
$$+H\hat{F}G = 17° \, 42' \, 21''$$
$$+F\hat{G}H = 27° \, 54' \, 35''$$

$$180° \, 00' \, 00''$$

Finally, determine distances FG and GH. As a check on the fix for H given by the intersection of the lines of sight of theodolite(s) at F and G, the distances, measured by EDM say, can be compared with the theoretical distances determined below.

Line FH

Easting difference $= 112.07$ m

Northing difference $= 43.68$ m.

Therefore $\qquad L_{FH} = \sqrt{(112.07^2 + 43.68^2)}$

$\qquad\qquad\qquad = 120.28$ m.

Check $\qquad \Delta E_{FH} = 120.28 \sin 291° \, 17' \, 37'' = -112.07$ m

Line GH

Easting difference $= 30.65$ m

Northing difference $= 71.89$ m.

Therefore $\qquad L_{GH} = \sqrt{(30.65^2 + 71.89^2)}$

$\qquad\qquad\qquad = 78.15$ m.

Check $\qquad \Delta N_{GH} = 78.15 \cos 156° \, 54' \, 33'' = -71.89$ m

Any differences between theoretical distances and measured distances should lie within limits specified for the work.

As an alternative the angles can be derived by standard formulae. The purpose of this section is to show how the general formulae given in the previous example can be used in this case. The formulae were derived on the assumption that A, B and C were laid down in a clockwise direction as shown in Fig. 7.2. Thus in Fig. 7.3, it is necessary to work in the direction G to F to H so that G effectively replaces A and F replaces B in the quoted expressions.

Hence for this example.

$$N_H = \frac{N_G \cot \phi_2 + N_F \cot \phi_1 - (E_F - E_G)}{\cot \phi_1 + \cot \phi_2}$$

$$E_H = \frac{E_G \cot \phi_2 + E_F \cot \phi_1 + (N_F - N_G)}{\cot \phi_1 + \cot \phi_2}.$$

Thus $1675.00 = \dfrac{1746.89 \cot \phi_2 + 1631.22 \cot \phi_1 - (3812.07 - 3669.35)}{\cot \phi_1 + \cot \phi_2}$

and $3700.00 = \dfrac{3669.35 \cot \phi_2 + 3812.07 \cot \phi_1 + (1631.32 - 1746.89)}{\cot \phi_1 + \cot \phi_2}.$

The reader can check that the values of $\phi_1 = 27° 54' 35''$ and $\phi_2 = 17° 42' 21''$ satisfy the above equations.

7.3 Setting out using the perpendicular

Two stations on a traverse conducted in an urban area have co-ordinates of 950.00 mE, 1200.00 mN and 983.50 mE, 1340.00 mN, respectively. A manhole is to be located at a point whose co-ordinates are 975.00 mE, 1266.00 mN. It is known that neither direct sightings nor measurements can be made from the two stations and accordingly the engineer decides to locate the manhole by linear measurements along and perpendicular to the traverse line. Determine those measurements.

Introduction. Use of the method of intersection, either by setting off angles ϕ_1 and ϕ_2 or by measuring distances AC and BC, is precluded by sighting conditions. Thus the approach indicated in Fig. 7.1(*b*) is to be adopted, D is located and perpendicular DC set out afterwards.

Solution. Initially, we shall derive standard expressions. In the previous examples two expressions of the form

$$N_C = \frac{N_A \cot \phi_2 + N_B \cot \phi_1 - (E_B - E_A)}{\cot \phi_1 + \cot \phi_2}$$

and $E_C = \dfrac{E_A \cot \phi_2 + E_B \cot \phi_1 + (N_B - N_A)}{\cot \phi_1 + \cot \phi_2}$

were stated; A, B and C being lettered in clockwise direction. These can be rearranged to eliminate ϕ_1 and ϕ_2.

From above $N_C(\cot \phi_1 + \cot \phi_2) = N_A \cot \phi_2 + N_B \cot \phi_1 - (E_B - E_A)$.
Thus $\cot \phi_1 (N_B - N_C) + \cot \phi_2 (N_A - N_C) = (E_B - E_A)$

but $\cot \phi_1 = \dfrac{AD}{CD}$

and $\cot \phi_2 = \dfrac{BD}{CD}$ in Fig. 7.2.

And so $\dfrac{AD}{CD}(N_B - N_C) + \dfrac{(AB - AD)}{CD}(N_A - N_C) = (E_B - E_A)$

or $\quad AD(N_B - N_A) + AB(N_A - N_C) = CD(E_B - E_A)$,

whence $\quad \dfrac{AD}{AB}(N_B - N_A) - \dfrac{CD}{AB}(E_B - E_A) = N_C - N_A.$

Similarly $\quad \dfrac{AD}{AB}(E_B - E_A) + \dfrac{CD}{AB}(N_B - N_A) = E_C - E_A.$

Thus AD and CD can be estimated provided that the three points have been co-ordinated.

Next calculate lengths AD and CD. In Fig. 7.2 we have

$$
\begin{aligned}
(E_B - E_A) &= 983.50 - 950.00 = 33.50 \text{ m} \\
(N_B - N_A) &= 1340.00 - 1200.00 = 140.00 \text{ m} \\
(E_C - E_A) &= 975.00 - 950.00 = 25.00 \text{ m} \\
\text{and} \quad (N_C - N_A) &= 1266.00 - 1200.00 = 66.00 \text{ m}.
\end{aligned}
$$

Also

$$
\begin{aligned}
AB &= \sqrt{[(E_B - E_A)^2 + (N_B - N_A)^2]} \\
&= 143.95 \text{ m}.
\end{aligned}
$$

Substituting in the two expressions

$$ AD\,\dfrac{140.00}{143.95} - CD\,\dfrac{33.50}{143.95} = 66.00 $$

and $\quad AD\,\dfrac{33.50}{143.95} + CD\,\dfrac{140.00}{143.95} = 25.00$

and so $\quad AD = \mathbf{70.01\ m}$

$CD = \mathbf{8.95\ m.}$

By setting out a length of 70.01 m from A along AB, and then setting off a perpendicular of length 8.95 m from that position the manhole is located.

An alternative approach is as follows:

Bearing of AB $= \tan^{-1}\dfrac{E_B - E_A}{N_B - N_A}$

$= 13°\ 27'\ 25.3''.$

Bearing of AC $= \tan^{-1}\dfrac{E_C - E_A}{N_C - N_A}$

$= 20°\ 44'\ 45.9''.$

Whence $\quad B\hat{A}C = 07°\ 17'\ 20.6'',$

also $\quad AC = \sqrt{[(E_C - E_A)^2 + (N_C - N_A)^2]}$

$= 70.58 \text{ m}.$

Therefore AD = 70.58 cos 07° 17′ 20.6″
            = 70.01 m.
And   DC = 70.58 sin 07° 17′ 20.6″
            = 8.95 m.

7.4 Resection

In order to set out the centre line of a bridge it is required to establish the point D which lies 115.00 m due south of the peg A, across the river. Stations B and C were established and referred to A by subtense tacheometry. Field notes were as in Table 7.2.

Table 7.2

From	To	Target bearings		
		Left	Centre	Right
A	B	269° 03′ 00″	270° 00′ 00″	270° 57′ 00″
	C	284° 13′ 30″	284° 30′ 00″	284° 46′ 30″

A peg was positioned at P which was thought to be close to D. At P the following angles were measured:

$$C\hat{P}B = 20° 54′ 00″, \quad B\hat{P}A = 28° 36′ 00″.$$

Calculate the bearing and distance PD.   [Bradford]

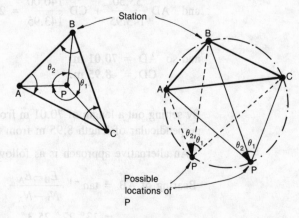

Figure 7.4

Introduction. In this example P is being located by resection. At least three control stations have to be fixed in position for the application of this method. Observations are then made on these stations so that there are at least two known horizontal angles θ_1 and θ_2 as in Fig. 7.4. It is essential that P does not lie on the circle circumscribing C, B and A or be near to it, because then θ_1 and

θ_2 would be subtended thereon by CB and BA, respectively, whatever the position of P, i.e. there would be a 'failure of fix'.

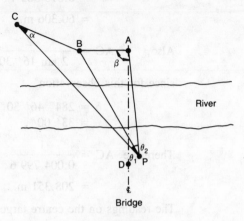

Figure 7.5

Solution. First derive a general expression. Consider triangles CBP and BPA in Fig. 7.5 in which CB, BA and CB̂A will be known. By the sine rule,

$$\frac{BC}{\sin \theta_1} = \frac{BP}{\sin \alpha} \quad \text{and} \quad \frac{BA}{\sin \theta_2} = \frac{BP}{\sin \beta}.$$

Therefore
$$BP = \frac{BC \sin \alpha}{\sin \theta_1} = \frac{BA \sin \beta}{\sin \theta_2}.$$

$$\sin \alpha = \frac{BA \sin \theta_1}{BC \sin \theta_2} \times \sin \beta.$$

Therefore
$$\frac{\sin \alpha}{\sin \beta} = \frac{BA}{BC} \times \frac{\sin \theta_1}{\sin \theta_2} = k, \text{ say,}$$

but
$$\alpha + \beta = 360° - C\hat{B}A - \theta_1 - \theta_2$$
$$= \phi, \text{ which is known.}$$

Therefore
$$\sin \alpha = \frac{BA}{BC} \times \frac{\sin \theta_1}{\sin \theta_2} \times \sin (\phi - \alpha).$$

This expression allows the determination of α, and hence β. There is now sufficient information to calculate lengths CP, BP and AP, allowing P to be fixed in position.

Calculation of lengths CB, BA and angle CB̂A. The lengths have been determined by subtense measurements. Assuming that a bar length 2 m was used (Fig. 2.4 refers)

$$AB = \frac{b}{2 \tan \delta/2} = \frac{2}{2 \tan 57'}$$

since
$$\delta = 270° \ 57' \ 00'' - 269° \ 03' \ 00''$$
$$= 1° \ 54' \ 00''.$$

$$AB = \frac{1}{0.016\ 582\ 1}$$

$$= 60.306 \text{ m}.$$

Also, $\quad AC = \dfrac{2}{2 \tan 16'\ 30''}$

since for this observation

$$\delta = 284°\ 46'\ 30'' - 284°\ 13'\ 30''$$
$$= 33'\ 00''.$$

Therefore $\quad AC = \dfrac{1}{0.004\ 799\ 6}$

$$= 208.351 \text{ m}.$$

The readings on the centre targets imply that

$$\hat{BAC} = 284°\ 30'\ 00'' - 270°\ 00'\ 00''$$
$$= 14°\ 30'\ 00''.$$

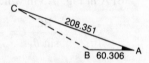

Figure 7.6

From triangle CBA (Fig. 7.6)

$$BC^2 = AC^2 + AB^2 - (2 \times AC \times AB \cos \hat{BAC}).$$
Therefore $\quad BC = 150.724 \text{ m},$

Also $\quad \dfrac{AC}{\sin \hat{CBA}} = \dfrac{CB}{\sin 14°\ 30'\ 00''}.$

Therefore $\sin \hat{CBA} = \dfrac{208.351}{150.724} \sin 14°\ 30'\ 00''$

$$\sin \hat{CBA} = 0.346\ 108\ 9$$
Therefore $\quad \hat{CBA} = 159°\ 45'\ 02''.$
External angle $\hat{CBA} = 200°\ 14'\ 58''$

Next calculate length PA. The relationship

$$\sin \alpha = \frac{BA}{BC} \times \frac{\sin \theta_1}{\sin \theta_2} \times \sin(\phi - \alpha)$$

has been derived in a previous section.

Now $\quad \phi = 360° - \hat{CBA} - \theta_1 - \theta_2$
$$= 360° - 200°\ 14'\ 58'' - 20°\ 54'\ 00'' - 28°\ 36'\ 00''$$
$$= 110°\ 15'\ 02''.$$

Therefore $\sin \alpha = \dfrac{60.306}{150.724} \times \dfrac{\sin 20° 54' 00''}{\sin 28° 36' 00''} \times \sin (110° 15' 02'' - \alpha)$

$\qquad = 0.298\,175\,1\,(\sin 110° 15' 02'' \cos \alpha - \cos 110° 15' 02'' \sin \alpha)$.

Therefore $\cot \alpha = 3.205\,762\,8$

$\qquad\qquad \alpha = 17° 19' 29''$,

whence $\qquad \beta = 92° 55' 33''$.

In triangle BAP

$$A\hat{B}P = 180° - 92° 55' 33'' - 28° 36' 00''$$
$$= 58° 28' 27''$$

and $\qquad \dfrac{AP}{\sin 58° 28' 27''} = \dfrac{BA}{\sin 28° 36' 00''}$.

Therefore $\qquad AP = 107.39$ m.

Finally calculate length and bearing of PA. With respect to A:

$$\text{Northing of P} = -107.39 \cos (92° 55' 33'' - 90°)$$
$$= -107.25 \text{ m},$$
$$\text{Easting of P} = 107.39 \sin (92° 55' 33'' - 90°)$$
$$= +5.48 \text{ m}.$$

But D lies 115.00 m south of A, and accordingly is 7.75 m south and 5.48 m west of P.

Therefore length $PD = \sqrt{(5.48^2 + 7.75^2)}$

$\qquad\qquad\qquad = \textbf{9.49 m.}$

$$\text{Bearing of PD} = \tan^{-1} \dfrac{-5.48}{-7.75}$$

$$= \textbf{215° 15' 50''}.$$

7.5 Resection – Tienstra's method	Point P was established within the triangle formed by control stations A (1020 mE, 2560 mN), B (1360 mE, 2520 mN), C (1150 mE, 2070 mN), such that angles APB = 128° 20' 20'' and BPC = 135° 47' 40''. What is the distance of point X (1200 mE, 2450 mN) from P and what is the WCB of XP? [Salford]

Solution. From the co-ordinate information,

$$AB = \sqrt{[(1360 - 1020)^2 + (2520 - 2560)^2]} = 342.345 \text{ m}$$
$$BC = \sqrt{[(1360 - 1150)^2 + (2520 - 2070)^2]} = 496.588 \text{ m}$$
$$AC = \sqrt{[(1150 - 1020)^2 + (2070 - 2560)^2]} = 506.952 \text{ m}$$

In addition, using the cosine rule in triangle ABC

$$B\hat{A}C = \cos^{-1} \left(\dfrac{AC^2 + AB^2 - BC^2}{2 \times AC \times AB} \right)$$

$$= \cos^{-1}\left(\frac{506.952^2 + 342.345^2 - 496.588^2}{2 \times 506.952 \times 342.345}\right)$$

$$= 68°\ 25'\ 54''$$

$$C\hat{B}A = \cos^{-1}\left(\frac{342.345^2 + 496.588^2 - 506.952^2}{2 \times 342.345 \times 496.588}\right)$$

$$= 71°\ 41'\ 34''$$

$$A\hat{C}B = \cos^{-1}\left(\frac{496.588^2 + 506.952^2 - 342.345^2}{2 \times 496.588 \times 506.952}\right)$$

$$= 39°\ 52'\ 32''$$

Check: $68°\ 25'\ 54'' + 71°\ 41'\ 34'' + 39°\ 52'\ 32'' = 180°\ 00'\ 00''$.

Also from the data

$$C\hat{P}A = 360° - 128°\ 20'\ 20'' - 135°\ 47'\ 40''$$
$$= 95°\ 52'\ 00''$$

Next determine the co-ordinates of P by means of Tienstra's Formulae which state

$$E_P = \frac{K_1 E_A + K_2 E_B + K_3 E_C}{K_1 + K_2 + K_3}$$

$$N_P = \frac{K_1 N_A + K_2 N_B + K_3 N_C}{K_1 + K_2 + K_3}$$

In the above

$$K_1 = \frac{1}{(\cot B\hat{A}C - \cot B\hat{P}C)}$$

$$= \frac{1}{(\cot 68°\ 25'\ 54'' - \cot 135°\ 47'\ 40'')} = 0.702\ 537$$

$$K_2 = \frac{1}{(\cot C\hat{B}A - \cot C\hat{P}A)}$$

$$= \frac{1}{(\cot 71°\ 41'\ 34'' - \cot 95°\ 52'\ 00'')} = 2.306\ 219$$

$$K_3 = \frac{1}{(\cot A\hat{C}B - \cot A\hat{P}B)}$$

$$= \frac{1}{(\cot 39°\ 52'\ 32'' - \cot 128°\ 20'\ 20'')} = 0.503\ 049$$

Whence $K_1 + K_2 + K_3 = 3.511\ 81$

Therefore

$$E_P = \frac{(0.702\,537 \times 1020) + (2.306\,219 \times 1360) + (0.503\,049 \times 1150)}{3.511\,81}$$

$$= 1261.900 \text{ m}$$

$$N_P = \frac{(0.702\,537 \times 2560) + (2.306\,219 \times 2520) + (0.503\,049 \times 2070)}{3.511\,81}$$

$$= 2463.538 \text{ m}$$

We can now calculate the required distance and bearing using the co-ordinates of X and P.

$$XP = \sqrt{[(1261.900 - 1200)^2 + (2463.538 - 2450)^2]}$$
$$= \textbf{63.363 m}$$

$$\text{Bearing} = \tan^{-1}\frac{(1261.900 - 1200)}{(2463.538 - 2450)}$$

$$= \textbf{77° 39' 47''}$$

Note. Angles are registered in a clockwise direction, as in Fig. 7.4, when being paired for the determination of K_1, etc. When P lies outside triangle ABC, as in Fig. 7.5, $A\hat{P}B = 360° - \theta_2$.

The co-ordinates of P can be calculated using the following computer program. The angles input must be the clockwise angles as indicated in Fig. 7.4.

Variables

A1	Clockwise angle between stations A and B	E2, N2	Co-ordinates of station B
A2	Clockwise angle between stations B and C	E3, N3	Co-ordinates of station C
		E4, N4	Co-ordinates of point P
A4, A6	Clockwise angles in triangle ABC	K1, K2, K3	Tienstra's coefficients, i.e. $1/(\cot C\hat{A}B - \cot C\hat{P}B)$
B1, B2, B3	Bearings of the lines in triangle ABC	K4	Sum of K1 + K2 + K3 for denominator
C1 to C6	cot of angles 1 to 6	M	Input angle, minutes
D	Input angle, degrees	S	Input angle, seconds
E1, N1	Co-ordinates of station A		

```
10 REM RESECTION
20 INPUT"INPUT ANGLE BETWEEN STATIONS A AND B ",D,M,S
30 A1=((D*3600)+(M*60)+S)/206264.8
40 C1=1/TAN(A1)
50 INPUT"INPUT ANGLE BETWEEN STATIONS B AND C ",D,M,S
60 A2=((D*3600)+(M*60)+S)/206264.8
70 C2=1/TAN(A2)
80 C3=1/TAN(3.14159-A1-A2)
90 INPUT"INPUT CO-ORDINATES OF A ",E1,N1
100 INPUT"INPUT CO-ORDINATES OF B ",E2,N2
110 INPUT"INPUT CO-ORDINATES OF C ",E3,N3
120 B1=ATN((E1-E3)/(N1-N3))
130 B2=ATN((E2-E3)/(N2-N3))
140 B3=ATN((E1-E2)/(N1-N2))
```

```
150 A4=B2-B1
155 C4=1/TAN(A4)
160 A6=B3-B2-3.14159
170 C6=1/TAN(A6)
180 C5=1/TAN(3.14159-A4-A6)
190 K1=1/(C6-C3)
200 K2=1/(C4-C1)
210 K3=1/(C5-C2)
220 K4=K1+K2+K3
230 E4=INT((((K1*E2)+(K2*E3)+(K3*E1))/K4)*100+0.5)/100
240 N4=INT((((K1*N2)+(K2*N3)+(K3*N1))/K4)*100+0.5)/100
250 PRINT"CO-ORDINATES ARE ",E4,N4
260 END
```

It is recommended that the reader evaluates the co-ordinates of P in Fig. 7.5, assuming those of A, B and C to be respectively 1026.600 mE, 2570.590 mN; 960.294 mE, 2570.590 mN; 818.886 mE, 2622.757 mN.

7.6 Weisbach triangle

Two surface stations P and Q, having co-ordinates 1250.00 mE, 1200.00 mN and 1300.00 mE, 1350.00 mN, were observed during the installation of plumb wires in a shaft. The readings in Table 7.3 were recorded by a theodolite set up at a surface station A near to line XY.

Table 7.3

Pointing on	Horizontal circle reading
P	273° 42′ 08″
Q	93° 43′ 54″
Plumbwire X	08° 00′ 50″
Plumbwire Y	07° 58′ 10″

Distances AP and AX were measured to be 78.855 m and 8.374 m, respectively, and XY was measured as 5.945 m. Determine the bearing of XY given that X was the nearer of the wires to A.

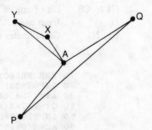

Figure 7.7

Introduction. This example illustrates the principles of the Weisbach triangle which may be used on the surface, as here, or below ground to establish the underground control system. In such a triangle a theodolite, reading directly to one second, is set up just off the line of the plumb wires X and Y (see Fig. 7.7) and angle XÂY measured as accurately as possible: this angle should be very small, i.e. not greater than a few minutes. Although distances XA, YA and XY should be measured, often the assumption is made, since XÂY is very small, that XY + XA = YA if only XY and XA have been measured, as in this example. PAQ also forms a Weisbach triangle when A lies near PQ since the angles at P and Q and the external angle at A will be small. Mention is made in the next worked example of the Weisbach triangle in an underground situation.

Solution. In Weisbach triangle PAQ the bearing of plumb plane XY will be derived from information obtained from the ground control points P and Q, station A tying the two Weisbach triangles together. In Fig. 7.7 we have the values in Table 7.4.

Table 7.4

Station	E (m)	N (m)
P	1250.00	1200.00
Q	1300.00	1350.00
Difference	50.00	150.00

Therefore length PQ $= \sqrt{(50.00^2 + 150.00^2)}$
$= 158.114$ m.

$$\text{Bearing of PQ} = \tan^{-1}\frac{50.00}{150.00}$$

$$= 18° \ 26' \ 05.8''.$$

Table 7.5

Pointing on	Horizontal circle readings
P	273° 42′ 08″
Q	93° 43′ 54″
Therefore clockwise angle QÂP = 179° 58′ 14″	

Also, from the circle readings at A we have the values in Table 7.5.

Length PQ $=$ 158.114 m (calculated)

PA $=$ 78.855 m (measured).

Therefore QA $=$ 79.259 m (assuming PA+QA = PQ).

Furthermore $\dfrac{\sin P\hat{A}Q}{158.114} = \dfrac{\sin P\hat{Q}A}{78.855}.$

Therefore $\sin \dfrac{179°\ 58'\ 14''}{158.114} = \dfrac{\sin 01'\ 46''}{158.114} = \dfrac{\sin \hat{PQA}}{78.85}$.

For small angles we can write

$$\sin 01'\ 46'' = \sin 106'' = 106'' \times \sin 1''$$

and

$$\sin \hat{PQA} = \hat{PQA}'' \times \sin 1''.$$

Whence

$$\dfrac{106'' \times \sin 1''}{158.114} = \dfrac{\hat{PQA}'' \times \sin 1''}{78.885}$$

$$\hat{PQA} = \dfrac{78.885 \times 106''}{158.114}$$

$$= 52.9''$$

and

$$\hat{QPA} = 106'' - 52.9''$$

$$= 53.1''.$$

Therefore   bearing of PA = bearing of $PQ - \hat{QPA}$

$$= 18°\ 26'\ 05.8'' - 53.1''$$

$$= 18°\ 25'\ 12.7''.$$

Weisbach triangle AXY.

From the circle readings at A we get the values in Table 7.6.

Table 7.6

Pointing on	Horizontal circle reading
X	08° 00′ 50″
Y	07° 58′ 10″
Therefore clockwise angle $X\hat{A}Y$ = 02′ 40″ = 160″	

This is an internal angle in triangle AXY.

Now

$$\dfrac{\sin X\hat{A}Y}{XY} = \dfrac{\sin X\hat{Y}A}{AX}$$

Therefore

$$\dfrac{\sin 160''}{5.945} = \dfrac{\sin X\hat{Y}A}{8.374}$$

Therefore

$$\dfrac{160'' \times \sin 1''}{5.945} = \dfrac{X\hat{Y}A'' \times \sin 1''}{8.374}$$

$$X\hat{Y}A = \dfrac{8.374}{5.945} \times 160''$$

$$= 225.4''.$$

Also, from the circle readings, we get the values in Table 7.7.

Table 7.7

Pointing on	Horizontal circle reading
X	8° 00′ 50″
P	273° 42′ 08″
Therefore clockwise angle PÂX = 94° 18′ 42″	

$$\begin{aligned}
\text{Now the bearing of PA} &= 18° \ 25′ \ 12.7″ \\
\text{Therefore the bearing of AP} &= 198° \ 25′ \ 12.7″ \\
\text{PÂX} &= \underline{94° \ 18′ \ 42″}
\end{aligned}$$

Therefore the bearing of AX = 292° 43′ 54.7″

But the external angle at X in triangle AXY is

$$\begin{aligned}
XŶA + XÂY &= 225.4″ + 160″ \\
&= 385.4″ = 6′ \ 25.4″.
\end{aligned}$$

Therefore	AXY = 179° 53′ 34.6″.
Now	bearing of AX = 292° 43′ 54.7″
add	179° 53′ 34.6″
	472° 37′ 29.3″
deduct	180° 00′ 00″
Bearing of XY	= 292° 37′ 29.3″
	= **292° 37′ 29″**, say.

7.7 Weisbach triangle and co-planing

A length of sewer CD is to be constructed in a tunnel at a bearing of 255° 00′ 00″ from C, whose co-ordinates are 1942.00 mE, 1281.80 mN. Two control points A and B are in the vicinity, and no site obstructions exist. Plumbwires X and Y, 3.640 m apart, were put down after the shaft at C had been excavated, the method of co-planing being adopted on the surface to achieve direction CD.

Theodolite station U was then established below ground, lying close to line YX produced and 5.295 m from nearer wire X. Horizontal circle readings for pointings on X and Y were 23° 20′ 12″ and 23° 15′ 35″ respectively.

Given that U is generally to the west of YX and that XC = CY, derive data for setting out:

(*a*) points on centre line CD 12.00 m and 25.00 m, respectively, from C;

(*b*) a reference station R, underground, having co-ordinates 1930.00 mE, 1280.20 mN

Co-ordinates of A and B are:

A 1850.75 mE, 1270.40 mN
B 1990.24 mE, 1310.66 mN

Introduction. This example illustrates the use of a Weisbach triangle in an underground situation, but it also allows the alternative method of co-planing to be reviewed. As mentioned previously, when co-planing the vertical plane formed by the plumb wires is placed in the line of sight of a theodolite (above ground) or the line of sight is placed in the vertical plane (below ground).

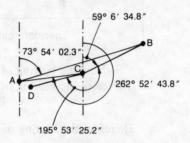

Figure 7.8

Solution. First consider the surface control (Fig. 7.8). The co-ordinates of control points A and B have been given, so a ground point corresponding to C could be located by methods previously discussed. For AB (Fig. 7.8) we have the values in Table 7.8.

Table 7.8

Station	E (m)	N (m)
A	1850.75	1270.40
B	1990.24	1310.66
Difference	139.49	40.26

Length of AB $= \sqrt{(139.49^2 + 40.26^2)}$
$= 145.184$ m.

Bearing of AB $= \tan^{-1} \dfrac{139.49}{40.26}$

$= 73° \ 54' \ 02.3''$.

For AC we get the values in Table 7.9.

Length of AC $= \sqrt{(91.25^2 + 11.40^2)}$
$= 91.959$ m.

Bearing of AC $= \tan^{-1} \dfrac{91.25}{11.40}$

$= 82° \ 52' \ 43.8''$.

Table 7.9

Station	E (m)	N (m)
A	1850.75	1270.40
C	1942.00	1281.80
Difference	91.25	11.40

For BC we obtain the values in Table 7.10.

Table 7.10

Station	E (m)	N (m)
B	1990.24	1310.66
C	1942.00	1281.80
Difference	48.24	28.86

$$\text{Length of BC} = \sqrt{(48.24^2 + 28.86^2)}$$
$$= 56.214 \text{ m.}$$

$$\text{Bearing of CB} = \tan^{-1} \frac{48.24}{28.86}$$

$$= 59° \ 06' \ 34.8''.$$

Thus $B\hat{A}C$ = Bearing of AC − Bearing of AB
$$= 82° \ 52' \ 43.8'' - 73° \ 54' \ 02.3''$$
$$= 08° \ 58' \ 41.5''.$$

And $A\hat{B}C$ = Bearing of BA − Bearing of BC
$$= 253° \ 54' \ 02.3'' - 239° \ 06' \ 34.8''$$
$$= 14° \ 47' \ 27.5''.$$

These are rather small angles for the purposes of intersection and it would be more satisfactory to locate C by a perpendicular offset from AB. This offset has a value of 91.959 sin 08° 58′ 41.5″, i.e. 14.351 m and it would be taken at a distance of $\sqrt{(91.959^2 - 14.351^2)}$ m, i.e. 90.832 m from A.

Setting out line CD on the surface. Initially line CD (Fig. 7.9) could be set out on the surface by 'turning off' clockwise angle BCD, with clockwise angle DCA as a check since A should be intersected.

Clockwise angle BCD = Bearing CD − Bearing of CB
$$= 255° \ 00' \ 00'' - 59° \ 06' \ 34.8''$$
$$= 195° \ 53' \ 25.2''.$$

Clockwise angle DCA = Bearing CA − Bearing of CD

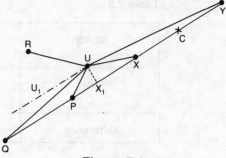

Figure 7.9

$$= 262° \ 52' \ 43.8'' - 255° \ 00' \ 00''$$
$$= 07° \ 52' \ 43.8''.$$

Stations could be established along CD at convenient locations, the nearest being just clear of the shaft sides to ensure no disturbance, say 7 m to 10 m from C. Pegs could also be positioned on DC produced. In this way the surface alignment CD can be recovered at anytime. When co-planing, the plumb wires X and Y are to be brought into this line. The greatest of care has to be exercised and observations should be made from both sides of the shaft to obtain the best positions.

Next establish the underground control. Figure 7.9 shows the disposition of the underground station U as it forms a Weisbach triangle with plumb wires X and Y, now suspended within the shaft. The relevant horizontal circle readings are given in Table 7.11.

Table 7.11

Pointing on	Circle reading
X	23° 20′ 12″
Y	23° 15′ 35″
Therefore YÛX = 04′ 37″ = 277″	

In triangle YUX

$$\frac{\sin Y\hat{U}X}{XY} = \frac{\sin U\hat{Y}X}{UX}$$

$$\frac{Y\hat{U}X \sin 1''}{3.640} = \frac{U\hat{Y}X \sin 1''}{5.295}$$

$$U\hat{Y}X = \frac{5.295 \times 277''}{3.640}$$

$$= 402.9''.$$

Therefore external angle at X $= 402.9'' + 277'' = 679.9''$

$$= 11'\ 19.9''$$

$$UX_1 = 5.295 \sin 11'\ 19.9''$$

$$= 0.0175 \text{ m, i.e. } 17.5 \text{ mm.}$$

Line UU_1 (parallel to YX produced) can be obtained by setting off an angle of $179°\ 48'\ 40.1''$ from UX. This could be used as a reference line if suitably 'pegged out'.

Also   Bearing of XU $=$ Bearing of YX $+$ External angle at X

$$= 255°\ 00'\ 00'' + 11'\ 19.9''$$

$$= 255°\ 11'\ 19.9''.$$

Now we find the co-ordinates of X $\left(\dfrac{3.64}{2} = 1.82 \text{ m from C}\right)$.

$$\text{Easting} = \text{Easting of C} + 1.82 \sin 255°\ 00'\ 00.''$$

$$= 1942.00 - 1.758$$

$$= 1940.242 \text{ m, say } 1940.24 \text{ m;}$$

$$\text{Northing} = \text{Northing of C} + 1.82 \cos 255°\ 00'\ 00''$$

$$= 1281.80 - 0.471$$

$$= 1281.329 \text{ m, say } 1281.33 \text{ m.}$$

The co-ordinates of U (which is 5.295 m from X) are:

$$\text{Easting} = \text{Easting of X} + 5.295 \sin 255°\ 11'\ 19.9''$$

$$= 1940.242 - 5.119$$

$$= 1935.123 \text{ m, say } 1935.12 \text{ m;}$$

$$\text{Northing} = \text{Northing of X} + 5.295 \cos 255°\ 11'\ 19.9''$$

$$= 1281.329 - 1.354$$

$$= 1279.975 \text{ m, say } 1279.98 \text{ m.}$$

The co-ordinate values have been 'rounded off' to be in accordance with those of the control stations.

Next locate the centre line stations at 12.00 m and 25.00 m from C. The first of these will be 10.18 m from X along the centre line. In triangle PUX

$$UX = 5.295 \text{ m}$$

$$PX = 10.180 \text{ m}$$

and $\qquad U\hat{X}P = 679.9''.$

Assume that UP $+$ UX $=$ PX

Therefore $\qquad UP = 10.180 - 5.295$

$$= 4.885 \text{ m,}$$

whence $\qquad \dfrac{\sin 679.9''}{UP} = \dfrac{\sin U\hat{P}X}{UX}$

$$\dfrac{679.9''\sin 1''}{4.885} = \dfrac{U\hat{P}X \sin 1''}{5.295}.$$

Therefore $\qquad U\hat{P}X = 737.0''$

$$= 12'\ 17''$$

and
$$P\hat{U}X = 180° - 12' 17'' - 11' 19.9''$$
$$= 179° 36' 23.1''.$$

Horizontal circle reading for pointing on P = Reading for pointing on X + 179° 36' 23.1"

$$= 23° 20' 12'' + 179° 36' 23.1''$$
$$= 202° 56' 35'', \text{ say.}$$

The co-ordinates of P are:

$$\text{Easting} = \text{Easting of } X + 10.18 \sin 255° 00' 00''$$
$$= 1940.242 - 9.833$$
$$= 1930.409 \text{ m, i.e. } 1930.41 \text{ m;}$$
$$\text{Northing} = 1281.329 + 10.18 \cos 255° 00' 00''$$
$$= 1281.329 - 2.635$$
$$= 1278.694 \text{ m, i.e. } 1278.69 \text{ m.}$$

Similarly, to locate station Q 25.00 m from C (or 23.18 m from X) a distance of (23.18 − 5.295) = 17.885 m could be set off from U at a horizontal circle reading of 203° 05′ 31″. The co-ordinates of Q are 1917.85 mE, 1275.33 mN. Having fixed stations U, P and Q the tunnel can be extended.

The second part of the question concerns the location of stations not on the centre line. Consider point R, co-ordinates 1930.00 mE, 1280.20 mN, in Table 7.12.

Table 7.12

Station	Easting (m)	Northing (m)
R	1930.000	1280.200
U	1935.123	1279.975
Difference	5.123	0.225

Hence length UR = $\sqrt{(5.123^2 + 0.225^2)}$
$$= 5.128 \text{ m.}$$

$$\text{Bearing UR} = \tan^{-1} \frac{-5.123}{0.225}$$

$$= 272° 30' 53.3''.$$

Since the bearing of XU = 255° 11′ 19.9″
clockwise angle X$\hat{U}$R = 197° 19′ 33.4″

But the horizontal circle reading for pointing on X is 23° 20′ 12″. Therefore the horizontal circle reading for pointing on R

$$= 23° 20' 12'' + 197° 19' 33.4''$$
$$= 220° 39' 45'', \text{ say.}$$

A series of points could be set out now in a similar manner, using directions UR, UP or PR, to establish a reference line having a bearing of 255° 00′ 00″ through R, which actually lies 1.5 m from the centre line.

7.8 Measuring depths down a shaft

Prove the following equation for the elongation of a steel measuring tape when used for measurements in a vertical plane.

$$s = \frac{gx}{AE} \left(M + \tfrac{1}{2} m \, (2l - x) \, - \, \frac{T_s}{g} \right)$$

where:

s is the elongation.
g is the gravitational acceleration.
x the length of suspended tape.
A the cross-sectional area of the tape.
E the modulus of elasticity.
M the attached mass.
m the mass of the tape per unit length.
l the total length of the tape.
T_s the standard tension.

A 1000 m mine shaft measuring tape has cross-sectional area of 10 mm², a mass of 0.07 kg/m and a modulus of elasticity of 2×10^5 N/mm². It is standardized as 1000.000 m at 180 N tension. Assuming gravitational acceleration to be 9.807 m/s² compute the correct depth of the mine shaft recorded as 750.52 m. Neglect the effects of temperature.   [Eng. Council]

Introduction.   When the shaft has been excavated it is essential that a temporary bench mark be established so that the construction can be carried out to correct levels as well as to correct line. Conventionally levels will be carried down from a known datum, at, say, the side of the shaft at the top, by means of a steel tape, hanging vertically and free of restrictions. It may be that, as in this case, a very long tape is available so that the operation is carried out in a single stage. Otherwise, the separate tape lengths will need to be marked out in descending order. The use of EDM to measure shaft depths is covered in Chapter 2.

Solution.   First derive the equation. Figure 7.10 shows the actual suspended length x of the tape, which hangs from a fixed point C. $(l - x)$ is the additional portion of the tape not required in the measurement but still contributing to the loading of the tape. Alternatively, the tape could be considered to be fully run out to length l and x to be a specific measurement or reading with respect to C. The tension sustained by the vertical tape due to self-loading is a maximum at C and varies with y, being a minimum at the longest point under consideration. Thus the extensions induced in the small elements of length dy are

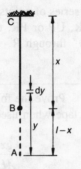

Figure 7.10

greater in magnitude in the upper regions of the hanging tape than in the lower regions but, naturally, all contribute to the overall elongation.

Consider an element of length dy at a point y from the 'free end' A.

$$\text{Load on that element} = mg \times y$$

$$\text{Therefore extension over length dy} = mg\, y\, \frac{dy}{AE}$$

$$\text{Therefore extension over length BC} = \int_{(l-x)}^{l} mg\, y\, \frac{dy}{AE}$$

$$= {}_{(l-x)}^{\quad l}\left[\frac{mg}{AE}\,\frac{y^2}{2}\right] + \text{constant.}$$

The extension is zero when $y = 0$, therefore the constant $= 0$.

$$\text{Extension over length BC} = \frac{mg}{2AE}\,[l^2 - (l-x)^2]$$

$$= \frac{mgx}{AE}\left(\frac{2l-x}{2}\right).$$

A mass may be attached to lower end A to ensure verticality and to minimize oscillation. It will have a uniform effect over the tape in so far as elongation is concerned, and in this context is analogous to the load applied to a 'horizontal' tape suspended in catenary between supports.

$$\text{Extension due to mass } M \text{ over length } x = Mg \times \frac{x}{AE}.$$

The tape has been standardized under tension T_s and this has to be allowed for in the same way as the standard tension in the pull correction mentioned in Chapter 2.

$$\text{Therefore elongation over length } x = \frac{x}{AE}\left[mg\,\frac{(2l-x)}{2} + Mg - T_s\right]$$

$$= \frac{gx}{AE}\left[\frac{m}{2}\,(2l-x) + M - \frac{T_s}{g}\right].$$

Now the depth of the shaft can be determined. It is stipulated that $T_s = 180$ N, $l = 1000.000$, $m = 0.07$ kg/m, $A = 10$ mm^2, $E = 2 \times 10^5$ N/mm^2 and $M = 0$. Hence the effective elongation when $x = 750.52$ m.

$$= \frac{9.807 \times 750.52}{10 \times 2 \times 10^5} \left[\frac{0.07}{2} (2000.000 - 750.52) + 0 - \frac{180}{9.807} \right]$$

$$= 0.093 \text{ m.}$$

Therefore corrected depth $= 750.52 + 0.093$
$$= \textbf{750.61 m,}$$

the measured depth being recorded to 0.01 m.

7.9 Rock plane

Table 7.13 gives data concerning the position of a rock plane stratum at three stations A, B and C. Determine the co-ordinates of the point at which the formation centre line of a cutting and tunnel constructed at a down gradient of 3 in 106 from A would find the stratum when driven in a north-easterly direction. At what depth will this point be below the surface, the cutting starting at ground level at A? If tunnelling commences when the formation is 18.0 m below ground level locate where the cutting finishes.

Table 7.13

Station	A	B	C
Ground level, AOD (m)	150.0	177.0	192.0
Depth to rock (m)	13.5	34.5	40.5
Co-ordinates (E,N)	(0,0)	(30,240)	(330,90)

[London]

Introduction. When considering the estimation of quantities, costs and the planning of this construction the engineer must have knowledge of such items as depths of superficial deposits, position of the rock plane in general and where it will be encountered as the work progresses. This example illustrates an analytical approach to the questions posed.

Solution. First reduce the data. Figure 7.11 shows the relative positions of stations A, B and C together with the centre line direction of the cutting and tunnel. It is necessary to locate E in plan and to establish rock level and ground level there before we can calculate intersection points, etc.

$$\text{Length BC} = \sqrt{[(330.0 - 30.0)^2 + (240.0 - 90.0)^2]}$$
$$= 335.41 \text{ m.}$$

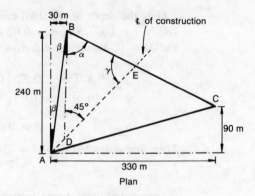

Figure 7.11

$$\text{Length AB} = \sqrt{(30.0^2 + 240.0^2)}$$
$$= 241.87 \text{ m.}$$

Also from the co-ordinate information

$$\tan \alpha = \frac{330.0 - 30.0}{240.0 - 90.0}.$$

Therefore $\qquad \alpha = 63° \, 26' \, 06'',$

and, drawing BD parallel to the northing direction at A,

$$\tan \beta = \frac{30.0 - 0.0}{240.0 - 0.0}.$$

Therefore $\qquad \beta = 07° \, 07' \, 30'',$
whence $\qquad \alpha + \beta = A\hat{B}C$
$$= 70° \, 33' \, 36''$$
and $\qquad \gamma = 180° - 63° \, 26' \, 06'' - 45° \, 00' \, 00''$
$$= 71° \, 33' \, 54''.$$

In triangle AEB

$$\frac{AE}{\sin (\alpha + \beta)} = \frac{AB}{\sin \gamma}.$$

Therefore $\qquad AE = \dfrac{AB \sin 70° \, 33' \, 36''}{\sin 71° \, 33' \, 54''}$

$$= 240.42 \text{ m, since AB} = 241.87 \text{ m,}$$

and $\qquad \dfrac{BE}{\sin (45 - \beta)} = \dfrac{BE}{\sin 37° \, 52' \, 30''} = \dfrac{AB}{\sin \gamma}.$

Therefore $\qquad BE = 241.87 \dfrac{\sin 37° \, 52' \, 30''}{\sin 71° \, 33' \, 54''}$

$$= 156.52 \text{ m.}$$

Also, from Table 7.13 we obtain the values in Table 7.14.

Table 7.14

Station	Ground level (m)	Rock depth (m)	Rock level (m AOD)
B	177.0	34.5	142.5
C	192.0	40.5	151.5

By proportion, assuming that ground surface is also plane,

$$\text{Rock level at E} = 142.5 + \frac{BE}{BC}(151.5 - 142.5)$$

$$= 142.5 + \frac{156.52}{335.41} \times 9.0$$

$$= 146.7 \text{ m AOD.}$$

Similarly, ground level at E $= 177.0 + \frac{156.52}{335.41}(192.0 - 177.0)$

$$= 184.0 \text{ m AOD.}$$

Next locate the intersection point. Figure 7.12 shows the longitudinal section along AE, indicating salient dimensions. I is the point of intersection of the

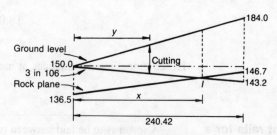

Figure 7.12

centre line of construction and the rock plane. The gradient of the formation of construction is 3 in 106 so that the fall over a distance of 240.42 m is

$$\tfrac{3}{106} \times 240.42 = 6.8 \text{ m.}$$

Therefore formation level at E $= 150.0 - 6.8$
$$= 143.2 \text{ m AOD.}$$

By similar triangles

$$\frac{150.0 - 136.5}{x} = \frac{146.7 - 143.2}{240.42 - x},$$

in which x is the distance in plan from A to I.

Therefore $\qquad\qquad x = 190.92 \text{ m.}$

Therefore the co-ordinates of I are

$$\text{Easting} = 190.92 \sin 45° + 0.0$$
$$= \textbf{135.0 m,}$$
$$\text{Northing} = 190.92 \cos 45° + 0.0$$
$$= \textbf{135.0 m.}$$

$$\text{Tunnel level at intersection point I} = 150.0 - \tfrac{3}{106} \times 190.92$$

$$= 144.6 \text{ m.}$$

But ground level at E = 184.0 m,

Therefore $$\text{Ground level at I} = 150.0 + (184.0 - 150.0) \times \frac{190.92}{240.42}$$

$$= 177.0 \text{ m AOD.}$$

$$\text{Depth of formation at I} = 177.0 - 144.6$$
$$= \textbf{32.4 m.}$$

For the location of the end of the cutting,

$$\text{Formation level at E} = 143.2 \text{ m AOD}$$
$$\text{Ground level at E} = 184.0 \text{ m AOD.}$$
Therefore depth of formation at E = 40.8 m.

If y be the distance in plan from A (Fig. 7.12) to the point at which formation is 18.0 m deep, then

$$\frac{y}{18.0} = \frac{240.42}{40.8}$$

$$y = 106.0 \text{ m.}$$

Hence tunnelling commences at a distance of **106.0 m** in plan from A.

7.10 Sight rails for a sewer

A sewer is to be laid between two points P and Q, 80 m apart. Levels were taken to establish the longitudinal profile, and the bookings in Table 7.15 obtained. Invert level at P is to be 120.750 m AOD and the sewer is to fall towards Q at a gradient of 1 in 120.

Table 7.15

Backsight	Intersight	Foresight	Distance	Remarks
0.633				OBM RL 124.82 m AOD
	2.925		0	GL at P
	2.697		20	
	2.560		40	
	2.915		60	
	3.020		80	GL at Q
		0.633		OBM

At the setting-out stage the level was set up close to its previous position and a backsight of 0.587 m was recorded on a staff held at the bench mark.

(*a*) Suggest a suitable length for the 'traveller'.

(*b*) Determine:

(*i*) the staff readings required for the fixing of sight rails at P and Q;
(*ii*) the heights of those rails above ground level at P and Q.

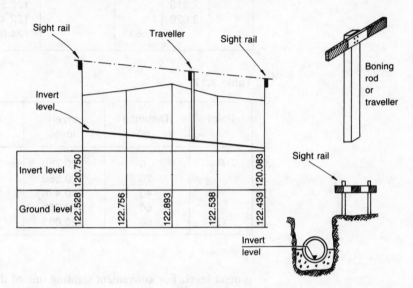

Figure 7.13

Introduction. The sight rails are positioned so that the line connecting their upper edges reflects the gradient of the trench bottom or the pipe invert, as applicable. A boning rod, or traveller, of correct length is held with the upper edge of its horizontal sight bar just in the line of sight given by the sight rails; in this position its lower end stands at the required level (Fig. 7.13). The horizontal sight rails are nailed to stout uprights, firmly installed on alternate sides of the trench. These uprights must be well clear of the sides of the trench. Frequent checking of their integrity is essential.

Solution. First determine the existing ground levels. The level bookings may be reduced as follows in Table 7.16 overleaf.

Next calculate the length of boning rod. Invert level at P is to be 120.750 m and the gradient of PQ is to be 1 in 120 falling towards Q, 80 m from P.

Therefore invert level at Q = $120.750 - \frac{80}{120}$

= 120.083 m AOD

whence we obtain the values in Table 7.17. The line of sight given by the sight rails is to have a gradient of 1 in 120 and it must have clearance above

Table 7.16

BS	IS	FS	Height of collimation	RL	Distance (m)	Remarks
0.633			125.453	124.820		BM
	2.925			122.528	0	Point P
	2.697			122.756	20	
	2.560			122.893	40	
	2.915			122.538	60	
	3.020			122.433	80	Point Q
		0.633		124.820		BM

Table 7.17

Point	Distance (m)	Invert level	Ground level	Difference (m)
P	0	120.750	122.528	1.778
	20	120.583	122.756	2.173
	40	120.416	122.893	2.477
	60	120.250	122.538	2.288
Q	80	120.083	122.433	2.350

ground level. For convenient sighting one of the rails should be about 1 m or so above ground level. Consideration of the above differences between invert and ground levels shows that a boning rod length of 3 m would be satisfactory. Hence the levels of the sight rails are to be

At P   120.750 + 3.000 = 123.750 m AOD
At Q   120.083 + 3.000 = 123.083 m AOD.

The staff readings for the positioning of the sight rails can now be calculated. Since the backsight observed by level on to the bench mark is 0.587 m when setting up the sight rails, the corresponding height of collimation

$$= 124.820 + 0.587$$
$$= 125.407 \text{ m AOD.}$$

Therefore staff reading for rail at P $= 125.407 - (120.750 + 3.000)$
$$= \textbf{1.657 m}$$
and staff reading for rail at Q $= 125.407 - (120.083 + 3.000)$
$$= \textbf{2.324 m.}$$

The staff could be moved up and down the uprights to give readings of 1.657 m and 2.324 m, respectively at P and Q, marks made thereon corresponding to the base of the staff. The sight rail is then nailed in position and checked. Alternatively, the tops of the uprights could be levelled and measurements made down the uprights to locate the 'finished' levels of 123.750 m and 123.083 m.

The heights of the sight rails above ground level are:

At P   rail height = 123.750 − 122.528
$\qquad\qquad\qquad$ = **1.222 m** above ground level;

At Q   rail height = 123.083 − 122.433
$\qquad\qquad\qquad$ = **0.650 m** above ground level.

Observations could be made quite conveniently from P towards Q.

7.11 Embankment profile boards

An embankment is to be constructed on ground having a transverse cross fall of 1 in 9. At a certain cross-section the formation level will be 267.50 m AOD, ground level at the centre line being 261.98 m AOD. Side slopes of 1 vertical in 2 horizontal have been specified together with a formation width of 20 m. How could profile boards be established to control the construction?

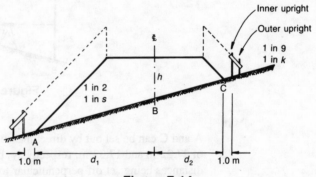

Figure 7.14

Introduction. The profile boards are nailed to two uprights which are firmly driven into the ground near the toes, A and C, of the embankment (Fig. 7.14). The inner uprights need to have clearances from the toes of the order of 1.0 m to prevent disturbance. The inner and outer uprights can be spaced up to 1.0 m apart, since the sloping boards reflecting the side gradients need to be of reasonable length for sighting purposes. A traveller is used in conjunction with the upper surface of the boards to achieve the gradients.

Various configurations are discussed in this example.

Solution. First establish the side widths and levels. As discussed in Chapter 5 the side widths, d_1 and d_2, can be determined from the expressions

$$d_1 = \left(\frac{b}{2} + sh\right)\left(\frac{k}{k-s}\right)$$

and $$d_2 = \left(\frac{b}{2} + sh\right)\left(\frac{k}{k+s}\right)$$

where b is the formation width; h is the height of embankment at centre line; s is the side slope of the embankment; and k is the transverse cross fall.

Now $h = 267.50 - 261.98 = 5.52$ m

$d_1 = (10 + 11.04) \left(\frac{9}{7}\right)$ $d_2 = (10 + 11.04) \left(\frac{9}{11}\right)$

$\quad = 27.05$ m $= 17.21$ m

The reduced level of B $= 261.98$ m AOD

and fall from B to A $= \dfrac{27.05}{9} = 3.01$ m.

Therefore reduced level of A $= 258.97$ m.

Also, rise from B to C $= \dfrac{17.21}{9} = 1.91$ m.

Therefore reduced level of C $= 263.89$ m.

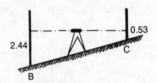

Figure 7.15

A and C can be set out by direct measurement from B, either sloping lengths of 27.22 m and 17.32 m, respectively, being measured or the above horizontal distances being set off perpendicular to the centre line. Alternatively, a trial method using a level can be adopted as in Fig. 7.15. Say that a reading of 2.44 m was obtained when the staff was held at B, the level being in any suitable position, not necessarily on the line of the cross-section. The staff man is now moved along BC until the reading of 0.53 m is attained, i.e. to realize the level difference of 1.91 m previously calculated. This position fixes C.

Next determine the profile board configuration. The two major considerations are the length of the traveller and the relationship between the profile board, existing ground and proposed slopes. In the first instance assume a traveller of length 1.25 m with the centre lines of the uprights at 1.00 m and 2.00 m, respectively, from toes A and C.

At A the sight line must be at an altitude of $1.25 + 258.97 = 260.22$ m. Thus at the uprights the corresponding altitudes of the sight line will be

$260.22 - 1.00 \times \frac{1}{2} = 259.72$ m

and $260.22 - 2.00 \times \frac{1}{2} = 259.22$ m.

The ground levels at the uprights will be

$258.97 - 1.00 \times \frac{1}{9} = 258.86$ m

and $258.97 - 2.00 \times \frac{1}{9} = 258.75$ m.

Table 7.18

	C	Inner upright	Outer upright
Altitude of sight line	265.14	264.64	264.14
Ground level	263.89	264.00	264.11
Difference (m)	1.25	0.64	0.03

Thus there is a reasonable height difference at each upright (0.86 m and 0.47 m, respectively) to allow the boards to be fixed. At C, the altitude of the sight line will be $1.25 + 263.89 = 265.14$ m. Thus for uprights spaced as at A we have the values in Table 7.18.

0.03 m at the outer upright is too small, but if the spacing of the uprights is reduced to 0.60 the difference at the outer upright becomes $264.34 - 264.07 = 0.27$ m, and this difference should be considered to be a minimum, allowing for the width of the board itself. Alternatively, if the traveller is made 1.5 m long, which is probably a maximum for convenience, the values in Table 7.19 apply when the uprights are 1.0 m apart.

Table 7.19

At A	A	Inner upright	Outer upright
Altitude of sight line	260.47	259.97	259.47
Ground level	258.97	258.86	258.75
Difference (m)	1.50	1.11	0.72
At C	**C**	**Inner upright**	**Outer upright**
Altitude of sight line	265.39	264.89	264.39
Ground level	263.89	264.00	264.11
Difference (m)	1.50	0.89	0.28

If the spacing of the uprights is reduced to 0.60 m the difference becomes 0.52 m at the outer upright at C. It will be seen that the setting-out engineer has some options open to him at this cross-section. However, several cross-sections may need consideration at a particular time and, for convenience, one length of traveller selected.

Problems

1 Boreholes were sunk at points X, Y and Z to locate a coal seam, and the information in Table 7.20 was obtained.

Table 7.20

Point	Co-ordinates		Ground level (m) AOD	Seam depth (m)
	E (m)	N (m)		
X	0	0	180.0	460.0
Y	30.0	1200.0	168.0	408.0
Z	780.0	150.0	192.0	528.0

Locate point P at which straight formations at an inclination of 1 in 18 can be driven upwards to Y and downwards to Z, respectively.
Answer   450.4 m from Z, bearing 227° 00′ 33″

2 The co-ordinates of two stations A and B are 434 762.19 mE, 376 592.83 mN and 435 476.80 mE, 377 404.35 mN, respectively. At A and B clockwise angles BAC and ABC are measured as 44° 29′ 35″ and 313° 32′ 43″, respectively. Determine the co-ordinates of C.
Answer   435 544.02 mE, 376 649.43 mN

3 Control station X lies inside triangle PQR such that $P\hat{X}Q = 128°$ 23′ 24″ and $Q\hat{X}R = 97° 10′ 11″$. Calculate the co-ordinates of point X if the co-ordinates of

P   are   1798.2 mE   7643.2 mN
Q   are   2534.8 mE   6327.1 mN
R   are   1531.7 mE   6215.6 mN

[Salford]

Answer   1916.7 mE, 6701.6 mN

4 A shaft has been sunk to a coal seam at point A and is subsequently to be deepened to a point Y. It is proposed to drive a cross-measure drift dipping at 1 in 200 from a point E to a point X which is 134 m due east of Y; YX is level. In Table 7.21 are the notes of a traverse made in the seam, from the centre of the shaft A to the point E.

Table 7.21

Line	Azimuth	Slope distances	Vertical angle
AB	270° 00′	127 m	Level
BC	184° 30′	550 m	Dipping 21°
CD	159° 15′	730 m	Dipping 18° 30′
DE	90° 00′	83 m	Level

Compute

(*a*) the azimuth and horizontal length of the drift EX, and (*b*) the amount by which it is necessary to deepen the shaft, i.e. the vertical distance AY. [CEI]

Answer (*a*) 358° 40′, 1159.58 m; (*b*) 434.53 m

5 (*a*) During the setting out of a tunnel the observations in Table 7.22 were made by theodolite at a surface station I near to a vertical shaft.

Table 7.22

Pointing	Horizontal circle reading	Distance from theodolite
Station A	193° 18′ 12″	446.35 m
Station B	13° 17′ 56″	170.60 m
Plumbwire P	106° 50′ 20″	7.290 m
Plumbwire Q	106° 50′ 29″	

The plumb wires down the shaft were 5.345 m apart, P being the nearer of the two to the theodolite at I. If the whole circle bearing of PQ was 307° 47′ 24″ deduce that of AB.

(*b*) Explain briefly two methods of using the gyro-theodolite to determine the azimuth of AB. [Salford]

Answer 214° 14′ 50″

6 An approach ramp of 1 vertical to 10 horizontal is to be supported by a single line of vertical columns based at 15 m intervals measured along a true curve plan of radius 30 m. The ramp starts at tangent point T on the flat and then rises with its centre line connecting columns A, B, C, D, etc.

(*a*) Calculate the data to set out the first three columns from T using a theodolite and steel tape.

(*b*) If the theodolite axis was set up 0.90 m vertically above T and the columns terminated at 0.6 m below the ramp surface, calculate the vertical angles required to check the stopping-off heights of the three column heads. [I. Struct. E.]

Answer Chords 14.84 m long, deflection angles 14° 19′ 26″, 28° 38′ 52″ and 42° 58′ 18″; vertical angles 00° 00′ 00″, 02° 59′ 06″, 04° 11′ 43″

7 Surveys connecting base lines of known azimuth with a vertical shaft plumb line (W) have been conducted on the surface and underground at a deep mine as part of a correlation survey. From the information given below calculate the National Grid co-ordinates of the survey stations which are all to the east generally of the shaft.

National Grid co-ordinates of shaft plumb line (W):
  E 460 257.664 N 350 258.130

Underground survey stations: U_1, U_2 and U_3
Surface survey stations: S_1, S_2 and S_3

National Grid azimuth U_3 to U_2 = 237° 20′ 00″
National Grid azimuth S_3 to S_2 = 308° 17′ 52″

Horizontal angles measured in a clockwise direction:

$$S_3 - \hat{S}_2 - S_1 = 136° \ 20' \ 21''$$
$$S_2 - \hat{S}_1 - W = 217° \ 15' \ 17''$$
$$U_3 - \hat{U}_2 - U_1 = 232° \ 49' \ 45''$$
$$U_2 - \hat{U}_1 - W = 199° \ 39' \ 40''$$

Horizontal distances:

S_3 to S_2	= 410.257 m	U_3 to U_2	= 700.157 m
S_2 to S_1	= 400.135 m	U_2 to U_1	= 405.316 m
S_1 to W	= 350.685 m	U_1 to W	= 690.334 m

[Eng. Council]

Answer U_3, 461 757.742 mE, 350 054.226 mN

8 The co-ordinates of three stations A, B and C are given in Table 7.23.

Table 7.23

Station	Easting (m)	Northing (m)
A	24 078.31	29 236.48
B	26 266.48	31 493.20
C	28 377.67	29 661.04

A point O is set up inside the triangle, and the observations in Table 7.24 are taken. Calculate the co-ordinates of station O.   [Leeds]

Table 7.24

Angle	Adjusted value
BOA	142° 48′ 32″
COB	92° 12′ 22″
AOC	124° 59′ 06″

Answer 25 960.42 mE, 30 547.89 mN

9 Observations were taken from a boat (X) to three shore stations A, B and C, lying to the west, in order to resect its position. If the co-ordinates of A, B and C were 5280.16 mE; 6213.40 mN, 3297.95 mE; 4502.75 mN and 5001.23 mE; 2987.46 mN, respectively and the angles on the boat were recorded as AXB = 50° 27′ 30″ and BXC = 46° 25′ 25″, show why this would produce a poor fix without attempting to calculate the co-ordinates of the boat.

Explain why modern hydrographic surveying has adopted automated techniques more rapidly than land surveying and why the use of electronics is more vital for hydrographic surveying.   [Leeds]

Answer $\hat{B} + \hat{X} = 179° \ 20' \ 00''$

10 Two stations X and Y on a street traverse and three manholes A, B, and C on a sewer, to be constructed in heading, have the co-ordinates given in Table 7.25.

Table 7.25

	E (m)	N (m)
X	1000.0	1000.0
Y	1300.0	1500.0
A	1100.0	1050.0
B	1180.0	1200.0
C	1325.0	1400.0

Determine (a) the co-ordinates of station D on XY such that DB is perpendicular to XY, and (b) angles DBA and DBC. Thence describe how lines BA and BC could be established underground.　　[CEI]
Answer (a) 1135.9 mE, 1226.6 mN; (b) 92° 54′, 94° 58′

11 A, B and C are three control points which lie in a straight line and have co-ordinates given in Table 7.26. The horizontal circle readings

Table 7.26

Station	A	B	C
E	550	610	690
N	550	700	900

on a theodolite stationed at T_1, a point to the east of the line AC, were:

From T_1: to A 360° 00′; to B 40° 00′; to C 70° 00′.

Calculate the co-ordinates of T_1.

T_1 is the tangent point on the leading end straight of a circular curve which has a bearing of 312° 15′. If the circular curve has a radius of 2000 m and a deviation angle of 13° 05′ to the right (+ve) calculate the co-ordinates of the second tangent point.　　[Bradford]
Answer 776.2 mE, 565.0 mN; 476.0 mE, 907.8 mN

12 The co-ordinates of three stations A, B and C are as given in Table 7.27.

Table 7.27

Station	Easting (m)	Northing (m)
A	11 264.69	21 422.30
B	12 142.38	21 714.98
C	12 907.49	21 538.66

Two unknown points X and Y lie to the southerly side of AC with Y on the easterly side of point X. The following angles are recorded:

$$AXB = 38° 15' 47''$$
$$BXC = 30° 38' 41''$$
$$CXY = 33° 52' 06''$$
$$CYX = 106° 22' 20''$$

Calculate the length and bearing of the line XY. [Leeds]

Answer 1013.27 m, 68° 10' 10''

13 Two surface reference stations X and Y having co-ordinates of 1000.00 mE, 1000.00 mN and 1300.00 mE, 1500.00 mN respectively were observed during a shaft plumbing exercise. A theodolite was set up at surface station A near to the line XY and the readings in Table 7.28 recorded.

Table 7.28

Pointing direction	Horizontal circle reading
X	273° 42' 24''
Y	93° 42' 08''
Plumb wire P	98° 00' 50''
Plumb wire Q	98° 00' 40''

The distances from the theodolite to X and P were 269.12 m and 8.374 m respectively, whilst P and Q were 5.945 m apart, P being nearer to A than Q. Estimate the bearing of PQ. [Salford]

Answer 35° 16' 00''

14 A and B are points on the centre line of a level mine roadway and C and D are points on the centre line of a lower roadway having a uniform gradient between C and D. It is proposed to connect the roadways by a drivage from point B on a bearing of 165° 35'. Given the data in Table 7.29, calculate

Table 7.29

Point	Northing (m)	Easting (m)	Reduced level (m)
A	2653	1321	462.5
B	2763	1418	462.5
C	2653	1321	418.2
D	2671	1498	441.8

(*i*) the actual length and gradient of the drivage, and
(*ii*) the co-ordinates of the point at which it meets the lower roadway. [C.E.I.]

Answer (*i*) 104.57 m, 1:3.6; (*ii*) 1443 mE, 2665 mN

15 During redevelopment in a city area, a block of flats rectangular in plan is to be built with the long elevation orientated due east—west. The building, which is to measure 86 m long by 21 m wide, is located initially by means of a peg at the south-east corner (point A, 1200 mE,

600 mN). The remainder of the building cannot be set out as existing properties have not yet been demolished and it is essential to check its clearance from a proposed new building line. The observations in Table 7.30 are therefore made.

Table 7.30

Line	Whole circle bearing	Horizontal distance (m)
AP	251° 00′	80.69
PQ	283° 00′	20.64

Point Q is located on the proposed building line which is straight with a WCB of 13° 30′. Calculate the minimum distance from the proposed development to the new building line. [Salford]
Answer 0.54 m

16 Three points P, Q and R have the following co-ordinates:

P   950.00 mE   1200.00 mN
Q   983.50 mE   1340.00 mN
R   1027.69 mE   1239.74 mN

Locate point S on PQ from which a perpendicular SR can be set out to establish R. What is the length of that perpendicular?
Answer PS = 56.73 m; SR = 66.31 m

17 State the advantages to be gained in correlating engineering surveys to the National Grid.

The National Grid co-ordinates of two survey stations A and B are as follows:

Station A   E 323 679.35   N 340 431.32
Station B   E 324 022.07   N 342 846.89

The grid distances from a third survey station C to the east generally of A and B are CA 1901.624 m and CB 1388.901 m. Compute the National Grid co-ordinates of station C. [Eng. Council]
Answer 324 967.91 mE, 341 829.82 mN

18 It is required to establish some planimetric control in an area where the only existing control visible is three intersected points, Stopworth Church Spire, the Drill Hall and Buckton Castle. A station is established at Rowley Mount, to the south of the line joining Stopworth Church Spire and Buckton Castle. The following angles are observed:

Stopworth Church Spire → Rowley Mount → The Drill Hall
27° 43′ 49″
The Drill Hall → Rowley Mount → Buckton Castle 26° 52′ 24″

The co-ordinates of the three intersected points are as in Table 7.31.

Table 7.31

	Easting (m)	Northing (m)
Stopworth Church Spire	1264.93	1421.26
The Drill Hall	1483.71	1390.29
Buckton Castle	1701.83	1430.02

Calculate the co-ordinates of Rowley Mount. [Leeds]
Answer 1480.00 mE, 1002.46 mN

19 A small control survey network contains four stations A, B, C and D. The stations C and D lie to the east of line AB. Given the following data calculate the co-ordinates of D. Check each stage of the calculation.

Co-ordinates: A 4763.252 E, 6372.156 N; B 2477.361 E, 1544.789 N.
Angles: CAB 49° 26′ 15″   BCA 65° 37′ 39″
Lengths: AD 4366.890, CD 3632.471, both reduced to horizontal.
                                                      [Bradford]

Answer 8777.247 mE, 4652.396 mN

20 The National Grid co-ordinates of three survey stations C, E and R are as follows:

Station C E 322 393.384 N 343 016.705
Station E E 323 048.449 N 343 362.094
Station R E 324 022.141 N 342 846.959

Horizontal angles observed from a fourth station G to the south generally of stations C, E and R are as follows:

CĜE = 27° 54′ 10″ and EĜR = 42° 52′ 29″

Compute the National Grid co-ordinates of station G and the azimuth of the line G to R.   [Eng. Council]
Answer 323 117.992 mE, 341 782.834 mN; 40° 21′ 12″

21 An embankment is to be constructed on ground which can be taken to have an effective transverse gradient of 1 in 12, the reduced ground level at the centre line being 49.53 m AOD. The side slopes of the embankment are to be 1 vertical in 2 horizontal and the reduced level of the formation, width 20 m, is to be 54.33 m. Profile boards are to be set up, the centre lines of the uprights nearest to the toes of the embankments being 1 m therefrom. The setting-out engineer is considering using a travelling rod 1.5 m long in conjunction with pairs of uprights 0.8 m apart at the lower toe C and 0.6 m apart at the higher toe A.
    Determine the difference between the heights of the sight rails and ground level at each of the four uprights. If a level be set up such that a staff reading of 1.02 m is given at the centre line what would be the staff readings at the uprights when positioning the sight rail near C?
Answer 0.75 m, 1.08 m at C; 0.92 m, 0.57 m at A; 2.06 m, 2.46 m

22 A steel tape of nominal length 200 m was used to measure a length down a shaft as 165.252 m. The mean recorded temperature was 3 °C and a plumb bob, mass 15 kg, had been attached. If the tape had been standardized to be 200.0012 m under a tension of 120 N at 20 °C determine the corrected length.

Cross-sectional area of tape	$= 10$ mm^2
Mass of tape	$= 0.07$ kg/m
Modulus of elasticity	$= 2 \times 10^5$ N/mm^2
Coefficient of linear expansion	$= 11 \times 10^{-6}/°C$
Acceleration due to gravity	$= 9.807$ m/s^2

Answer 165.231 m

23 During levelling operations carried out in connection with the construction of a proposed sewer, the consecutive staff readings given in Table 7.32 were obtained.

Table 7.32

Staff reading (m)	Distance (m)	Remarks
1.40		OBM 34.20 m AOD
1.90	0	Ground level at outfall
1.19	40	Ground level
0.84	80	Ground level
1.15	120	Ground level Change point
2.96	120	Ground level Change point
1.49	160	Ground level
1.26	200	Ground level
1.68	—	TBM 35.73 m AOD

Details of the proposed sewer are given in Table 7.33.

Table 7.33

Chainage	Manhole	Invert level (m AOD)	Remarks
0	Outfall	31.98	Gradient of outfall to MH 'A' to be 1 in 140
100	'A'	?	
150	'B'	33.50	Outside diameter of sewer is 0.400 m
200	'C'	35.40	

(*a*) Book and reduce the levels carrying out appropriate arithmetical checks.

(*b*) Draw a longitudinal section of the ground.

(*c*) Calculate the invert level of MH 'A' and the gradients of sewer runs from MH 'A' to MH 'B' and MH 'B' to MH 'C', and show this information on the section. Determine the minimum cover to the sewer.

(*d*) A traveller 3.0 m long was used when the trench was excavated. Calculate the staff readings adopted when setting out the sight rails at the manholes and the outfall, and state from which instrument position the readings were obtained. Assume that the level was set up exactly as it was in (a).

Answer (c) 32.69, 1 in 61.7, 1 in 26.3, 1.04 m say; (d) 0.62 m at outfall from first position, 1.72 m at A, 0.91 m at B, −0.99 m at C from second position

24 A section of sewer is to be laid between two manholes, 12 and 13, which are 100 m apart. Sight rails have been established, based on the staff readings given in Table 7.34. The distance refers to the lengths of sight from the instrument.

Table 7.34

BS	IS	FS	Distance (m)	Remarks
1.480			50	BM RL 48.290
	0.855		75	Sight rail at MH 12
	2.105		40	Sight rail at MH 13
		1.480	50	BM RL 48.290

If the sewer is to be constructed using a travelling rod of length 3.25 m, determine the invert levels at the manholes and the gradient of the sewer as designed.

At a later date it was found that the tilting level was out of adjustment and the gradient of the sewer was actually 1 in 78. What was the error in collimation of the level?

Answer 45.655 m, 44.415 m; 0.0009 rad (upwards)

25 The plan centre line P Q R S of a bridge over a river lies on a 300 m radius curve. The bridge has three spans PQ, QR, and RS, P and S being the centres of the faces of the abutments and Q and R the centres of the piers. The faces of the piers and the abutments are parallel and the length of each span, measured square to those faces, from abutment to pier, pier to pier, and pier to abutment respectively is 40.500 m.

The chainage of P, the centre of the face of the first abutment, is 8539.800 m and the angle of skew between that face and the tangent to the curve at P is 74° 00′ 00″. Calculate (i) the angles of skew at Q, R, and S, (ii) the chainages of Q, R, and S.

If the chainages of the tangent points A, near P, and B, near S, on the main straights are 8525.650 m and 8675.250 m respectively, determine (iii) length AQ and clockwise angle BAQ to locate Q, and (iv) length BR and clockwise angle ABR to locate R.   [London]

Answer (i) 81° 54′ 55″, 89° 40′ 37″, 97° 25′ 58″; (ii) 8581.244 m, 8621.884 m, 8662.406 m; (iii) 55.515 m, 351° 01′ 23″; (iv) 53.296 m, 9° 11′ 23″

26 During the setting out of a tunnel, the observations given in Table 7.35 were made at a shaft, the instrument stations being established eastwards of the plumb wires and with X the nearer in each case. The

Table 7.35

Theodolite at surface station		
Pointing	Circle reading	Distance from theodolite (m)
Reference station P	288° 31′ 10″	140.32
Reference station Q	108° 31′ 19″	476.24
Plumb wire X	196° 08′ 18″	8.825
Plumb wire Y	196° 08′ 35″	

Theodolite at underground station		
Pointing	Circle reading	Distance from theodolite (m)
Plumb wire X	54° 31′ 30″	7.263
Plumb wire Y	54° 31′ 54″	
Reference station R	224° 34′ 34″	27.68
Reference station S	44° 34′ 17″	24.35

plumb wires were 6.250 m apart and the whole circle bearing of PQ was 210° 17′ 22″. Determine the bearing of SR.   [Salford]

Answer 107° 57′ 08″

27 You are employed as a surveyor by a company engaged in the construction of a tunnel about 3 km long, with a depth between 30 m and 80 m. Vertical shafts are being sunk along the line of the tunnel and it is intended to connect the bottom of each shaft by tunnelling outwards in both directions.

Discuss:

(*a*) the correlation of the surface and underground surveys;

(*b*) the problems of maintaining plan and height control underground.

[RICS]

28 State the factors to be considered in designing the survey stations and the conditions to be satisfied in setting out control networks for precise deformation surveys.

Two fixed and stable stations, A and B have been set out to facilitate the location of two stations, W9 and W46, on top of a diaphragm retaining wall. The measurements in Table 7.36 were recorded, before and

Table 7.36

Before excavations commenced			
Reduced horizontal distances		Horizontal angles	
A to B	= 123.9065 m	B − Â − W9	= 29° 04′ 40.4″
A to W9	= 138.7447 m	W9 − Â − W46	= 42° 18′ 50.4″
A to W46	= 56.0792 m		

After excavations completed			
Reduced horizontal distances		Horizontal angles	
A to B	= 123.9065 m	B − Â − W9	= 29° 04′ 27.6″
A to W9	= 138.7486 m	W9 − Â − W46	= 42° 18′ 22.9″
A to W46	= 56.0895 m		

after deep excavations were made adjacent to the diaphragm wall, in order to detect any movement of the wall.

Calculate the horizontal movement of the stations W9 and W46. List the equipment and instruments required to measure the vertical movement of the stations on the diaphragm wall. [Eng. Council]

Answer 9.5 mm, 15.0 mm

8

Observations and adjustments

Error types

A constant error, such as an incorrectly standardized tape, has the same influence throughout the programme of observations, whilst a systematic error like the influence of field temperature on the length of that measuring tape, can be of varying magnitude and changing sense. Accidental errors, such as instrument imperfections, are of a random nature and could be said to be second-order errors when compared to the types just mentioned.

In this chapter a number of examples are solved on the assumption that the observations are free of all errors other than accidental errors. In other words, the work has been carried out carefully, with as many checks as possible to eradicate mistakes and, in addition, corrections have been applied to eliminate any constant errors and systematic errors.

Accidental errors

Accidental errors are unavoidable in so far as the observer is concerned. They usually occur with positive and negative values of the same size and frequency, the larger errors occurring less frequently than the small errors. If a number of measurements of the length of a line are repeatedly made under similar conditions a spread of values will be obtained and the difference between the largest and the smallest values is defined by the term 'range'. The values can be placed into groups or classes having specified limits, the number of measures in each group being termed the class frequency. If these numbers are expressed as a percentage of the total then proportional frequencies are obtained. Alternatively, the numbers themselves, taken in conjunction with their respective classes, give the frequency distribution of the whole set of measurements.

Probability distribution

If a very large number of measurements has been taken the frequency distribution could be considered to be the probability distribution. The normal distribution is usually assumed for survey observations, being expressed by the equation,

$$\mathrm{d}p = \frac{1}{\sigma \sqrt{(2\pi)}} \, e^{-(x_1 - \mu)^2 / 2\sigma^2} \, \mathrm{d}x$$

where $\mathrm{d}p$ is the probability that value x will lie between the limits of x_1 and $(x_1 + \mathrm{d}x)$; μ is the true mean of the population; and σ is the standard deviation.

Most probable value

In practice, a limited number of observations of a quantity, such as the length of a line or the magnitude of an angle, will be taken and this set is effectively a sample; μ and σ are unknown. Accordingly, we calculate the most probable value of the quantity from that sample and estimate the standard deviation of the population. If the measurements are of equal reliability, or precision, the arithmetic mean ($\bar{x}$) of a set of observations is taken to be the most probable value of that quantity. The difference between the value (x) of one of the measurements and the most probable value of the quantity, i.e. ($x - \bar{x}$), is termed the residual (or the residual error). Had we known μ, the true error would be known to be ($x - \mu$).

Standard deviation

Standard deviation is a measure of the spread of a distribution and for the population, assuming the readings are of equal reliability,

$$\sigma = \pm \sqrt{\left[\frac{\Sigma(x-\mu)^2}{n}\right]}.$$

However, μ cannot be determined from a small sample of readings. Instead, the arithmetic mean ($\bar{x}$) is accepted as the most probable value and the population standard deviation is estimated as

$$s = \pm \sqrt{\left[\frac{\Sigma(x-\bar{x})^2}{n-1}\right]},$$

in which n is the number of observations and

$$\bar{x} = \frac{\Sigma x}{n}.$$

Confidence limits

Having established the sample mean as an estimate of the true value of the quantity, the observer might wish to state the range of values within which that true value should lie for a given probability. This range is termed the confidence interval, its bounds being the confidence limits. Having calculated the standard deviation for a set of observations, confidence limits can be established for that stated probability, statistics tables being available for this purpose. Frequently, a figure of 95% is chosen, and this implies that nineteen times out of twenty the true value will lie within the computed limits. If normally distributed small random errors obtain, the presence of a very large residual suggests an occurrence to the contrary. Such an observation can be rejected if its residual error is larger than three times the standard deviation.

Standard error

The standard error of the mean ($s_{\bar{x}}$) is given by

$$s_{\bar{x}} = \pm \frac{s}{\sqrt{n}}$$

and hence the precision of the mean is enhanced with respect to that of a single observation. There are n deviations (or residuals) from the mean of the sample and their sum will be zero. Thus, knowing ($n-1$) deviations the surveyor could deduce the remaining deviation and it may be said that there are ($n-1$) degrees of freedom. This number is used when estimating the population standard deviation.

Weight

When the measurements are not of equal reliability weights (w) are applied to the individual measures to reduce them to one standard. The most probable value is then the weighted mean.

The relationship $w_1 s_1{}^2 = w_2 s_2{}^2 = \ldots = s^2$ applies, in which w_1 and s_1 are the weight and standard error for observations 1, w_2 and s_2 are the weight and standard error for observations 2, whilst s is the standard error for the observations having unit weight. Hence

$$w_1 = \frac{s^2}{s_1{}^2}, \quad w_2 = \frac{s^2}{s_2{}^2} \text{ etc.,}$$

and $\quad \dfrac{w_1}{w_2} = \dfrac{s_2{}^2}{s_1{}^2}$ etc.

Thus the weight of an observation can be taken to be inversely proportional to the square of its standard error. The most probable value of the weighted mean $(\bar{x}_w)$ of weighted observations is

$$\bar{x}_w = \frac{\Sigma w_1 x_1}{\Sigma w_1},$$

in which x_1 is observation 1, etc.

The other formulae given above are modified as follows:

(a) Standard deviation of an observation of unit weight

$$= \pm \sqrt{\left[\frac{\Sigma w_1 (x_1 - \bar{x}_w)^2}{(n-1)}\right]}.$$

(b) Standard deviation of an observation of weight w_n

$$= \pm \sqrt{\left[\frac{\Sigma w_1 (x - \bar{x}_w)^2}{w_n (n-1)}\right]}.$$

(c) Standard error of the weighted mean

$$= \pm \sqrt{\left[\frac{\Sigma w_1 (x_1 - \bar{x}_w)^2}{(n-1)\Sigma w_1}\right]}.$$

Method of least squares

This is a widely used method of determining the most probable values of observed quantities, assuming that only accidental errors are present. It states that the sum of the weighted residuals squared will be a minimum, i.e. $\Sigma[\text{weight} \times (\text{residual})^2]$ is to be a minimum.

For n observations

$$w_1 r_1{}^2 + w_2 r_2{}^2 + w_3 r_3{}^2 + \ldots + w_n r_n{}^2$$

is to be a minimum. Thus

$$[r_1 \; r_2 \; r_3 \; \ldots \; r_n] \begin{bmatrix} w_1 & 0 & 0 & \ldots & 0 \\ 0 & w_2 & 0 & \ldots & 0 \\ \cdot & \cdot & \cdot & & \cdot \\ \cdot & \cdot & \cdot & & \cdot \\ 0 & 0 & 0 & \ldots & w_n \end{bmatrix} \begin{bmatrix} r_1 \\ r_2 \\ \cdot \\ \cdot \\ r_n \end{bmatrix}$$

or $r^T W r$ is to be a minimum. Note that the alternative form of $\Sigma(\sqrt{w_1}\, r_1)^2$ can be minimized.

8.1 The distribution of a set of readings

The following readings were repeated on the same angle

$73° 40'$ plus $12''$, $15''$, $09''$, $14''$, $10''$, $18''$, $16''$, $13''$, $15''$ and $18''$.

Determine:

(a) the most probable value of the angle,
(b) the range,
(c) the standard deviation,
(d) the standard error of the mean, and
(e) the 95% confidence limits.

Solution. To calculate the mean value of the sample of 10 observations it is convenient to tabulate the data as in Table 8.1.

Table 8.1

Reading	$(x - \bar{x})$	$(x - \bar{x})^2$
$73° 40'$ $12''$	-2	4
$15''$	$+1$	1
$09''$	-5	25
$14''$	0	0
$10''$	-4	16
$18''$	$+4$	16
$16''$	$+2$	4
$13''$	-1	1
$15''$	$+1$	1
$18''$	$+4$	16
$\Sigma = 140''$	$\Sigma = 0$	$\Sigma = 84$

$$\frac{\Sigma x}{n} = 73° 40' + \frac{140''}{10}$$

(a) Most probable value $= \bar{x} = $ **73° 40′ 14″**.
(b) Range $= 73° 40' 18'' - 73° 40' 09''$
$= $ **9″**.

(c) Standard deviation $= \pm \sqrt{\left[\dfrac{\Sigma(x - \bar{x})^2}{n-1} \right]}$

$$= \pm \sqrt{\frac{84}{9}}$$

$$= \pm\ 3.1''.$$

(d) Standard error of the mean $= \pm \dfrac{s}{\sqrt{n}}$

$$= \pm \dfrac{3.1}{\sqrt{10}}$$

$$= \pm\ 1.0''.$$

(e) 95% confidence limits:

The lower confidence limit $= \bar{x} - \dfrac{t \times s}{\sqrt{n}}$

The upper confidence limit $= \bar{x} + \dfrac{t \times s}{\sqrt{n}}$

t is selected from statistical tables for a given value of n. When $n = 10$, $t = 2.26$ and so

$$\frac{t \times s}{\sqrt{n}} = 2.26 \times \frac{3.1}{\sqrt{10}}$$

$$= 2.2''.$$

The 95% confidence limits are **73° 40′ 14.0″ ± 2.2″**.

8.2 Relative precision

Prove that the mean value of a set of observations satisfies the least squares condition.

The length of a line was measured using two different EDM instruments A and B with the following results:

Table 8.2

EDM A	EDM B
785.546	785.545
785.538	785.549
785.535	785.543
785.550	785.546
785.542	785.545
785.547	785.550
	785.547
	785.543

Assuming that conditions were identical for both sets of readings, determine the relative precision of the two instruments. What is the most probable length of the line based upon the results? Work in millimetres.

[Bradford]

Solution. (*a*) Let $\bar{x}$ be the arithmetic mean of a set of measurements x_1, x_2, ..., x_n made on a certain quantity, and let a be any other value assigned to that quantity. Residuals of $(x_1-\bar{x})$, $(x_2-\bar{x})$, etc., and (x_1-a), (x_2-a), etc., respectively, then arise.

$$\Sigma(x-a)^2 - \Sigma(x-\bar{x})^2 = \Sigma x^2 - 2a\Sigma x + \Sigma a^2 - \Sigma x^2 + 2\bar{x}\Sigma x - \Sigma(\bar{x})^2$$

$$= -2a\Sigma x + na^2 + 2\frac{\Sigma x}{n} \times \Sigma x - n\left(\frac{\Sigma x}{n}\right)^2$$

(for n measures)

$$= n\left(a - \frac{\Sigma x}{n}\right)^2,$$

which is always positive. Hence the sum of the squared residuals based on the arithmetic mean is always less (i.e. a minimum) than the equivalent based on any other value.

(*b*) Consider A and B in turn.

Table 8.3

A	Measurement	$(x-\bar{x})$ (mm)	$(x-\bar{x})^2$
	785.546	+3	9
	785.538	−5	25
	785.535	−8	64
	785.550	+7	49
	785.542	−1	1
	785.547	+4	16
Mean $\bar{x}_A$ = 785.543			164

The standard deviation of the measures by A

$$= s_A = \pm\sqrt{\left[\frac{\Sigma(x-\bar{x})^2}{n-1}\right]} = \pm\sqrt{\left(\frac{164}{6-1}\right)}$$

$$= \pm 5.73 \text{ mm.}$$

Standard error of the mean

$$= s_{\bar{A}} = \pm\frac{5.73}{\sqrt{6}} = \pm 2.34 \text{ mm.}$$

Similarly standard deviation S_B

$$s_B = \pm\sqrt{\left[\frac{\Sigma(x-\bar{x})^2}{n-1}\right]} = \pm\sqrt{\left(\frac{46}{8-1}\right)}$$

$$= \pm 2.56 \text{ mm.}$$

Table 8.4

B	Measurement	$(x - \bar{x})$ (mm)	$(x - \bar{x})^2$
	785.545	−1	1
	785.549	+3	9
	785.543	−3	9
	785.546	0	0
	785.545	−1	1
	785.550	+4	16
	785.547	+1	1
	785.543	−3	9
Mean $\bar{x}_B$ = 785.546			46

Standard error of the mean

$$= s_{\bar{B}} = \pm \frac{2.56}{\sqrt{8}}$$

$$= \pm\ 0.91 \text{ mm}.$$

Next calculate the relative precision. The length of the line given by EDM A is 785.543 m with a standard error of 2.34 mm, whilst by EDM B it is 785.546 m with a standard error of 0.91 mm. Now

$$\frac{w_A}{w_B} = \frac{s_{\bar{B}}^2}{s_{\bar{A}}^2} = \frac{0.91^2}{2.34^2}$$

$$= \frac{1}{6.6}.$$

This ratio is a measure of the relative precision of the instruments.

Finally calculate the most probable length of the line. The most probable length of the line is the weighted mean of the two observed lengths. Now

$$\bar{x}_w = \frac{\Sigma w_1\, x_1}{\Sigma w_1}.$$

In this example

$$\bar{x}_w = \frac{w_A\, L_A + w_B\, L_B}{w_A + w_B},$$

in which L_A and L_B are the mean lengths recorded by EDM A and EDM B, respectively.

Hence $\bar{x}_w = \dfrac{w_A \times 785.543 + w_B \times 785.546}{w_A + w_B}$

$$= \frac{\dfrac{w_B}{6.6} \times 785.543 + w_B \times 785.546}{\dfrac{w_B}{6.6} + w_B}$$

$$= \textbf{785.5456 m.}$$

$\bar{x}_w$ could be written as 785.546 m to the nearest millimetre, in accordance with the observed measurements. In practice, the arithmetic mean and standard deviation or error may be expressed to one decimal place further than that to which the measurements had been made.

8.3 The distribution of weighted readings

An angle was measured as follows

73° 40′ 12″	weight 2
73° 40′ 15″	weight 3
73° 40′ 09″	weight 1
73° 40′ 14″	weight 4
73° 40′ 10″	weight 1
73° 40′ 18″	weight 1
73° 40′ 16″	weight 2
73° 40′ 13″	weight 3

Determine

(a) its most probable value;
(b) the standard deviation of an observation of unit weight;
(c) the standard deviation of an observation of weight 3; and
(d) the standard error of the weighted mean.

Table 8.5

Measure	x	Weight (w)	wx	Residual $r = (x - \bar{x}_w)$	wr^2
73° 40′ 12″	12	2	24	−1.7″	5.78
73° 40′ 15″	15	3	45	+1.3″	5.07
73° 40′ 09″	9	1	9	−4.7″	22.09
73° 40′ 14″	14	4	56	+0.3″	0.36
73° 40′ 10″	10	1	10	−3.7″	13.69
73° 40′ 18″	18	1	18	+4.3″	18.49
73° 40′ 16″	16	2	32	+2.3″	10.58
73° 40′ 13″	13	3	39	−0.7″	1.47
		$\Sigma w_1 = 17$	233		77.53

Solution. Tabulating the data and weighted residuals working from a datum of 73° 40′ we get the values in Table 8.5.

(a) The most probable value $= \bar{x}_w = $ Datum $+ \dfrac{\Sigma w_1 x_1}{\Sigma w_1}$

$$= 73° \ 40' + \frac{233}{17}$$

$$= \mathbf{73° \ 40' \ 13.7''}.$$

(b) Standard deviation (s) of an observation of unit weight

$$s = \pm \sqrt{\left[\frac{\Sigma w_1 (x_1 - \bar{x}_w)^2}{n-1}\right]}$$

$$s = \pm \sqrt{\left[\frac{77.53}{(8-1)}\right]} \text{ (since } n = 8)$$

$$= \pm \mathbf{3.33''}.$$

(c) Standard deviation of an observation of weight 3.

Since $\quad w_1 s_1{}^2 = w_2 s_2{}^2 = \ldots = s^2$

if $\quad\quad w_3 = 3$

$$3 s_3{}^2 = s^2 = 3.33^2.$$

Therefore $\quad s_3 = \pm \dfrac{3.33}{\sqrt{3}}$

$$= \pm \mathbf{1.92''}.$$

(d) Find the standard error of the weighted mean.

The weight of this mean $= \Sigma w_1 = 17.$

Therefore standard error $= \pm \sqrt{\left[\dfrac{\Sigma w_1 (x_1 - \bar{x}_w)^2}{(n-1) \ \Sigma w_1}\right]}$

$$= \pm \sqrt{\left[\frac{77.53}{(8-1) \times 17}\right]}$$

$$= \pm \mathbf{0.81''},$$

or, since $\quad w_m s_m{}^2 = s^2,$

$$s_m = \frac{s}{\sqrt{w_m}} = \pm \frac{3.33}{\sqrt{17}}$$

$$= \pm \mathbf{0.81''}.$$

8.4 Probability distribution

Taking the standard deviation σ of a single measure in Example 8.1 to be $\pm 3''$ calculate:

(a) the magnitude of the deviation likely to occur once in every two measurements;

(b) the probability that a single measurement may deviate from the true value by $\pm 6''$;

(c) the probability that the mean of nine measurements may deviate from the true value by $\pm 1.5''$.

Introduction. The expression for the normal distribution is

$$dp = \frac{1}{\sigma\sqrt{2\pi}} \, e^{-(x_1 - \mu)^2 / 2\sigma^2} \, dx.$$

On writing $\quad u = \dfrac{x - \mu}{\sigma}\quad$ the expression becomes

$$dp = \frac{1}{\sqrt{2\pi}} \, e^{-u^2/2} \, du.$$

This is the standardized form of the above expression, and the relationship between dp/du and u is illustrated in Fig. 8.1.

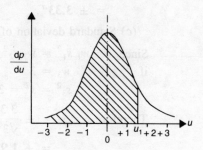

Figure 8.1

The curve is symmetrical and its total area is 1, the two parts about $u = 0$ having areas of 0.5. The shaded area has the value

$$\int_{-\infty}^{+u_1} \frac{1}{\sqrt{2\pi}} \, e^{-u^2/2} \, du$$

and it gives the probability that u lies between $-\infty$ and $+u_1$, i.e. less or equal to u_1. The unshaded area gives the probability that u will be larger than $+u_1$. Since the curve is symmetrical, the probability that u takes up a value outside the range $+u_1$ to $-u_1$ is given by the two areas indicated in Fig. 8.2.

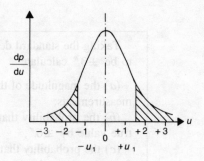

Figure 8.2

The values of the ordinates of the standardized form, and the corresponding definite integrals, have been determined for a wide range of u. These have been tabulated and are available in various publications. Typical values used in this example are given in Table 8.6.

Table 8.6

u	$\dfrac{dp}{du}$	$\displaystyle\int_{-\infty}^{+u}$
0.0	0.3989	0.5000
0.6	0.3332	0.7257
0.7	0.3123	0.7580
1.5	0.1295	0.9332
2.0	0.0540	0.9772

Solution. (a) For a deviation to occur once in every two measurements a probability of 50% is implied. Thus in Fig. 8.2 the two shaded parts have areas of 0.25 each and the shaded area is 0.5.

Bearing in mind that $\int_{-\infty}^{+u_1}$ is the shaded area in Fig. 8.1, we require a value of u in the table such that

$$1 - \int_{-\infty}^{+u_1} = 0.25, \quad \text{i.e.} \quad \int_{-\infty}^{+u_1} = 0.75.$$

By inspection, u lies between 0.6 and 0.7, for which the integral has values of 0.7257 and 0.7580, respectively. Hence the value u for the integral to be 0.75 is 0.6745.

Now $\qquad u = \dfrac{x - \mu}{\sigma} = 0.6745.$

Therefore deviation $(x - \mu) = 0.6745\sigma$
$$= \pm\, 2.0''$$

since σ is given as $3''$.

The term probable error may be encountered, rather than standard deviation; for measurements of equal reliability it is estimated as

$$0.6745 \sqrt{\left[\frac{(x - \bar{x})^2}{n - 1}\right]}.$$

(b) For a deviation $(x - \mu)$ of $\pm 6''$ for a single measure

$$u = \frac{x - \mu}{\sigma} = \frac{6}{3}$$

$$= 2.0''.$$

For $u + 2.0$

$$\int_{-\infty}^{+2.0} = 0.9772.$$

Hence $1 - \int_{-\infty}^{+2.0} = 1 - 0.9772$

$$= 0.0228.$$

For the deviation to lie at the limits of, or outside, the range $+6''$ to $-6''$ the probability is $2 \times 0.0228 = 0.0456$, or **4.6%**.

(c) The standard deviation of the mean of nine observations

$$\sigma_{\bar{x}} = \pm \frac{\sigma}{\sqrt{9}} = \pm \frac{3.0}{3}$$

$$= \pm 1.0''.$$

For a deviation of $\pm 1.5''$

$$u = \frac{x - \mu}{\sigma} = \frac{1.5}{1.0}$$

$$= 1.5'',$$

whence $\int_{-\infty}^{1.5} = 0.9332$

$$1 - \int_{-\infty}^{1.5} = 0.0668.$$

Therefore the probability of assuming a deviation of $\pm 1.5''$

$$= 2 \times 0.0668$$

$$= 0.1336, \quad \text{or } \textbf{13.4\%}.$$

8.5 Combinations of errors

The co-ordinates of two surveying stations A and B are given below, together with their standard deviations. Calculate the length and the standard deviation of AB determined from these values.

Station A Easting   456.961 m $\pm$ 20 mm,
           Northing   573.237 m $\pm$ 30 mm
Station B Easting   724.616 m $\pm$ 40 mm,
           Northing   702.443 m $\pm$ 50 mm

If an independent measurement by taping gives the length of AB to be 297.426 m $\pm$ 70 mm and separate determination by EDM to be 297.155 $\pm$ 15 mm, calculate the most probable length of the line using all the available information. What is the standard deviation of this value?

[Bradford]

Introduction. The surveyor might require to know the precision of a quantity which has been determined from other quantities of known precision, i.e. the horizontal length of a line calculated from measured slope length and angle of slope. There are various rules allowing the necessary combinations to be effected.

Let

$$a = f(b,c,d, \ldots).$$

Then

$$\sigma_a{}^2 = \left(\frac{\partial f}{\partial b}\right)^2 \sigma_b{}^2 + \left(\frac{\partial f}{\partial c}\right)^2 \sigma_c{}^2 + \left(\frac{\partial f}{\partial d}\right)^2 \sigma_d{}^2 \ldots$$

In practice

$$s_a{}^2 = \left(\frac{\partial f}{\partial b}\right)^2 s_b{}^2 + \left(\frac{\partial f}{\partial c}\right)^2 s_c{}^2 + \left(\frac{\partial f}{\partial d}\right)^2 s_d{}^2 \ldots, \qquad [8.1]$$

in which s_a, s_b, s_c and s_d etc. are the standard deviations (or the standard errors) of a, b, c and d, respectively. Similarly, if $a = b + c + d + \ldots$

$$s_a{}^2 = s_b{}^2 + s_c{}^2 + s_d{}^2 + \ldots, \text{ since } \left(\frac{\partial f}{\partial b}\right) \text{ etc.} = 1.$$
$$[8.2]$$

And if $\qquad a = Kb$, in which K is free of error
$$s_a = Ks_b,$$

since $\left(\dfrac{\partial f}{\partial b}\right) = K.$ $\qquad\qquad\qquad\qquad\qquad\qquad$ [8.3]

In the above relationships it is assumed that $b, c, d, \ldots$ are independent, implying that the probability of any single one having a certain value does not depend on the values observed for the others.

Solution. First determine length AB from the co-ordinates (see Table 8.7).

Table 8.7

Station	Easting (m)	Northing (m)
B	724.616	702.443
A	456.961	573.237
Difference	x = 267.655	y = 129.206

Therefore length of AB $= \sqrt{(267.655^2 + 129.206^2)}$
$\qquad\qquad\qquad\qquad\quad = 297.209$ m.

Next calculate the standard deviations of the easting and northing differences. The easting difference was calculated as $(724.616 - 456.961)$ m. Thus the standard deviation (s_x) of this difference is given by

$$s_x{}^2 = s_{eB}{}^2 + s_{eA}{}^2, \text{ as in eq. [8.2]}$$
$$s_x{}^2 = 40^2 + 20^2$$

$$s_x = \sqrt{2000}$$
$$= 44.7 \text{ mm}.$$

Similarly, the standard deviation (s_n) of the northing difference is given by

$$s_y{}^2 = s_{nB}{}^2 + s_{nA}{}^2$$
$$= 50^2 + 30^2$$
$$s_y = \sqrt{3400}$$
$$= 58.3 \text{ mm}.$$

Now calculate the standard deviation of the calculated length AB.

$$L_{AB}{}^2 = x^2 + y^2$$
$$L_{AB} = \sqrt{(x^2 + y^2)}.$$

Whence, from eq. [8.1],

$$s_{AB}{}^2 = \left(\frac{\partial L_{AB}}{\partial x}\right)^2 s_x{}^2 + \left(\frac{\partial L_{AB}}{\partial y}\right)^2 s_y{}^2,$$

in which $s_x = 44.7$ mm and $s_y = 58.3$ mm.

Now $\quad \dfrac{\partial L_{AB}}{\partial x} = \frac{1}{2} \times 2x(x^2 + y^2)^{-1/2} = \dfrac{x}{\sqrt{(x^2 + y^2)}} = \dfrac{x}{L_{AB}}$

and $\quad \dfrac{\partial L_{AB}}{\partial y} = \frac{1}{2} \times 2y(x^2 + y^2)^{-1/2} = \dfrac{y}{\sqrt{(x^2 + y^2)}} = \dfrac{y}{L_{AB}}.$

Whence $\quad s_{AB}{}^2 = \left(\dfrac{267.655}{297.209}\right)^2 \times 44.7^2 + \left(\dfrac{129.206}{297.209}\right)^2 \times 58.3^2$

$$= 1622 + 642 = 2264 \text{ mm}^2.$$

Therefore $\quad s_{AB} = 47.58$ mm,

i.e. the standard deviation of length AB determined from co-ordinate information is 47.6 mm.

We can now calculate the most probable length of AB. Three lengths (measured or calculated) are known:

(a) by tape — 297.426 ± 70 mm;
(b) by EDM — 297.155 ± 15 mm;
(c) by calculation — 297.209 ± 47.6 mm.

Weighting inversely to the standard deviations squared we have the values in Table 8.8.

Table 8.8

Measure	l	w (mm^{-2})
1	297.426	1/4900
2	297.155	1/ 225
3	297.209	1/2264

The most probable length $= \dfrac{w_1 \, l_1 + w_2 \, l_2 + w_3 \, l_3}{w_1 + w_2 + w_3}$

$$= \frac{\left(\dfrac{297.426}{4900}\right) + \left(\dfrac{297.155}{225}\right) + \left(\dfrac{297.209}{2264}\right)}{\left(\dfrac{1}{4900}\right) + \left(\dfrac{1}{225}\right) + \left(\dfrac{1}{2264}\right)}$$

$$= \mathbf{297.171 \ m.}$$

Now $\qquad w_{\bar{x}} = \Sigma w = \dfrac{1}{4900} + \dfrac{1}{225} + \dfrac{1}{2264}$

$$= 0.005 \ 09 \ \text{mm}^{-2}.$$

Since weights were applied as the inverse of the squares of the standard deviations

$$w_{\bar{x}} = \frac{1}{s_{\bar{x}}^2}.$$

Therefore $\qquad s_{\bar{x}}^2 = \dfrac{1}{0.005 \ 09}$

and $\qquad s_{\bar{x}} = \pm \ \mathbf{14.0 \ mm.}$

Standard deviation of length 297.171 m is $\pm$ 14.0 mm.

8.6 Accuracy of auxiliary base subtense tacheometry

The length of a line PQRST was measured by establishing auxiliary base lines at Q and S, respectively. The lengths of these lines, which were perpendicular to PQRST, were determined using a 2 m subtense bar, positioned such that the angles subtended by the auxiliary bases at P, R and T were equal to those subtended by the subtense bars at Q and S. Given that PQ = QR = RS = ST, and assuming that the standard error of angular measurement was $\pm 1''$, determine from first principles the error in the deduced length of 2000 m for PT. [Salford]

Introduction. This is a further example of the combination of errors. Figure 8.3 shows the method of measurement of PT, the auxiliary bases QU and SV

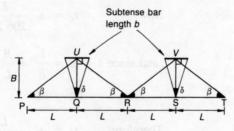

Figure 8.3

being of length B. Angles δ and β were measured by theodolite, and optimum results occur when $\delta = \beta$.

Solution. First derive standard formulae. If the length of the subtense bar be b

$$B = \frac{b}{2 \tan \delta/2} = \frac{b}{\delta}, \text{ when } \delta \text{ is small.}$$

Therefore $\quad dB = \dfrac{-b}{\delta^2}\, d\delta = \dfrac{-B^2}{b}\, d\delta$ ($d\delta$ in radians).

Also, $\quad L = \dfrac{B}{\tan \beta} = \dfrac{B}{\beta}$, when β is small

and $\quad dL = \dfrac{-B}{\beta^2}\, d\beta = \dfrac{-L^2}{B}\, d\beta$ ($d\beta$ in radians).

Now $\quad s_L{}^2 = \left(\dfrac{\partial L}{\partial \beta}\right)^2 s_\beta{}^2 + \left(\dfrac{\partial L}{\partial B}\right)^2 s_B{}^2,$

in which s_L, s_β and s_B are the respective standard errors.

Therefore $\quad s_L{}^2 = \left(\dfrac{-L^2}{B}\right)^2 s_\beta{}^2 + \left(\dfrac{1}{\beta}\right)^2 s_B{}^2$

and $\quad s_B = -\dfrac{B^2}{b}\, s_\delta$

$$s_L{}^2 = \frac{L^4}{B^2}\, s_\beta{}^2 + \frac{L^2}{B^2}\, \frac{B^4}{b^2}\, s_\delta{}^2.$$

Therefore $\quad s_L = \pm \dfrac{L}{206\,265} \sqrt{\left(\dfrac{L^2}{B^2}\, s_\beta{}^2 + \dfrac{B^2}{b^2}\, s_\delta{}^2\right)},$

when s_L, s_δ and s_β are expressed in seconds of arc.
Next determine the standard error in PT. Since

$$PT = 2000 \text{ m}$$
$$PQ = QR = RS = ST = L = 500 \text{ m}.$$

Now $\quad s_L = \pm \dfrac{L}{206\,265} \sqrt{\left(\dfrac{L^2}{B^2}\, s_\beta{}^2 + \dfrac{B^2}{b^2}\, s_\delta{}^2\right)}$

and since $\delta = \beta = \dfrac{b}{B} = \dfrac{B}{L}$

$$B^2 = bL.$$

Therefore $\quad s_L = \pm \dfrac{L}{206\,265} \sqrt{\left(\dfrac{L^2}{bL}\, s_\beta{}^2 + \dfrac{bL}{b^2}\, s_\delta{}^2\right)}$

$$= \pm \frac{L}{206\ 265} \sqrt{\left(\frac{2L}{b}\right)} \text{ (when } s_\delta = s_\beta = \pm 1'')$$

$$= \pm \frac{500}{206\ 265} \sqrt{\left(\frac{2 \times 500}{2}\right)}$$

$$= \pm 0.054 \text{ m.}$$

Also $\quad s_{PT}^2 = \pm \sqrt{(s_{PQ}^2 + s_{QR}^2 + s_{RS}^2 + s_{ST}^2)}$

$$= \pm \sqrt{(4s_L^2)} \text{ (since PQ = QR = RS = ST).}$$

Therefore $\quad s_{PT}^2 = \pm \sqrt{[4 \times (0.054)^2]}$

and $\qquad\qquad s_{PT} = \pm \textbf{0.108 m.}$

8.7 Accuracy of levelling

Two vertical levelling staves were securely fixed in position at stations A and B, 93.00 m apart. Ten instrument stations were then located between A and B as shown in Fig. 8.4 and the differences in height between A and B were then measured by a tilting level set up in turn at the ten stations (see Table 8.9).

Table 8.9

Level position	Difference in height (m)
1	0.146
2	0.140
3	0.136
4	0.133
5	0.128
6	0.121
7	0.115
8	0.111
9	0.106
10	0.102

Determine the most probable values of the height differences between A and B and the inclination of the line of sight of the level when the bubble is centred.

Introduction. The relative positions of the instrument stations in this example (Fig. 8.4) are very similar to those adopted when estimating the accuracy of a level by the Princeton standard test. (See The Princeton Standard Test for Estimating the Accuracy of a Level by P. Kissam *Surveying and Mapping*, March 1963.) In this particular example a set of ten differences in

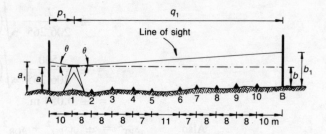

Figure 8.4

height have been established, whereas in the Princeton test a further three sets are required. The stations are set out on fairly level ground, and in that test all stadia lines are read and adjustments are made empirically for curvature and refraction. The principle of least squares is used to establish the average slope of the line of sight and the standard error is then calculated as a measure of accuracy of the level. In this example, as in the Princeton test, a regression analysis is carried out to fit a linear model to the data plotted in Fig. 8.5.

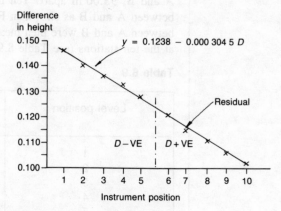

Figure 8.5

Solution. First derive a general expression. In Fig. 8.4 it has been assumed that the line of sight is inclined upwards when the bubble is central.

Apparent difference in level between A and B $= a_1 - b_1$.
True difference in level between A and B $\quad = a - b$.

Now
$$a - b = (a_1 - p_1 \, \theta) - (b_1 - q_1 \, \theta)$$
$$= (a_1 - b_1) - \theta \, (p_1 - q_1)$$
$$= (a_1 - b_1) - D_1 \, \theta \text{ (putting } D_1 = (p_1 - q_1)),$$
or $\quad (a_1 - b_1) = (a - b) + D_1 \, \theta.$

If t be the best, i.e. most probable, value for $(a-b)$ and θ be the best value for the inclination of the line of sight, the best value for the difference in level (Y) obtained by observations from any instrument station will be $Y = t + D_1 \, \theta$, and this is the equation of the regression line to fit the measurements.

Thus for an observed or apparent difference of y_1 a residual of value $y_1 - (t + D_1 \theta)$ results. Applying the principle of least squares the values of t and θ which minimize the sum of the square of the residuals (or deviations) between the observed measurements y_1 and Y_1 are found. $\Sigma[y_1 - (t + D_1\theta)]^2$ is to be a minimum. Writing this sum as E

$$E = \Sigma(y_1{}^2 - 2y_1(t + D_1\ \theta) + t^2 + 2t\ D_1\ \theta + D_1{}^2\ \theta^2).$$

For a minimum $\dfrac{\partial E}{\partial t} = 0$ and $\dfrac{\partial E}{\partial \theta} = 0.$

Therefore $\qquad -\Sigma 2y_1 + \Sigma 2t + \Sigma 2D_1\ \theta = 0$

and $\qquad -\Sigma 2y_1\ D_1 + \Sigma 2tD_1 + \Sigma 2D_1{}^2\ \theta = 0.$

By symmetry $\Sigma D_1 = 0$ because $(p_1 - q_1)$ is negative for stations 1 to 5, and positive for stations 6 to 10, with corresponding sight differences.

Therefore $\quad t = \dfrac{\Sigma y_1}{n}$ ($n = 10$ in this example),

i.e. the best value for t is the mean of the ten differences measured and

$$\theta = \frac{\Sigma y_1 D_1}{\Sigma D_1{}^2}.$$

The data can be tabulated as in Table 8.10.

Table 8.10

Station	y_1	p_1	q_1	D_1	$y_1 D_1$	$D_1{}^2$	r
1	0.146	10.00	83.00	−73.00	−10.658	5329	0
2	0.140	18.00	75.00	−57.00	− 7.890	3249	−0.0012
3	0.136	26.00	67.00	−41.00	− 5.576	1681	−0.0013
4	0.133	34.00	59.00	−25.00	− 3.325	625	+0.0016
5	0.128	41.00	52.00	−11.00	− 1.408	121	+0.0009
6	0.121	52.00	41.00	+11.00	+ 1.331	121	+0.0006
7	0.115	59.00	34.00	+25.00	+ 2.875	625	−0.0012
8	0.111	67.00	26.00	+41.00	+ 4.551	1681	−0.0003
9	0.106	75.00	18.00	+57.00	+ 6.042	3249	−0.0004
10	0.102	83.00	10.00	+73.00	+ 7.446	5329	+0.0004
Σ	1.238			0.00	− 6.702	22 010	

Most probable difference in height

$$t = \frac{\Sigma y_1}{10}$$

$$= \frac{1.238}{10}$$

$= 0.1238$ m, say **0.124 m**.

Most probable value of the inclination of the line of sight

$$\theta = - \frac{\Sigma y_1 D_1}{\Sigma D_1{}^2}$$

$$= - \frac{6.702}{22010}$$

$= - 0.000\ 304\ 5$ radian, i.e. **0.000 30 radian**
(inclined downwards).

Note that inclination θ is an instrumental error; in this example we have estimated its most probable value. Accidental errors are present in the observations, Fig. 8.5 indicating residuals (or deviation) from the regression line. These are included in the table and are based on $r = y_1 - (t + D_1\theta)$, where $t = 0.1238$ and $\theta = 0.000\ 304\ 5$ rad.

8.8 Least squares — the normal equations

A series of level networks are run, and the results are as follows:

	Weight
Height of A − Height of B = 18.614	3
Height of B − Height of C = 16.264	2
Height of C − Height of D = 22.385	2
Height of A − Height of D = 57.247	1

Calculate the most probable differences in height between A and C and between B and D.　　　　　　　　　　　　　　　　[Leeds]

Introduction. When using the method of least squares to solve such problems as this two approaches present themselves:

(*a*) reduction to a minimum number of unknowns; and
(*b*) the method of correlates (Lagrange's method of undetermined multipliers).

The former method which produces equations, known as normal equations, is used in this example. Two solutions will be given, the first assuming the observations were of equal weight and the second accepting the weights stipulated in the question.

Solution. (*a*) **Assume equal weighting**. Examination of the data shows that D is the lowest of the four stations and for convenience it will be treated

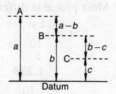

Figure 8.6

as the datum. All height differences are now positive, as in Fig. 8.6, and the four observations have been arranged to give three unknowns. Let the most probable heights of A, B and C above D be a, b and c, respectively. Thus the most probable differences in height between A and B, B and C will be $(a-b)$ and $(b-c)$, respectively. Individual residuals can now be deduced as

$$r_1 = 57.247 - a$$
$$r_2 = 18.614 - (a-b)$$
$$r_3 = 16.264 - (b-c)$$
$$r_4 = 22.385 - c.$$

Σr_1^2 is to be a minimum, i.e.

$$R = (57.247-a)^2 + [18.614-(a-b)]^2 + [16.264-(b-c)]^2$$
$$+ (22.385-c)^2, \text{ is to be a minimum.}$$

Hence $\dfrac{\partial R}{\partial a} = 0, \ \dfrac{\partial R}{\partial b} = 0 \text{ and } \dfrac{\partial R}{\partial c} = 0,$

and so with respect to a

$$(-2 \times 57.247) + 2a - (2 \times 18.614) + 2a - 2b = 0$$

with respect to b

$$(+2 \times 18.614) - 2a + 2b - (2 \times 16.264) + 2b - 2c = 0$$

with respect to c

$$(+2 \times 16.264) - 2b + 2c - (2 \times 22.385) + 2c = 0.$$

Dividing throughout by 2 results in the normal equations

$$2a - b - 75.861 = 0$$
$$-a + 2b - c + 2.350 = 0$$
$$-b + 2c - 6.121 = 0$$

Whence $a = 57.251$ m, $b = 38.641$ m, $c = 22.381$ m, i.e. the most probable height difference between A and C is **34.870 m** and between B and D is **38.641 m.**

(b) **Adopt the assigned weights**. The residuals r_1, etc., maintain the same values but they do not have the same reliability or precision. Hence $\Sigma w_1 r_1^2$ is to be a minimum in which

$$r_1 = 57.247 - a \quad \text{(weight 1)},$$
$$r_2 = 18.614 - (a-b) \quad \text{(weight 3)},$$
$$r_3 = 16.264 - (b-c) \quad \text{(weight 2)},$$
$$r_4 = 22.385 - c \quad \text{(weight 2)},$$

and $R = (57.247-a)^2 + 3[18.614-(a-b)]^2 + 2[16.264-(b-c)]^2$
$$+ 2(22.385-c)^2.$$

Again $\dfrac{\partial R}{\partial a} = 0, \ \dfrac{\partial R}{\partial b} = 0 \text{ and } \dfrac{\partial R}{\partial c} = 0,$ and so with respect to a

$$(-2 \times 57.247) + 2a - (6 \times 18.614) + 6a - 6b = 0$$

with respect to b

$$(+6 \times 18.614) - 6a + 6b - (4 \times 16.264) + 4b - 4c = 0$$

with respect to c

$$(+4 \times 16.264) - 4b + 4c - (4 \times 22.385) + 4c = 0.$$

Dividing throughout by 2 results in the normal equations

$$4a - 3b - 113.089 = 0$$
$$-3a + 5b - 2c + 23.314 = 0$$
$$- 2b + 4c - 12.242 = 0.$$

Whence $a = 57.254$ m, $b = 38.642$ m, $c = 22.381$ m. Therefore the most probable difference in height between A and C is **34.873 m** and the most probable difference in height between B and D is **38.642 m.**

Alternative solution. There is a more convenient method of solving this problem which involves tabulating the coefficients from the various residuals as in Table 8.11. Now multiply all terms on each line by the weight of that line and the coefficient of the unknown (a, b or c) under consideration.

Table 8.11

Weight w	a	b	c	N
1	−1			+57.247
3	−1	+1		+18.614
2		−1	+1	+16.264
2			−1	+22.385

For a, which appears on two lines:

for the first line, $(w) (c_a) (\text{term } a) + (w) (c_a) (\text{term } N)$,

i.e. $(1)(-1) \ (-1)a + (1)(-1) (57.247)$;

and for the second line,

$$(3)(-1)(-1)a + (3)(-1)(+1)b + (3)(-1) (18.614).$$

These summarize as

$$a - 57.247 + 3a - 3b - 55.842,$$

and so we obtain the normal equation

$$4a - 3b - 113.089 = 0.$$

For b, which appears on two lines

$$(3)(+1)(-1)a + (3)(+1)(+1)b + (3)(+1)(18.614)$$
$$(2)(-1)(-1)b + (2)(-1)(+1)c + (2)(-1)(16.264)$$

which give:

$$-3a + 3b + 55.842 + 2b - 2c - 32.528$$
$$\text{or} \quad -3a + 5b - 2c + 23.314 = 0.$$

For c, which appears on two lines

$$(2)(+1)(-1)b + (2)(+1)(+1)c + (2)(+1)(16.264)$$
$$(2)(-1)(-1)c + (2)(-1)(22.385),$$

which give:

$$-2b + 2c + 32.528 + 2c - 44.770$$
$$\text{or} \quad -2b + 4c - 12.242 = 0.$$

It will be realised that we have 'automatically' carried out the partial differentiations of R demanded by the principle of least squares.

This problem can be solved by the following computer program using the matrix equation $X = (A^T W A)^{-1} (A^T W B)$, where

$$X = \begin{bmatrix} a \\ b \\ c \end{bmatrix}, \quad A = \begin{bmatrix} -1 & 0 & 0 \\ -1 & +1 & 0 \\ 0 & -1 & +1 \\ 0 & 0 & -1 \end{bmatrix}, \quad W = \begin{bmatrix} 1 & 0 & 0 & 0 \\ 0 & 3 & 0 & 0 \\ 0 & 0 & 2 & 0 \\ 0 & 0 & 0 & 2 \end{bmatrix}, \quad B = \begin{bmatrix} 57.247 \\ 18.614 \\ 16.264 \\ 22.385 \end{bmatrix}.$$

Further details on the matrix method of solving this type of problem are given in Example 8.10.

The DIM statement in line 20 is set for a problem with 10 unknowns in 10 equations. For readers with computers with limited memory the minimum values for the array variables are: A(N2, N1); B(N2); D(N1, 2 * N1); E(N1); T(N1, N2); W(N2).

Variables

A(i,j) = Coefficient matrix A

B(i) = Product matrix B

D(i,j) = Values to N1,N1 are $[A^T W A]$, values from N1, N1+1 are $[A^T W A]^{-1}$

E(i) = Matrix $[A^T W B]$

F = An element of the output matrix

I,J,K = Loop counters

M = Line divisor in matrix inversion

N1 = Number of unknowns

N2 = Number of equations

T(i,j) = Transpose matrix $[A^T W]$

W(i) = Weight matrix W

X = Divisor of each line in matrix inversion

```
10 REM LEAST SQUARES MATRIX SOLUTION
20 DIM A(10,10),B(10),D(10,20),E(10),T(10,10),W(10)
30 INPUT"NUMBER OF UNKNOWNS ";N1
40 INPUT"NUMBER OF EQUATIONS ";N2
50 FOR I=1 TO N2
60 PRINT"FOR EQUATION";I
70 FOR J=1 TO N1
80 PRINT"INPUT COEFFICIENT OF VARIABLE";J
90 INPUT A(I,J)
100 NEXT J
110 INPUT"PRODUCT OF THIS EQUATION ";B(I)
120 INPUT"WEIGHT  OF THIS EQUATION ";W(I)
130 NEXT I
140 FOR I=1 TO N1
150 FOR J=1 TO N2
160 T(I,J)=A(J,I)*W(J)
170 NEXT J
180 NEXT I
```

```
190 FOR K=1 TO N1
200 FOR I=1 TO N1
210 FOR J=1 TO N2
220 D(K,I)=D(K,I)+(T(K,J)*A(J,I))
230 NEXT J
240 NEXT I
250 FOR I=N1+1 TO 2*N1
260 IF N1+K=I THEN 290
270 D(K,I)=0
280 GOTO 300
290 D(K,I)=1
300 NEXT I
310 FOR J=1 TO N2
320 E(K)=E(K)+(T(K,J)*B(J))
330 NEXT J
340 NEXT K
350 FOR K=1 TO N1
360 FOR J=1 TO N1
370 IF J=K THEN 420
380 M=D(J,K)
390 FOR I=1 TO 2*N1
400 D(J,I)=D(J,I)*D(K,K)-D(K,I)*M
410 NEXT I
420 NEXT J
430 NEXT K
440 FOR J=1 TO N1
450 X=D(J,J)
460 F=0
470 FOR I=N1+1 TO 2*N1
480 D(J,I)=D(J,I)/X
490 F=F+D(J,I)*E(I-N1)
500 NEXT I
510 F=INT(F*10000+0.5)/10000
520 PRINT"VARIABLE";J;"=";F
530 NEXT J
540 END
```

8.9 Least squares — method of correlates

In a round of levels the results in Table 8.12 were obtained. Determine the most probable values of the levels of B, C and D above A, given that the height difference value for the line A to B is the mean of two runs, all other lines being observed once only.

Table 8.12

Level line	Length (km)	Rise (+) Fall (−) (m)
A to B	20	+ 42.285
B to C	10	− 12.016
B to D	25	− 20.240
C to D	20	− 8.114
D to A	30	− 22.224

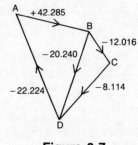

Figure 8.7

[Salford]

Introduction. This example will be solved by the Lagrange method of undetermined multipliers (method of correlates) where the errors (or their corrections) in each line are considered directly. The difference in height of two terminals, between which a run of flying levels has been conducted, is the algebraic sum of the rises and falls between the change points. Thus, assuming equal distances for backsights and foresights, accidental errors should be proportional to $\sqrt{}$(number of instrument stations) and hence proportional to $\sqrt{}$(length of line). Accordingly, weighting, which is assumed to be inversely proportional to the square of the errors, can be applied as the reciprocal of the length of line.

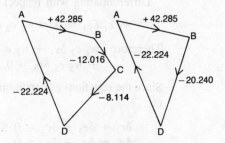

Figure 8.8

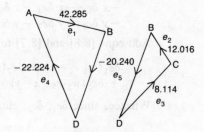

Figure 8.9

Solution. The network of levels together with the directions of the rises and falls are shown in Fig. 8.7. A study of this figure shows that there are three closed circuits of level runs, namely ABCDA, ABDA and BCDB. Two out of the three are needed in the analysis, but whichever are chosen the 'directions' of levelling should be compatible. For instance, in Fig. 8.8, ABCDA and ABDA are compatible. Choosing ABDA and BCDB requires some adjustment for conformity along BD and the falls along BC and CD are transformed into rises to satisfy the changes in direction in Fig. 8.9. Such transformations should minimize any confusion in respect of signs. In each of the closed circuits it is required that the sum of the 'rises' will equal the sum of the 'falls'. Accordingly there is one condition equation for each circuit.

Let the configurations of Fig. 8.9 be adopted in the analysis and let corrections (equal and opposite to the corresponding errors) e_1, e_2, e_3, e_4 and e_5 be required for the corresponding adjustment.

In circuit ABDA the error $= +42.285 - (20.240 + 22.224)$
$$= -0.179 \text{ m.}$$
In circuit BCDB the error $= +(12.016 + 8.114) - 20.240$
$$= -0.110 \text{ m.}$$

The corrections are to be such that the two condition equations will be satisfied and these reduce to

$$e_1 + e_4 + e_5 = +0.179 \text{ m}$$
$$e_2 + e_3 + e_5 = +0.110 \text{ m}.$$

A third condition has to be satisfied, and this is that $\Sigma w_1 \, e_1{}^2$ is to be a minimum.

Differentiating with respect to e_1, e_2, etc.,

$$w_1 \, e_1 \, \delta e_1 = 0 = w_2 \, e_2 \, \delta e_2, \text{ etc.}$$

Therefore $w_1 \, e_1 \, \delta e_1 + w_2 \, e_2 \, \delta e_2 + w_3 \, e_3 \, \delta e_3 + w_4 \, e_4 \, \delta e_4$
$$+ \, w_5 \, e_5 \, \delta e_5 = 0. \qquad [8.4]$$

Since the equations of condition must remain unaltered by increments δe_1, δe_2, etc.,

$$\delta e_1 + \delta e_4 + \delta e_5 = 0 \qquad [8.5]$$
$$\delta e_2 + \delta e_3 + \delta e_5 = 0. \qquad [8.6]$$

If the equations are multiplied by factors $-x$ and $-y$, respectively,

$$-x \, \delta e_1 - x \, \delta e_4 - x \, \delta e_5 = 0 \qquad [8.7]$$
$$-y \, \delta e_2 - y \, \delta e_3 - y \, \delta e_5 = 0. \qquad [8.8]$$

Add eqns [8.8] and [8.7] to eqn [8.4]

$$\delta e_1(w_1 \, e_1 - x) + \delta e_2(w_2 \, e_2 - y) + \delta e_3(w_3 \, e_3 - y) + \delta e_4(w_4 \, e_4 - x)$$
$$+ \, \delta e_5(w_5 \, e_5 - x - y) = 0.$$

Whence, since δe_1, δe_2, etc., are independent quantities

$$e_1 = \frac{x}{w_1}, \, e_2 = \frac{y}{w_2}, \, e_3 = \frac{y}{w_3}, \, e_4 = \frac{x}{w_4}, \, e_5 = \frac{x+y}{w_5}.$$

Now weights can be applied inversely to the length of run, and so
$$w_2 = \tfrac{1}{10}, \, w_3 = \tfrac{1}{20}, \, w_4 = \tfrac{1}{30} \text{ and } w_5 = \tfrac{1}{25}.$$

Two runs were made on line AB and the mean value of level difference has been used above.

It has been mentioned in Example 8.2 that

$$s_{\bar{x}} = \frac{s_x}{\sqrt{n}}$$

and $w_{\bar{x}} \, s_{\bar{x}}{}^2 = w_x \, s_x{}^2$.

Therefore $w_{\bar{x}} = n \, w_x$,

which relates the weight of the mean of n observations to that of the single observation. Hence in this example

$$w_1 = \frac{n}{l_1} = \frac{2}{20}, \text{ since } n = 2$$

$$= \frac{1}{10}$$

Therefore $e_1 = 10x$, $e_2 = 10y$, $e_3 = 20y$, $e_4 = 30x$ and
$$e_5 = 25x + 25y.$$

Substituting in the equations of condition, which have to be satisfied

$$10x + 30x + (25x + 25y) = 0.179$$
$$10y + 20y + (25x + 25y) = 0.111$$

and so $65x + 25y = +0.179$
$$25x + 55y = +0.110,$$

whence $y = 0.000\ 91$, $x = 0.002\ 41$,

$$e_1 = 10x \qquad = +0.024 \text{ m}$$
$$e_2 = 10y \qquad = +0.009 \text{ m}$$
$$e_3 = 20y \qquad = +0.018 \text{ m}$$
$$e_4 = 30x \qquad = +0.072 \text{ m}$$
$$e_5 = 25x + 25y = +0.083 \text{ m}$$

The most probable values of the levellings are:

A to B $= +42.285 + 0.024 = +42.309$ m
C to B $= +12.016 + 0.009 = +12.025$ m
D to C $= +\ 8.114 + 0.018 = +\ 8.132$ m
D to A $= -22.224 + 0.072 = -22.152$ m
B to D $= -20.240 + 0.083 = -20.157$ m

The most probable levels above A are:

B $\qquad\qquad\qquad\quad$ = **42.309 m**
C $\quad 42.309 - 12.025$ = **30.284 m**
D $\qquad\qquad\qquad\quad$ = **22.152 m**

The computer program listed with Example 8.8 can be used to solve this problem.

8.10 Least squares — angle observations

From an observation station O four trigometrical stations A, B, C and D can be seen. The readings in Table 8.13 were taken in order to determine the most likely values of the included angles.

Table 8.13

Angle	Reading	Weight
AÔB	25° 16′ 27″	1
BÔC	37° 15′ 33″	1
CÔD	47° 22′ 15″	1
AÔC	62° 32′ 05″	2
BÔD	84° 37′ 41″	2
AÔD	109° 54′ 21″	3

By the method of least squares using either observation equations or condition equations determine the most likely values of angles AÔB, BÔC and CÔD. [Bradford]

Solution. (*a*) **By condition equations**. The most probable values should satisfy the condition equations

$$A\hat{O}B + B\hat{O}C = A\hat{O}C$$
$$B\hat{O}C + C\hat{O}D = B\hat{O}D$$
$$A\hat{O}B + B\hat{O}C + C\hat{O}D = A\hat{O}D$$

Let corrections of e_1, e_2, e_3, e_4, e_5, e_6 be applied to the observed values $A\hat{O}B$, $B\hat{O}C$, $C\hat{O}D$, $A\hat{O}C$, $B\hat{O}D$ and $A\hat{O}D$, respectively. Then the most probable value of $A\hat{O}B$ will be $25°\ 16'\ 27'' + e_1$, and similarly for the remaining five angles.

Thus $25°\ 16'\ 27'' + e_1 + 37°\ 15'\ 33'' + e_2 = 62°\ 32'\ 05'' + e_4$
$37°\ 15'\ 33'' + e_2 + 47°\ 22'\ 15'' + e_3 = 84°\ 37'\ 41'' + e_5$

and $25°\ 16'\ 27'' + e_1 + 37°\ 15'\ 33'' + e_2 + 47°\ 22'\ 15'' + e_3$
$= 109°\ 54'\ 21'' + e_6.$

Whence the condition equations reduce to

$$e_1 + e_2 - e_4 = +05''\ \text{[8.9]}$$
$$e_2 + e_3 - e_5 = -07''\ \text{[8.10]}$$
$$e_1 + e_2 + e_3 - e_6 = +06''\ \text{[8.11]}$$

and $\Sigma w_1\, e_1{}^2$ is to be a minimum to satisfy the principle of least squares. On differentiation for the minimum condition

$$w_1\, e_1\, \delta e_1 = 0,\ \text{etc.,}$$

and so

$$w_1\, e_1\, \delta e_1 + w_2\, e_2\delta e_2 + w_3\, e_3\, \delta e_3 + w_4\, e_4\, \delta e_4 + w_5\, e_5\, \delta e_5$$
$$+ w_6\, e_6\, \delta e_6 = 0.\ \text{[8.12]}$$

Also $\quad\delta e_1 + \delta e_2 - \delta e_4 = 0\ \text{[8.13]}$
$\quad\quad\quad\delta e_2 + \delta e_3 - \delta e_5 = 0\ \text{[8.14]}$
$\quad\quad\quad\delta e_1 + \delta e_2 + \delta e_3 - \delta e_6 = 0.\ \text{[8.15]}$

Multiplying eqns [8.13], [8.14] and [8.15] by $-x$, $-y$ and $-z$, respectively, and then adding to eqn [8.12]

$$\delta e_1(w_1 e_1 - x - z) + \delta e_2(w_2 e_2 - x - y - z) + \delta e_3(w_3 e_3 - y - z)$$
$$+ \delta e_4(w_4 e_4 + x) + \delta e_5(w_5 e_5 + y) + \delta e_6(w_6 e_6 + z) = 0.$$

Hence $\quad e_1 = \dfrac{x+z}{w_1},\ e_2 = \dfrac{x+y+z}{w_2},\ e_3 = \dfrac{y+z}{w_3},\ e_4 = \dfrac{-x}{w_4},\ e_5 = \dfrac{-y}{w_5},$

$$e_6 = \dfrac{-z}{w_6}.$$

Inserting the values for the weights and incorporating the above in eqns [8.9], [8.10] and [8.11] gives

$$(x+z) + (x+y+z) + \dfrac{x}{2} = +05''$$

$$(x+y+z) + (y+z) + \frac{y}{2} = -07''$$

$$(x+z) + (x+y+z) + (y+z) + \frac{z}{3} = +06''$$

and so
$$2.5x + y + 2z = +05''$$
$$x + 2.5y + 2z = -07''$$

$$2x + 2y + \frac{10z}{3} = +06'',$$

whence $x = -0.181''$, $y = -8.182''$ and $z = +6.818''$.

Therefore the most probable values are

$$\text{A}\hat{\text{O}}\text{B} = 25°\ 16'\ 27'' - 0.18'' + 6.82'' \qquad = \mathbf{25°\ 16'\ 33.6''}$$
$$\text{B}\hat{\text{O}}\text{C} = 37°\ 15'\ 33'' - 0.18'' - 8.18'' + 6.82'' = \mathbf{37°\ 15'\ 31.5''}$$
$$\text{C}\hat{\text{O}}\text{D} = 47°\ 22'\ 15'' - 8.18'' + 6.82'' \qquad = \underline{\mathbf{47°\ 22'\ 13.6''}}$$

Total $\qquad\qquad\qquad\qquad\qquad\qquad\qquad\qquad = 109°\ 54'\ 18.7''$
Check $\qquad \text{A}\hat{\text{O}}\text{D} = 109°\ 54'\ 21'' - 2.3'' = 109°\ 54'\ 18.7''$.

Alternative solution. (b) **By observation equations**. Let measurements b_1, b_2, ..., b_n be related to unknowns x_1, x_2, ..., x_i by observation equations of the type

$$a_{11} x_1 + a_{12} x_2 + \ldots + a_{1i} x_i = b_1$$
$$a_{21} x_1 + a_{22} x_2 + \ldots + a_{2i} x_i = b_2$$
$$\vdots \qquad\qquad \vdots \qquad\qquad \vdots \qquad\qquad \vdots$$
$$a_{n1} x_1 + a_{n2} x_2 + \ldots + a_{ni} x_i = b_n,$$

in which $n > i$. If $x_1\ x_2\ \ldots\ x_n$ are the best values, residuals can be written as

$$r_1 = -(a_{11} x_1 + a_{12} x_2 + \ldots + a_{1i} x_i) + b_1$$
$$r_2 = -(a_{21} x_1 + a_{22} x_2 + \ldots + a_{2i} x_i) + b_2$$
$$\vdots \qquad \vdots \qquad \vdots \qquad\qquad \vdots \qquad\qquad \vdots$$
$$r_n = -(a_{n1} x_1 + a_{n2} x_2 + \ldots + a_{ni} x_i) + b_n$$

and expressed as $r = -Ax + b$.

Allowing for weighted observations the principle of least squares requires

$$(-Ax + b)^{\text{T}}\ W(-Ax + b) = x^{\text{T}} A^{\text{T}}WAx - b^{\text{T}}WAx$$
$$- x^{\text{T}}A^{\text{T}}Wb + b^{\text{T}}Wb$$

to be a minimum,

whence $\quad 2A^{\text{T}}WAx - A^{\text{T}}Wb - A^{\text{T}}Wb = 0$
or $\qquad\qquad\qquad\qquad\qquad A^{\text{T}}WAx = A^{\text{T}}Wb$
and $\qquad\qquad\qquad\qquad\qquad\quad x = (A^{\text{T}}WA)^{-1}\ A^{\text{T}}Wb.$

Now let $\text{A}\hat{\text{O}}\text{B} = a$, $\text{B}\hat{\text{O}}\text{C} = b$ and $\text{C}\hat{\text{O}}\text{D} = c$. In the example six observation equations can be established.

$$
\begin{aligned}
a &= 25^\circ\ 16'\ 27'' = 0.441\ 117\ 4 \text{ radian (weight 1)} \\
b &= 37^\circ\ 15'\ 33'' = 0.650\ 295\ 1 \text{ radian (weight 1)} \\
c &= 47^\circ\ 22'\ 15'' = 0.826\ 777\ 0 \text{ radian (weight 1)} \\
a+b &= 62^\circ\ 32'\ 05'' = 1.091\ 436\ 8 \text{ radian (weight 2)} \\
b+c &= 84^\circ\ 37'\ 41'' = 1.477\ 038\ 2 \text{ radian (weight 2)} \\
a+b+c &= 109^\circ\ 54'\ 21'' = 1.918\ 218\ 6 \text{ radian (weight 3)}
\end{aligned}
$$

Hence

$$
A = \begin{bmatrix} +1 & 0 & 0 \\ 0 & +1 & 0 \\ 0 & 0 & +1 \\ +1 & +1 & 0 \\ 0 & +1 & +1 \\ +1 & +1 & +1 \end{bmatrix}
\quad
W = \begin{bmatrix} 1 & 0 & 0 & 0 & 0 & 0 \\ 0 & 1 & 0 & 0 & 0 & 0 \\ 0 & 0 & 1 & 0 & 0 & 0 \\ 0 & 0 & 0 & 2 & 0 & 0 \\ 0 & 0 & 0 & 0 & 2 & 0 \\ 0 & 0 & 0 & 0 & 0 & 3 \end{bmatrix}
\quad \text{and} \quad
b = \begin{bmatrix} 0.441\ 117\ 4 \\ 0.650\ 295\ 1 \\ 0.826\ 777\ 0 \\ 1.091\ 436\ 8 \\ 1.477\ 038\ 2 \\ 1.918\ 218\ 6 \end{bmatrix}
$$

$$
A^{T}W = \begin{bmatrix} +1 & 0 & 0 & +1 & 0 & +1 \\ 0 & +1 & 0 & +1 & +1 & +1 \\ 0 & 0 & +1 & 0 & +1 & +1 \end{bmatrix}
\begin{bmatrix} 1 & 0 & 0 & 0 & 0 & 0 \\ 0 & 1 & 0 & 0 & 0 & 0 \\ 0 & 0 & 1 & 0 & 0 & 0 \\ 0 & 0 & 0 & 2 & 0 & 0 \\ 0 & 0 & 0 & 0 & 2 & 0 \\ 0 & 0 & 0 & 0 & 0 & 3 \end{bmatrix}
$$

$$
= \begin{bmatrix} 1 & 0 & 0 & 2 & 0 & 3 \\ 0 & 1 & 0 & 2 & 2 & 3 \\ 0 & 0 & 1 & 0 & 2 & 3 \end{bmatrix}
$$

$$
A^{T}WA = \begin{bmatrix} 6 & 5 & 3 \\ 5 & 8 & 5 \\ 3 & 5 & 6 \end{bmatrix}
$$

$$
A^{T}Wb = \begin{bmatrix} 1 & 0 & 0 & 2 & 0 & 3 \\ 0 & 1 & 0 & 2 & 2 & 3 \\ 0 & 0 & 1 & 0 & 2 & 3 \end{bmatrix}
\begin{bmatrix} 0.441\ 117\ 4 \\ 0.650\ 295\ 1 \\ 0.826\ 777\ 0 \\ 1.091\ 436\ 8 \\ 1.477\ 038\ 2 \\ 1.918\ 218\ 6 \end{bmatrix}
= \begin{bmatrix} 8.378\ 646\ 8 \\ 11.541\ 900\ 9 \\ 9.535\ 509\ 2 \end{bmatrix}
$$

Therefore
$$
\begin{bmatrix} a \\ b \\ c \end{bmatrix} = (A^{T}WA)^{-1}\ A^{T}Wb
$$

$$
= \begin{bmatrix} +\frac{23}{66} & -\frac{15}{66} & +\frac{1}{66} \\ -\frac{15}{66} & +\frac{27}{66} & -\frac{15}{66} \\ +\frac{1}{66} & -\frac{15}{66} & +\frac{23}{66} \end{bmatrix}
\begin{bmatrix} 8.378\ 646\ 8 \\ 11.541\ 900\ 9 \\ 9.535\ 509\ 2 \end{bmatrix}
$$

$$
\begin{aligned}
a &= 0.441\ 149\ 6 \text{ radian} = \mathbf{25^\circ\ 16'\ 33.6''} \\
b &= 0.650\ 287\ 6 \text{ radian} = \mathbf{37^\circ\ 15'\ 31.5''} \\
c &= 0.826\ 770\ 4 \text{ radian} = \mathbf{47^\circ\ 22'\ 13.6''}
\end{aligned}
$$

Note: The reader will realize that having determined the six residuals from the observations, i.e. $25° 16' 17'' - a$, $37° 15' 33'' - b$, etc., a solution could be derived using the method given in Example 8.8. The computer program listed with Example 8.8 will solve this problem.

8.11 Braced quadrilateral by least squares and equal shifts

In order to determine the co-ordinates of the two ends A and B of the centre-line of a bridge across a wide river, a braced quadrilateral ABCD was set out. The mean observed angles and their log sines are given in Table 8.14. Adjust the angles.

Table 8.14

	Mean observed angle	Log sine	Log sine difference for 1″ (x 10⁻⁷)
CAB	43° 48′ 22″	$\bar{1}$.840 244 2	22
ABD	38° 36′ 57″	$\bar{1}$.795 251 1	26
DBC	33° 52′ 55″	$\bar{1}$.746 232 2	31
BCA	63° 41′ 24″	$\bar{1}$.952 506 5	10
ACD	49° 20′ 43″	$\bar{1}$.880 041 1	17
CDB	33° 04′ 56″	$\bar{1}$.737 067 2	32
BDA	50° 10′ 43″	$\bar{1}$.885 386 7	18
DAC	47° 23′ 28″	$\bar{1}$.866 872 9	19

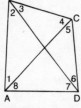

Figure 8.10

[ICE]

Solution. (*a*) **By least squares**. The layout of the triangulation is shown in Fig. 8.10. The most probable values of the eight angles must satisfy four condition equations, three being

$$1+2+3+4+5+6+7+8 = 360°$$
$$1+2 = 5+6 \quad \text{(Vertically opposite angles being equal in the}$$
$$3+4 = 7+8 \quad \text{respective triangles.)}$$

The fourth arises from a side relationship, as follows

$$AB = BC \frac{\sin4}{\sin1} = CD \frac{\sin6 \, \sin4}{\sin3 \, \sin1} = AD \frac{\sin8 \, \sin6 \, \sin4}{\sin5 \, \sin3 \, \sin1}$$

$$= AB \frac{\sin2 \, \sin8 \, \sin6 \, \sin4}{\sin7 \, \sin5 \, \sin3 \, \sin1}.$$

Therefore $\sin1 \times \sin3 \times \sin5 \times \sin7 = \sin2 \times \sin4 \times \sin6 \times \sin8$

or Σ log sin 'odd' = Σ log sin 'even'.

The values of the measured angles give

$$1+2+3+4+5+6+7+8 = 359° 59' 28''$$
$$1+2 = 82° 25' 19''$$
$$3+4 = 97° 34' 19''$$

$$5+6 = 82°\ 25'\ 39''$$
$$7+8 = 97°\ 34'\ 11''$$

Let corrections of e_1, e_2, ..., e_8 be applied to angles 1, 2, ..., 8, respectively. Then

$$e_1 + e_2 + e_3 + e_4 + e_5 + e_6 + e_7 + e_8 = 32''$$
$$82°\ 25'\ 19'' + e_1 + e_2 = 82°\ 25'\ 39'' + e_5 + e_6$$
$$97°\ 34'\ 19'' + e_3 + e_4 = 97°\ 34'\ 11'' + e_7 + e_8,$$

and so
$$e_1 + e_2 - e_5 - e_6 = +20''$$
$$e_3 + e_4 - e_7 - e_8 = -08''.$$

The corrections also influence the log sine values. Before adjustment the sums are

log sin 1 = $\bar{1}.840\ 244\ 2$		log sin 2 = $\bar{1}.795\ 251\ 1$
3 = $\bar{1}.746\ 232\ 2$		4 = $\bar{1}.952\ 506\ 5$
5 = $\bar{1}.880\ 041\ 1$		6 = $\bar{1}.737\ 067\ 2$
7 = $\bar{1}.885\ 386\ 7$		8 = $\bar{1}.866\ 872\ 9$
$\bar{1}.351\ 904\ 2$		$\bar{1}.351\ 697\ 7$

Denoting the differences of the log sines as v_1, v_2, ..., v_8, i.e. 22×10^{-7}, 26×10^{-7}, ..., 19×10^{-7}, respectively, we have,

$$\bar{1}.351\ 904\ 2 + e_1 v_1 + e_3 v_3 + e_5 v_5 + e_7 v_7$$
$$= \bar{1}.351\ 697\ 7 + e_2 v_2 + e_4 v_4 + e_6 v_6 + e_8 v_8.$$

Hence $\qquad \Sigma e_1 v_1 = \Sigma e_2 v_2 - 0.000\ 206\ 5$

Therefore $\quad \Sigma e_1 v_1 - \Sigma e_2 v_2 = -2065 \times 10^{-7}$.

Multiplying throughout by 10^7 gives

$$\Sigma 22 e_1 - \Sigma 26 e_2 = -2065.$$

The condition equations have therefore reduced to

$$e_1 + e_2 + e_3 + e_4 + e_5 + e_6 + e_7 + e_8 = +32'' \qquad [8.16]$$
$$e_1 + e_2 - e_5 - e_6 = +20'' \qquad [8.17]$$
$$e_3 + e_4 - e_7 - e_8 = -08'' \qquad [8.18]$$
$$e_1 v_1 + e_3 v_3 + e_5 v_5 + e_7 v_7 - e_2 v_2 - e_4 v_4 - e_6 v_6 - e_8 v_8$$
$$= -2065. \qquad [8.19]$$

Now, by 'least squares', $\Sigma e_1{}^2$ is to be a minimum. On differentiation

$$e_1 \delta e_1 + e_2 \delta e_2 + e_3 \delta e_3 + e_4 \delta e_4 + e_5 \delta e_5 + e_6 \delta e_6 + e_7 \delta e_7$$
$$+ e_8 \delta e_8 = 0 \qquad [8.20]$$
$$\delta e_1 + \delta e_2 + \delta e_3 + \delta e_4 + \delta e_5 + \delta e_6 + \delta e_7 + \delta e_8 = 0 \qquad [8.21]$$
$$\delta e_1 + \delta e_2 - \delta e_5 - \delta e_6 = 0 \qquad [8.22]$$
$$\delta e_3 + \delta e_4 - \delta e_7 - \delta e_8 = 0 \qquad [8.23]$$

and $v_1 \delta e_1 + v_3 \delta e_3 + v_5 \delta e_5 + v_7 \delta e_7 - v_2 \delta e_2 - v_4 \delta e_4 - v_6 \delta e_6$
$$- v_8 \delta e_8 = 0. \qquad [8.24]$$

Multiply equations [8.21], [8.22], [8.23], [8.24] by $-a$, $-b$, $-c$, $-d$, respectively, and add to equation [8.20] to obtain

$$\delta e_1(e_1-a-b-dv_1) + \delta e_2(e_2-a-b+dv_2) + \delta e_3(e_3-a-c-dv_3)$$
$$+ \delta e_4(e_4-a-c+dv_4) + \delta e_5(e_5-a+b-dv_5)$$
$$+ \delta e_6(e_6-a+b+dv_6) + \delta e_7(e_7-a+c-dv_7)$$
$$+ \delta e_8(e_8-a+c+dv_8) = 0.$$

Since δe_1, δe_2, etc., are independent quantities

$$e_1 = a+b+dv_1, \qquad e_2 = a+b-dv_2, \qquad e_3 = a+c+dv_3,$$
$$e_4 = a+c-dv_4, \qquad e_5 = a-b+dv_5, \qquad e_6 = a-b-dv_6,$$
$$e_7 = a-c+dv_7, \qquad e_8 = a-c-dv_8.$$

Substituting in equations [8.16], [8.17], [8.18] and [8.19] and remembering that all v values have been multiplied by 10^7

$$8a + d(v_1 - v_2 + v_3 - v_4 + v_5 - v_6 + v_7 - v_8) = +32$$
$$4b + d(v_1 - v_2 - v_5 + v_6) = +20$$
$$4c + d(v_3 - v_4 - v_7 + v_8) = -8$$
$$a(v_1 - v_2 + v_3 - v_4 + v_5 - v_6 + v_7 - v_8)$$
$$+ b(v_1 - v_2 - v_5 + v_6) + c(v_3 - v_4 - v_7 + v_8)$$
$$+ d \, \Sigma v_1^2 = -2065.$$

Whence
$$8a + d = +32$$
$$4b + 11d = +20$$
$$4c + 22d = -8$$
$$a + 11b + 22c + 4219d = -2065.$$

Which solve for $a = +4.063$, $b = +6.386$, $c = +0.772$ and $d = -0.504$.
Thus

$$
\begin{array}{llll}
e_1 = 4.063 + 6.386 - 11.088 = & -0.639, & \text{i.e.} & -0.7'' \\
e_2 = 4.063 + 6.386 + 13.104 = & +23.553 & & +23.6'' \\
e_3 = 4.063 + 0.772 - 15.624 = & -10.789 & & -10.8'' \\
e_4 = 4.063 + 0.772 + 5.040 = & +9.875 & & +9.9'' \\
e_5 = 4.063 - 6.386 - 8.568 = & -10.891 & & -10.9'' \\
e_6 = 4.063 - 6.386 + 16.128 = & +13.805 & & +13.8'' \\
e_7 = 4.063 - 0.772 - 9.072 = & -5.781 & & -5.8'' \\
e_8 = 4.063 - 0.772 + 9.576 = & +12.867 & & +12.9''
\end{array}
$$

$$\text{Check} \quad +32.000 \qquad +32.0''$$

Adjusted values (which could be rounded off to the nearest second)

CÂB = 43° 48′ 21.3″
AB̂D = 38° 37′ 20.6″
DB̂C = 33° 52′ 44.2″
BĈA = 63° 41′ 33.9″
AĈD = 49° 20′ 32.1″
CD̂B = 33° 05′ 09.8″
BD̂A = 50° 10′ 37.2″
DÂC = 47° 23′ 40.9″

360° 00′ 00″

Alternative solution. (*b*) **By 'equal shifts'.** In this method the four condition equations have to be established as in (*a*) but the corrections are derived without necessarily satisfying the principle of least squares.

Consider eqns [8.16], [8.17] and [8.18].

Step 1: eqn [8.16] is stated as

$$e_1 + e_2 + e_3 + e_4 + e_5 + e_6 + e_7 + e_8 = 32''.$$

The total correction is shared equally to each of the eight angles, i.e.

$$e = +4''.$$

Step 2: eqn [8.17] is stated as

$$e_1 + e_2 - e_5 - e_6 = +20''.$$

Now share out the corrections so that $e_1 = e_2 = +5''$ and $e_5 = e_6 = -5''$. In this way the total correction becomes $+20''$ but eqn [8.16] still holds, i.e.

$$\Sigma e_1 = +32''.$$

Step 3: eqn [8.18] stated that

$$e_3 + e_4 - e_7 - e_8 = -8''.$$

Again share out the corrections so that $e_3 = e_4 = -2''$ and $e_7 = e_8 = +2''$. The total correction is $-8''$ but eqn [8.16] still holds, i.e.

$$\Sigma e_1 = +32''.$$

At this stage the corrections in Table 8.15 will satisfy the requirements of eqns [8.16], [8.17] and [8.18].

Table 8.15

Angle	Step 1	Step 2	Step 3	Net correction
1	$+4''$	$+5''$		$+9''$
2	$+4''$	$+5''$		$+9''$
3	$+4''$		$-2''$	$+2''$
4	$+4''$		$-2''$	$+2''$
5	$+4''$	$-5''$		$-1''$
6	$+4''$	$-5''$		$-1''$
7	$+4''$		$+2''$	$+6''$
8	$+4''$		$+2''$	$+6''$
Σ	$+32''$	$0''$	$0''$	$+32''$

Step 4: the log sine condition determined in (*a*) stated

$$\bar{1}.351\ 904\ 2 + e_1 v_1 + e_3 v_3 + e_5 v_5 + e_7 v_7$$
$$= \bar{1}.351\ 697\ 7 + e_2 v_2 + e_4 v_4 + e_6 v_6 + e_8 v_8$$

or $\quad 0.000\ 260\ 5 + e_1 v_1 + e_3 v_3 + e_5 v_5 + e_7 v_7$
$$= e_2 v_2 + e_4 v_4 + e_6 v_6 + e_8 v_8.$$

To eliminate the difference of 0.000 206 5 it is necessary to reduce the sum of the log sines of angles 1, 3, 5 and 7 and to increase the sum of the log sines of angles 2, 4, 6 and 8.

Let a further correction of $\pm x''$ be applied to the net corrections determined in Steps 1 to 3, being positive in the case of angles 2, 4, 6 and 8, and negative in respect of angles 1, 3, 5 and 7. By this means angles 2, 4, 6 and 8 will be increased and angles 1, 3, 5 and 7 will be decreased (see Table 8.16), but there will be no alteration to

$$\Sigma e_1 = 32''.$$

Table 8.16

Angle	Net correction (steps 1–3)	$\pm x$	Final correction	$v \times 10^7$
1	$+9''$	$-x$	$+9-x$	22
2	$+9''$	$+x$	$+9+x$	26
3	$+2''$	$-x$	$+2-x$	31
4	$+2''$	$+x$	$+2+x$	10
5	$-1''$	$-x$	$-1-x$	17
6	$-1''$	$+x$	$-1+x$	32
7	$+6''$	$-x$	$+6-x$	18
8	$+6''$	$+x$	$+6+x$	19
Σ	$32''$	0	$32''$	

Whence $2065 + 22(9-x) + 31(2-x) + 17(-1-x) + 18(6-x)$
$$= 26(9+x) + 10(2+x) + 32(-1+x) + 19(6+x).$$

Therefore $x = \dfrac{2080}{175} = 11.9''$.

Adjusted values (rounded off to the nearest second) are given in Table 8.17.

Table 8.17

Angle	Correction	Adjusted value
1	$-2.9''$	43° 48′ 19″
2	$+20.9''$	38° 37′ 18″
3	$-9.9''$	33° 52′ 45″
4	$+13.9''$	63° 41′ 38″
5	$-12.9''$	49° 20′ 30″
6	$+10.9''$	33° 05′ 07″
7	$-5.9''$	50° 10′ 37″
8	$+17.9''$	47° 23′ 46″
Σ	$+32''$	360° 00′ 00″

This problem can be solved by the following computer program, which will also compute the corrections to any polygon with a central station. Readers with programmable calculators may find that this program is too large for their machine. This can be overcome by making the program problem-specific. For the braced quadrilateral case of this example lines 40−290 are not required, but a new line 40 GOTO 870 should be inserted, or the program should be rearranged. For the central polygon (Example 8.12) lines 40, 60 and 870−1030 can be left out.

The DIM statement in line 20 is set for a six-sided figure. The minimum requirement is $A(3*N)$; $C(3*N)$; $E(N+1)$.

The output is formatted into a table and readers using computers with limited display will need to adjust lines 300, 420, 470, 610, 620, 790 and 850. The angles must be input in the correct order, working logically around the outside of the figure, with 1 and 2 in the same triangle and then going to the centre angles as shown in Fig. 8.10 and Fig. 8.11.

Variables

A(i)	= Angle i	N3	= $3*N$
B	= $\pi/180°$ conversion to radians	O	= Sum of odd log sines
		Q\$	= Printing advance control
C(i)	= Correction to angle (i)	S	= Input/output, seconds
D	= Input/Output, degrees	S1	= Sum of differences
E	= Sum of even log sines	S2	= Sum of correction × difference
E(i)	= Error in triangle i		
E(N+1)	= Error at centre angle	T	= Total correction at centre angle due to triangles
F	= Difference for 1″		
G	= Log sin error	T2	= Total correction for side angles
H	= ±sign indicator		
I,J	= Loop counters	T3	= Total correction for centre angles
M	= Input/Output, minutes		
N	= Number of sides to the polygon	V	= Log sin correction per angle
		X	= Log sin of angle
N2	= $2*N$	Y	= Log sin of angle plus 1″

```
10 REM EQUAL SHIFTS
20 DIM A(18),C(18),E(7)
30 B=3.1415962 /648000
40 PRINT"FOR A BRACED QUADRILATERAL INPUT ZERO"
50 INPUT"INPUT NUMBER OF SIDES TO POLYGON ";N
60 IF N=0 THEN 880
70 IF N<3 THEN PRINT"INPUT ERROR":GOTO 40
80 N2=N*2
90 N3=N*3
100 GOSUB 1090
110 FOR I=1 TO N
120 E(I)=648000 -A(2*I-1)-A(2*I)-A(I+N2)
130 PRINT"CORRECTION TO TRIANGLE";I;"   =";E(I)
140 T=T+E(I)
150 NEXT I
160 E(N+1)=1296000
170 FOR I=N2+1 TO N3
180 E(N+1)=E(N+1)-A(I)
190 NEXT I
200 PRINT"CORRECTION TO CENTRE ANGLES =";E(N+1)
210 T2=T-(3*E(N+1))
220 T3=(3*E(N+1))-T
230 FOR I=1 TO N
240 C(2*I-1)=(T2+N2*E(I))/(6*N)
250 NEXT I
260 FOR I=N2+1 TO N3
```

```
270 C(I)=(T3+N*E(I-N2))/(3*N)
280 A(I)=A(I)+C(I)
290 NEXT I
300 PRINT "ANGLE         LOGSIN EVEN     LOGSIN ODD       DIFFERENCE"
310 PRINT
320 FOR I=1 TO N2 STEP 2
330 C(I+1)=C(I)
340 J=I
350 X=LOG10(SIN(A(J)*B))
360 Y=LOG10(SIN((A(J)+1)*B))
370 F=ABS(X-Y)
380 S1=S1+F
390 A=A(J)
400 GOSUB 1050
410 IF J<>I THEN 470
420 PRINT D;TAB(6);M;TAB(10);S;TAB(30);X;TAB(45);F
430 O=O+X
440 S2=S2+(C(J)*F)
450 J=I+1
460 GOTO 350
470 PRINT D;TAB(6);M;TAB(10);S;TAB(15);X;TAB(45);F
480 E=E+X
490 S2=S2-(C(J)*F)
500 NEXT I
510 G=E-O
520 IF G<0 THEN 550
530 H=1
540 GOTO 560
550 H=-1
560 V=ABS((G-S2)/S1)
570 FOR I=1 TO N2
580 A(I)=A(I)+C(I)+(H*V)
590 H=H*(-1)
600 NEXT I
610 PRINT SPC(14);"----------     -----------"
620 PRINT TAB(15);E;TAB(30);O
630 PRINT
640 V=INT(V*100+0.5)/100
650 PRINT"LOGSIN CORRECTION =";V;"SEC. PER ANGLE"
660 PRINT
670 PRINT"PRESS [RETURN] TO CONTINUE OUTPUT"
680 PRINT
690 INPUT Q$
700 PRINT"CORRECTED ANGLES"
710 PRINT
720 PRINT"REF   ANG.CORRN.     LOGSIN CORRN.         ANGLE"
730 PRINT
740 FOR I=1 TO N2
750 V=V*H
760 C(I)=INT(C(I)*100+0.5)/100
770 A=A(I)
780 GOSUB 1050
790 PRINT I;TAB(9);C(I);TAB(25);V;TAB(39);D;TAB(44);M;TAB(49);S
800 NEXT I
810 FOR I=N2+1 TO N3
820 C(I)=INT(C(I)*100+0.5)/100
830 A=A(I)
840 GOSUB 1050
850 PRINT I;TAB(9);C(I);TAB(39);D;TAB(44);M;TAB(49);S
860 NEXT I
870 END
880 REM BRACED QUADRILATERAL
890 N3=8
900 N=3
910 GOSUB 1090
920 E(1)=648000 -(A(1)+A(2)+A(3)+A(4))
930 E(2)=648000 -(A(3)+A(4)+A(5)+A(6))
940 E(3)=648000 -(A(5)+A(6)+A(7)+A(8))
950 E(4)=2*E(2)
960 FOR I=1 TO N
970 "CORRECTION TO TRIANGLE";I;"     =";E(I)
980 NEXT I
990 C(1)=-(E(4)-(3*E(1))-E(3))/8
1000 C(3)=-(E(3)-E(1)-E(4))/8
1010 C(5)=-(E(1)-E(3)-E(4))/8
1020 C(7)=-(E(4)-(3*E(3))-E(1))/8
1030 N2=8
1040 GOTO 300
1050 D=INT(A/3600)
```

```
1060 M=INT((A-(D*3600))/60)
1070 S=INT(A-(D*3600)-(M*60)+0.5)
1080 RETURN
1090 FOR I=1 TO N3
1100 PRINT"INPUT ANGLE";I;"IN DEG,MIN,SEC"
1110 INPUT D,M,S
1120 A(I)=(D*3600)+(M*60)+S
1130 NEXT I
1140 RETURN
```

8.12 Polygon with centre station by equal shifts

The observed angles of a polygon ABCDEFA with central point P are set out in the table together with their log sine difference for 1″. Adjust the polygon angles to the nearest second by the method of equal shifts. All angles are of equal weight. The left-hand log sine sum minus right-hand log sine sum = 0.000 146 0.

Table 8.18

Triangle	Observed central angle	LH angles observed	Diff. 1″	RH angles observed	Diff. 1″
APB	59° 15′ 25″	67° 34′ 20″	9	53° 10′ 15″	16
BPC	53° 56′ 25″	53° 34′ 27″	15	72° 29′ 08″	6
CPD	65° 10′ 03″	50° 50′ 53″	17	63° 59′ 07″	10
DPE	75° 05′ 02″	48° 46′ 42″	18	56° 08′ 09″	14
EPF	60° 03′ 05″	64° 38′ 20″	10	55° 18′ 40″	15
FPA	46° 30′ 00″	78° 26′ 35″	4	55° 03′ 34″	15

[London]

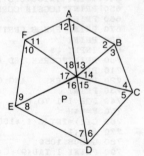

Figure 8.11

Introduction. There are seven condition equations relating directly to the angular observations since there are six triangles with a common centre point. Thus six of the condition equations demand that the sum of the respective angles total 180°, whilst the seventh demands that the angles at P have a total of 360°. In addition there is an eighth equation specifying that the sum of L.H. log sines will equal the sum of the R.H. log sines.

Solution. First correct the six triangles. The observations imply the values in Table 8.19.

Table 8.19

Triangle	Sum of angles	Correction required
APB	180° 00′ 00″	00″
BPC	180° 00′ 00″	00″
CPD	180° 00′ 03″	−03″
DPE	179° 59′ 53″	+07″
EPF	180° 00′ 05″	−05″
FPA	180° 00′ 09″	−09″
		Total −10″

The corrections are applied equally (the angles being of equal weight) to the three angles within the respective triangles as in Table 8.20.

Table 8.20

Triangle	Centre	L.H.	R.H.	Correction
APB	0″	0″	0″	0″
BPC	0″	0″	0″	0″
CPD	−1″	−1″	−1″	−3″
DPE	$+\frac{7}{3}''$	$+\frac{7}{3}''$	$+\frac{7}{3}''$	+7″
EPF	$-\frac{5}{3}''$	$-\frac{5}{3}''$	$-\frac{5}{3}''$	−5″
FPA	−3″	−3″	−3″	−9″
Totals	$-\frac{10}{3}''$	$-\frac{10}{3}''$	$-\frac{10}{3}''$	−10″

Next correct the central angles at P. The sum of the observed angles is 360° 00′ 00″ but the corrections applied to the individual triangles have resulted in an 'overcorrection' of $-10/3''$. This must be balanced by $+10/3''$ to restore the sum of the angles to 360° but at the same time equal and opposite corrections must be applied to the L.H. and R.H. angles so that the overall corrections applied to the triangles are not disturbed, as in Table 8.21.

Finally adjust the log sine condition. This condition demands that

$$\Sigma \log \sin \text{L.H.} = \Sigma \log \sin \text{R.H.}$$

after the adjustments have been effected. Let further corrections of $\pm x''$ be applied to the angles. Thus

$$\Sigma \log \sin \text{L.H.} + \Sigma e_L v_L = \Sigma \log \sin \text{R.H.} + \Sigma e_R v_R,$$

where v relates to the log sine differences for the individual angles. In the table these have been multiplied by 10^7. Hence, since

$$\Sigma \log \sin \text{L.H.} - \Sigma \log \sin \text{R.H.} = 0.000\ 146\ 0 = 1460 \times 10^{-7}$$

Table 8.21

Triangle	Centre	L.H.	R.H.	Correction
APB	$+\frac{10}{18}{''}$	$-\frac{5}{18}{''}$	$-\frac{5}{18}{''}$	$0''$
BPC	$+\frac{10}{18}{''}$	$-\frac{5}{18}{''}$	$-\frac{5}{18}{''}$	$0''$
CPD	$-1''+\frac{10}{18}{''}$	$-1''-\frac{5}{18}{''}$	$-1''-\frac{5}{18}{''}$	$-3''$
DPE	$+\frac{7}{3}{''}+\frac{10}{18}{''}$	$+\frac{7}{3}{''}-\frac{5}{18}{''}$	$+\frac{7}{3}{''}-\frac{5}{18}{''}$	$+7''$
EPF	$-\frac{5}{3}{''}+\frac{10}{18}{''}$	$-\frac{5}{3}{''}-\frac{5}{18}{''}$	$-\frac{5}{3}{''}-\frac{5}{18}{''}$	$-5''$
FPA	$-3''+\frac{10}{18}{''}$	$-3''-\frac{5}{18}{''}$	$-3''-\frac{5}{18}{''}$	$-9''$
	$0''$	$-\frac{90}{18}$	$-\frac{90}{18}$	$-10''$

$$1460 + 9(-\tfrac{5}{18}-x) + 15(-\tfrac{5}{18}-x) + 17(-\tfrac{23}{18}-x) + 18(\tfrac{37}{18}-x)$$
$$+ 10(-\tfrac{35}{18}-x) + 4(-\tfrac{59}{18}-x) = 16(-\tfrac{5}{18}+x) + 6(-\tfrac{5}{18}+x)$$
$$+ 10(-\tfrac{23}{18}+x) + 14(+\tfrac{37}{18}+x) + 15(-\tfrac{35}{18}+x) + 15(-\tfrac{59}{18}+x).$$

Therefore $x = 10.10''$.

Note that x has been applied to reduce all left-hand angles and to increase

Table 8.22

Triangle	Centre	L.H.	R.H.	Correction
APB	$+0.56$	-10.38	$+9.82$	$0''$
BPC	$+0.56$	-10.38	$+9.82$	$0''$
CPD	-0.44	-11.38	$+8.82$	$-3''$
DPE	$+2.89$	-8.04	$+12.16$	$+7''$
EPF	-1.11	-12.04	$+8.16$	$-5''$
FPA	-2.44	-13.38	$+6.82$	$-9''$
	$+0.02$			

Table 8.23

Triangle	APB	BPC	CPD	DPE	EPF	FPA
Centre	$0''$	$0''$	$0''$	$+3''$	$-1''$	$-2''$
L.H.	$-10''$	$-10''$	$-11''$	$-8''$	$-12''$	$-13''$
R.H.	$+10''$	$+10''$	$+8''$	$+12''$	$+8''$	$+7''$
	$0''$	$0''$	$-3''$	$+7''$	$-5''$	$-8''$

the right-hand angles since, initially,

$$\Sigma \log \sin \text{L.H.} > \Sigma \log \sin \text{R.H.}$$

Hence the corrections will be as in Table 8.22. Rounded off to 1″ the corrections are as in Table 8.23. The program listed with Example 8.11 can be used to solve this problem. The angles must be input in accordance with the number system of Fig. 8.11.

8.13 Variation of co-ordinates

In Fig. 8.12 two control stations A and B have the co-ordinates given in Table 8.24.

Table 8.24

	E (m)	N (m)
A	5210.15	12267.92
B	8785.62	9686.45

Observations were taken from these stations to determine the co-ordinates of a third station G, as follows:

Bearing AG = 76° 06′ 29″ (standard error ± 4.0″)
Bearing BG = 14° 24′ 27″ (standard error ± 4.0″)
Length AG = 4663.08 m (standard error ± 0.05 m)
Length BG = 3821.21 m (standard error ± 0.05 m)
Angle BĜA = 61° 41′ 57″ (standard error ± 5.6″)

By means of the method of variation of co-ordinates compute the co-ordinates of G, which have been accepted provisionally as 9736.54 mE, 13387.68 mN.

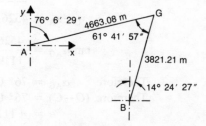

Figure 8.12

Introduction. In Fig. 8.12

$$l_{AG}^{2} = (x_G - x_A)^2 + (y_G - y_A)^2.$$

If small displacements dx_G, dx_A, dy_G and dy_A be applied to those co-ordinates, the change in length AG is given by differentiation, as

$$dl_{AG} = -\frac{(x_G - x_A)}{l_{AG}} dx_A - \frac{(y_G - y_A)}{l_{AG}} dy_A + \frac{(x_G - x_A)}{l_{AG}} dx_G$$

$$+ \frac{(y_G - y_A)}{l_{AG}} dy_G.$$

If α_{AG} be the bearing of AG

$$\tan \alpha_{AG} = \frac{x_G - x_A}{y_G - y_A}$$

and the change in that bearing due to the changes dx_A, dy_A, etc., is

$$d\alpha_{AG} = -\frac{(y_G - y_A)}{l_{AG}{}^2} dx_A + \frac{(x_G - x_A)}{l_{AG}{}^2} dy_A + \frac{(y_G - y_A)}{l_{AG}{}^2} dx_G$$

$$-\frac{(x_G - x_A)}{l_{AG}{}^2} dy_G.$$

Similar expressions can be derived for BG, whilst the change in angle $B\hat{G}A$ can be related to $d\alpha_{AG}$ and $d\alpha_{BG}$. In the method of variation of co-ordinates provisional co-ordinates are allocated to points requiring adjustment. These points are subjected to displacements dx and dy duly determined by a 'least-squares' analysis. Residuals (r) can be derived in the form

$$r = O - C - d\gamma,$$

where O is the measured value of a length, bearing or angle, C is the value of that quantity as calculated from relevant co-ordinates, and $d\gamma$ is the change in that quantity due to the displacements of the respective points. $(C + d\gamma)$ is the best value for the quantity.

Solution. (a) Calculate values of $(O - C)$ for bearings α_{AG} and α_{BG}.

$x_G = 9736.54$	$y_G = 13\ 387.68$ (provisional)
$x_A = \underline{5210.15}$	$y_A = \underline{12\ 267.92}$ (fixed)
$x_G - x_A = 4526.39$	$y_G - y_A = 1119.76$

$$\tan \alpha_{AG} = \frac{4526.39}{1119.76}$$

Therefore $\quad \alpha_{AG} = 76° \ 06' \ 17.5'' = C$
Therefore $(O - C) = 76° \ 06' \ 29'' - 76° \ 06' \ 17.5''$
$$= +11.5''$$

$x_G = 9736.54$	$y_G = 13\ 387.68$ (provisional)
$x_B = \underline{8785.62}$	$y_B = \underline{9686.45}$ (fixed)
$x_G - x_B = 950.92$	$y_G - y_B = 3701.23$

$$\tan \alpha_{BG} = \frac{950.92}{3701.23}$$

$$\alpha_{BG} = 14° \ 24' \ 31.7'' = C$$

Therefore $(O-C) = 14° 24' 27'' - 14° 24' 31.7''$
$$= -4.7''$$

(b) Calculate values of $(O-C)$ for l_{AG} and l_{BG}.

$$l_{AG} = \sqrt{(4526.39^2 + 1119.76^2)}$$
$$= 4662.84 \text{ m} = C.$$

Therefore $(O-C) = 4663.08 - 4662.84$
$$= +0.24 \text{ m.}$$
$$l_{BG} = \sqrt{(950.92^2 + 3701.23^2)}$$
$$= 3821.43 \text{ m} = C.$$
$$(O-C) = 3821.21 - 3821.43$$
$$= -0.22 \text{ m.}$$

(c) Calculate the value of $(O-C)$ for $B\hat{G}A$, measuring clockwise.

$$B\hat{G}A = 76° 06' 17.5'' - 14° 24' 31.7''$$
$$= 61° 41' 45.8'' = C.$$

Therefore $(O-C) = 61° 41' 57'' - 61° 41' 45.8''$
$$= +11.2''.$$

We can now determine the residuals. Since A and B are fixed control points $dx_A = dy_A = dx_B = dy_B = 0$.

Thus
$$d\alpha_{AG} = \frac{(y_G - y_A)}{l_{AG}^2} dx_G - \frac{(x_G - x_A)}{l_{AG}^2} dy_G$$

$$d\alpha_{AG} = \frac{1119.76}{4662.84^2} dx_G - \frac{4526.39}{4662.84^2} dy_G$$
$$= 0.000\ 051\ 5\ dx_G - 0.000\ 208\ 2\ dy_G.$$

and
$$d\alpha_{BG} = \frac{3701.23}{3821.43^2} dx_G - \frac{950.92}{3821.43^2} dy_G$$
$$= 0.000\ 253\ 4\ dx_G - 0.000\ 065\ 1\ dy_G.$$

$$dl_{AG} = \frac{(x_G - x_A)}{l_{AG}} dx_G + \frac{(y_G - y_A)}{l_{AG}} dy_G$$

$$= \frac{4526.39}{4662.84} dx_G + \frac{1119.76}{4662.84} dy_G$$
$$= 0.970\ 736\ 7\ dx_G + 0.240\ 145\ 5\ dy_G$$

and
$$dl_{BG} = \frac{950.92}{3821.43} dx_G + \frac{3701.23}{3821.43} dy_G$$

$$= 0.248\ 838\ 8\ dx_G + 0.968\ 545\ 8\ dy_G.$$

Also $(d\alpha_{AG} - d\alpha_{BG}) = -0.000\ 201\ 9\ dx_G - 0.000\ 143\ 1\ dy_G.$

Whence, expressing angles in radians, for α_{AG}

$$r = (O-C) - d\alpha_{AG}$$

$$= + \frac{11.5}{206\ 265} - 0.000\ 051\ 5\ dx_G + 0.000\ 208\ 2\ dy_G.$$

For α_{BG}

$$r = -\frac{4.7}{206\ 265} - 0.000\ 253\ 4\ dx_G + 0.000\ 065\ 1\ dy_G,$$

For l_{AG}

$$r = (O-C) - dl_{AG}$$
$$= +0.24 - 0.970\ 736\ 7\ dx_G - 0.240\ 145\ 5\ dy_G.$$

For l_{BG}

$$r = -0.22 - 0.248\ 838\ 8\ dx_G - 0.968\ 545\ 8\ dy_G.$$

For $B\hat{G}A$

$$r = 61° 41' 57'' - (\alpha_{AG} + d\alpha_{AG} - \alpha_{BG} - d\alpha_{BG})$$
$$= (O-C) - (d\alpha_{AG} - d\alpha_{BG})$$

$$= +\frac{11.2}{206\ 265} + 0.000\ 201\ 9\ dx_G + 0.000\ 143\ 1\ dy_G.$$

Now apply weightings. Weights will be applied which are inversely proportional to the squares of standard errors, as follows, adopting radian measure in respect of bearings and angles.

Table 8.25

Quantity	Weight x	$\sqrt{w}$
Bearing	$\left(\dfrac{206\ 265}{4.0}\right)^2$	51 566.25
Length	$\left(\dfrac{1}{0.05}\right)^2$	20.00
Angle	$\left(\dfrac{206\ 265}{5.6}\right)^2$	36 833.04

(d) Normal equations can now be derived. The method discussed in Example 8.8 will be adopted to minimize $\Sigma[\sqrt{w_1}\ r_1]^2$ with respect to the two unknowns dx_G and dy_G. In Table 8.26 the five residuals for α_{AG}, α_{BG}, l_{AG}, l_{BG} and angle

Table 8.26

dx_G	dy_G	N
− 2.655 661 9	+ 10.736 093 2	+ 2.877 396 7
− 13.066 887 7	+ 3.356 962 9	− 1.175 710 5
− 19.414 734 0	− 4.802 910 0	+ 4.800 000 0
− 4.976 776 0	− 19.370 916 0	− 4.400 000 0
+ 7.436 590 8	+ 5.270 808 0	+ 2.000 034 1

BĜA have been multiplied by the respective values of $\sqrt{w}$ to obtain the specific values within the columns. The two normal equations are obtained from the above table as

$$634.799\ 172\ 6\ dx_G + 156.472\ 281\ 3\ dy_G = 48.697\ 989\ 3$$
$$156.472\ 281\ 3\ dx_G + 552.614\ 645\ 3\ dy_G = -99.665\ 040\ 7$$

and these solve for

$$dx_G = +0.130 \text{ m, i.e. } +0.13 \text{ m}$$
$$dy_G = -0.217 \text{ m, i.e. } -0.22 \text{ m.}$$

Hence the most probable value of the co-ordinates of G are

$$9736.54 + 0.13 = 9736.67 \text{ mE}$$
$$\text{and} \quad 13\ 387.68 - 0.22 = 13\ 387.46 \text{ mN.}$$

Problems

1 A 'total station' instrument was set up at A and used to make observations towards two other stations B and C.

The following data were obtained together with their corresponding standard deviations:

Co-ordinates of A: 4372.651 m ± 8 mm E,
6751.322 m ± 12 mm N

Vertical angles: A to B 27° 15′ 27″ ± 9″,
A to C 15° 31′ 56″ ± 6″

Slant distances: A to B 2365.228 m ± 16 mm,
A to C 1496.314 ± 27 mm

Fixed bearings: A to B 64° 15′ 12″ ± 14″,
A to C 107° 42′ 36″ ± 20″

Calculate the eastings and northings of B together with their standard deviations.

What is the standard deviation of the difference in level between B and C calculated from the above data?   [Bradford]
Answer 6266.500 m ± 77 mm, 7664.671 m ± 131 mm, ± 101 mm

2 (*a*) List the various errors introduced in distance measurement using wires in catenary, indicating how their effects could be minimized or eliminated.

(*b*) From a marked point, another point is set out by using a theodolite and a tape for bearing and distance. If the standard deviations of the measured bearing and distance are ± σ_α and ± σ_l respectively, prove that the sum $(\sigma_E^2 + \sigma_N^2)$ is independent of the bearing, where ± σ_E and ± σ_N are the standard deviations of the computed co-ordinates of the other point.

(*c*) In part (*b*) above if $(\sigma_E^2 + \sigma^2_N)$ is not to exceed 25 cm², *l* is 400 m and σ_l is ± 4 cm, calculate the required maximum standard deviation of the observed bearing in seconds of arc. (Assume cosec $1''$ $= 20.6 \times 10^4$.)   [London]

Answer $\pm 15.45''$

3 (*a*) State the law of propagation of random error and name the limitation on its application. Hence show that if σ is the standard error of each of a set of n observations, the standard error of the mean equals $\sigma/\sqrt{n}$.

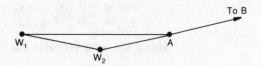

Figure 8.13

(*b*) Figure 8.13 represents a Weisbach triangle and the first leg of an underground traverse. W_1W_2 was measured to be 2.965 m and AW_2 2.097 m with standard errors 3 mm and 2 mm respectively. Angle W_1AB is known with a standard error of $4''$. The approximate value of angle W_2AW_1 is $0° 06' 50''$.

If the standard error of a single measurement of angle W_2AW_1 is estimated to be $10''$, calculate how many times this angle must be measured in order that the standard error of the bearing AB is not to exceed $5''$. (Assume that bearing W_1W_2 is known without error.)

[London]

Answer 6

4 An important line in setting out a major engineering project has been measured by an electromagnetic distance measuring instrument as 225.626 m ± 2 mm. A three-bay subtense-bar check measurement on the same line, using a first-order theodolite reading direct to 0.5 of a second of arc and a 2 metre invar subtense bar, gave the results listed in Table 8.27.

Calculate the three individual bay lengths, the total length of the survey line and the associated standard deviations of the means. Compare the subtense bar and electromagnetic distance measurements and state, with

Table 8.27

Bay 1			Bay 2			Bay 3		
°	′	″	°	′	″	°	′	″
1	28	20.0	1	26	50.3	1	40	13.0
1	28	15.8	1	26	52.7	1	40	17.3
1	28	14.1	1	26	48.8	1	40	14.1
1	28	14.8	1	26	51.8	1	40	14.0
1	28	17.3	1	26	53.4	1	40	12.7
1	28	18.5	1	26	50.2	1	40	14.4
1	28	18.3	1	26	50.5	1	40	14.9
1	28	17.2	1	26	47.9	1	40	15.0
1	28	16.1	1	26	52.2	1	40	14.0
1	28	17.0	1	26	52.9	1	40	16.7

reasons, whether you consider the agreement between the measurements to be within expected limits. Describe how the accuracy of both the subtense-bar and the electromagnetic distance measurements could be increased.

[Eng. Council]

Answer Bay 1 77.877 m ± 8.2 mm; Bay 2 79.160 m ± 8.8 mm; Bay 3 68.583 m ± 5.2 mm; Total 225.620 m ± 13 mm

5 The mean values of four rounds of each of the three angles of the triangle ABC, using three different theodolites with the same observer, are as follows:

Using a theodolite reading to 20 seconds:
ABC = 64° 13′ 50″ BCA = 61° 39′ 25″ CAB = 54° 06′ 45″
Using a theodolite reading to 5 seconds:
ABC = 64° 13′ 43″ BCA = 61° 39′ 27″ CAB = 54° 06′ 47″
Using a theodolite reading to 1 second:
ABC = 64° 13′ 45″ BCA = 61° 39′ 29″ CAB = 54° 06′ 49″

If the co-ordinates of A and C are 24 261.46 mE, 19 015.83 mN and 26 472.88 mE, 20 838.05 mN respectively and B lies to the north of the line AC, calculate the most probable value of the co-ordinates of station B.

[Leeds]

Answer 24 085.52 mE, 21 810.82 mN

6 Two different electronic distance measuring devices were used to make repeated determinations of the length of a line. The readings are given in Table 8.28 in metres.

Table 8.28

Instrument A	Instrument B
1056.429	1056.435
1056.435	1056.435
1056.440	1056.432
1056.432	1056.436
1056.433	1056.435
1056.436	1056.436
	1056.434
	1056.435

Assuming identical conditions for each set of determinations, calculate the relative precision of the two instruments based upon these observations.

What is the most probable length of the line based upon these readings? Determine the standard deviation of this value.

Work in millimetres and set out the calculation clearly step by step.

[Bradford]

Answer 1:11.5, 1056.453 m ± 0.4 mm

7 The three interior angles of a triangle ABC, where A, B and C appear in a clockwise direction, were measured as

ABC 63° 41′ 08″ standard error ± 3″
BCA 56° 55′ 21″ standard error ± 4″
CAB 59° 23′ 27″ standard error ± 2″

Assuming the weights to be inversely proportional to the squares of the standard errors, adjust the angles of the triangle for a 180° closure. Given the whole circle bearing of AB to be 118° 20′ 15″, state the quadrant bearings of AB, BC and CA.   [CEI]
Answer AB S 61° 39′ 45″ E, BC S 54° 39′ 06″ W, CA N 02° 16′ 17.4″ W

8 (*a*) A distance has been measured repeatedly on three different occasions. The mean values recorded and their standard errors are listed below.

Measurement 1   112.125 m ± 0.007 m
Measurement 2   112.130 m ± 0.002 m
Measurement 3   112.128 m ± 0.005 m

Determine the weighted mean of the three sets of measurements.
   (*b*) A survey of a conical tip at a mine has shown the diameter of the base to be 150 m with a standard error of ± 0.5 m and the height to be 50 m with a standard error of ± 1.0 m. What is the volume of the cone and the standard error of the volume?   [CEI]
Answer (*a*) 112.129 m; (*b*) 294 524.3 m^3 ± 6209.1 m^3

9 The standard deviation of a single measurement of an angle has been calculated as 2.75″. How many measurements should be made assuming similar conditions so that the standard deviation of the mean of a set of angles is 1.0″?
Answer 8

10 Determine the most probable value of a quantity which was measured nine times by a method having a standard deviation of four units giving an average value of 900 units, and four times by a method having a standard deviation of six units then giving an average value of 904 units.
Answer 900.7 units

11 Length *a* has been determined using the equation
$$a^2 = b^2 + c^2 - 2bc \cos A$$
in which $b = 146.00$ m, $c = 168.00$ m and $A = 58° 00′$. If the standard errors of the three measurements are ± 0.73 m, ± 0.84 m and ± 20″, respectively, determine the standard error of *a*.
Answer ± 0.57 m

12 Explain the terms 'most probable value' and 'weighting'.
   Angles A, B and C were observed as part of a programme of measurement at one station, and the data in Table 8.29 obtained. Use the method

Table 8.29

Angle	Observed value	Weight
A	30° 24′ 29.8″	1
B	44° 35′ 14.3″	1
A + B	74° 59′ 45.2″	2
B + C	110° 55′ 48.6″	1
A + B + C	141° 20′ 17.4″	2

of correlates to determine the most probable values of angles A, B and C. [CEI]

Answer A 30° 24′ 29.8″; B 44° 35′ 15.0″; C 66° 20′ 32.9″

13 In order to establish the area of a tract of land ABC, the sides AB and AC were measured by subtense bar. The shorter side AB was measured by sub-division into four equal lengths, whilst the longer side AC was measured using an auxiliary base of optimum length 42.94 m at right angles to CA. Given that the length of AB and the value of angle BAC were found to be 641.60 m and 46° 28′ 54″, respectively, and assuming that:

(*a*) standard errors of ± 1″ apply to all subtense measurements; and

(*b*) the standard error applicable to angle BAC is ± 10″.

Determine the area of ABC and its standard error.

Answer 21.45 hectare ± 53 m^2

14 A tape of length 30 m, standardized on the flat under a tension of 49 N, was used in catenary under a tension of 147 N. Determine the standard deviation of the combined pull correction and sag correction over a 30 m length if the standard deviation of the field tension is ± 3 N.

Cross-sectional area of tape = 0.406 cm^2
Young's modulus = 207000 N/mm^2
Mass of tape = 0.27 gm/cm
Acceleration due to gravity = 9.806 m/s^2

Answer ± 15 mm

15 A slope length of 178.741 m was measured with a standard error of ± 5 mm. To determine the corresponding horizontal length an angle of depression of 05° 14′ 25″ was measured with a standard error of ± 10″. Calculate the horizontal length and its standard error.

Answer 177.994 m ± 5 mm

16 A polygon ABCDEA with a central station O forms part of a triangulation scheme. The angles in each of the figures which form the complete network are being adjusted, and in this case the angles in each of the triangles DOE and EOA have already been adjusted and need no further correction.

Making use of the information given in Table 8.30, use the method

Table 8.30

Triangle	Angle	Observed value	Log sine	Log sine diff. for 1″ (×10⁻⁷)
AOB	OAB	40° 17′ 57″	1.810 755 7	25
	OBA	64° 11′ 20″	1̄.954 355 6	10
	AOB	75° 30′ 52″		
BOC	OBC	37° 22′ 27″	1̄.783 201 4	28
	OCB	71° 10′ 50″	1̄.976 139 0	7
	BOC	71° 26′ 22″		
COD	OCD	24° 51′ 25″	1̄.623 615 4	46
	ODC	51° 48′ 47″	1̄.895 421 4	17
	COD	103° 19′ 33″		
		Adjusted value		
DOE	ODE	67° 18′ 59″	1̄.965 036 2	
	OED	51° 02′ 00″	1̄.890 707 1	
	DOE	61° 39′ 01″		
EOA	OEA	116° 47′ 40″	1̄.950 671 4	
	OAE	15° 08′ 02″	1̄.416 766 2	
	EOA	48° 04′ 18″		

of equal shifts to determine the correction that must be applied to each of the remaining angles.   [ICE]

Answer   OB̂C + 14.6″, BĈO + 4.4″, BÔC + 2″

17   A line of length D was set out and then divided into n bays of length $d_1, d_2, \ldots, d_n$. Each individual length D, d_1 etc., was then measured by an EDM instrument. Show that:

(*a*) the additive constant of the device can be calculated from the expression

$$D - \frac{\Sigma d_n}{n-1};$$

(*b*) the standard deviation of that constant is

$$s\,\frac{\sqrt{(n+1)}}{n-1},$$

s being the standard deviation of a single measurement of distance.

18   The internal angles of a closed polygon ABCDEFA have been measured by theodolite and are tabulated (see Table 8.31). As it was

Table 8.31

Measured angle		Weighting
FAB	98° 13′ 13″	2
ABC	101° 08′ 54″	3
BCD	112° 41′ 20″	4
CDE	113° 48′ 12″	2
DEF	126° 28′ 02″	1
EFA	167° 40′ 37″	3
BCF	73° 12′ 54″	1
CFA	87° 24′ 47″	1

found possible to make sightings along a line CF, the angles BCF and CFA were also measured. 'Weightings' have been assigned to each of the measurements based on local conditions at the time of observation.

Using the method of correlates or otherwise, determine the most probable corrections (to the nearest 0.1″) to be applied to the angles.

[ICE]

Answer −0.7″, −0.5″, −2.0″, −4.0″, −8.0″, −2.7″, +6.6″, +6.6.″

19 Distances AB, BC, CD and DE along a straight base line AE have been measured with an electromagnetic distance measuring instrument as 100.241, 100.358, 100.303 and 100.338 m respectively. It is known from accurate catenary measurements that AD is 300.9000 m and that BE is 301.0000 m.

Calculate the most probable values of the electromagnetic distance measurements AB, BC, CD and DE by the method of least squares. The most probable values must conform with the catenary measurements.

[CEI]

Answer AB 100.2394 m; BC 100.3578 m; CD 100.3028 m; DE 100.3394 m

20 The observed angles of the braced quadrilateral ABCD are as follows:

AD̂B = 42° 40′ 40″   CB̂D = 52° 46′ 49″
DÂC = 62° 43′ 13″   AĈB = 52° 37′ 00″
CÂB = 44° 02′ 55″   AĈD = 36° 20′ 10″
AB̂D = 30° 33′ 28″   BD̂C = 38° 16′ 09″

Adjust the angles of the quadrilateral using the method of equal shifts.

[Leeds]

Answer AD̂B, −4.2″; AB̂D, −3.6″; BD̂C, −1.7″

21 Repeat Q. 20 using the method of least squares.
Answer AD̂B, −4.2″; AB̂D, −3.7″; BD̂C, −1.7″

22 D and G are two fixed triangulation stations. A point M, which lies north of the line D−G, is fixed by observing all three angles of the triangle. Using the data in Table 8.32 calculate the co-ordinates of M.

[London]

Answer 828.00 mE, 578.72 mN

Table 8.32

	Angles		Co-ordinates	
			E (m)	N (m)
MGD	46° 35′ 22″			
DMG	89° 12′ 54″	D	543.56	245.32
GDM	44° 11′ 35″	G	1144.13	301.38

23 Three survey stations A, B and C form the core of the control network on a site which is to be used to house equipment for detecting intercontinental ballistic missiles. They form an accurate equilateral triangle with 1000 m sides. C lies due east of A. B is north of the line AC.

In order to locate the new equipment it is necessary to locate the centroid of the triangle ABC to within ± 1 mm. A trial point T has been established which is known to be within 10 cm of the required position.

The following values are the mean of six rounds of angles taken with a half-second theodolite set up over T and should be accurate to the first place of decimals of a second quoted. How far from T and in what direction is the required centroid?

Reading towards A 00° 00′ 00.0″, towards B 119° 59′ 53.0″, towards C 240° 00′ 30.8″.　　　　　　　　　　　　　　[Bradford]

Answer 65.0 mm, 99° 20′ 43″ from pointing on C

24 (*a*) A horizontal angle between two fixed stations is observed at a third station. Deduce the observation equation for the angle as used in the method of variation of co-ordinates.

(*b*) Using the data set out below compute the co-ordinates of P by the method of variation of co-ordinates.

The observed horizontal angle at P between A and B = 53° 18′ 00″ and between B and C = 101° 18′ 28″.

The co-ordinates of A, B, C and P′, an estimated position of P, are given in Table 8.33. The joins have been calculated in Table 8.34.

[London]

Table 8.33

	E (m)	N (m)
A	1797.47	312.66
B	1606.21	347.96
C	1594.23	676.67
P′	1735.00	520.00

Table 8.34

	(m)	
P′A	216.55	163° 13′ 58″
P′B	214.91	216° 49′ 07″
P′C	210.62	318° 03′ 36″

Answer 1734.79 mE, 521.23 mN

25 During a topographical survey a series of runs of levels was made as follows:

A to B +10.72 m,	B to C +6.58 m,	C to D +4.97 m,
D to E −15.32 m,	E to A −6.99 m,	A to F +8.65 m,
F to G + 6.34 m,	G to D +7.28 m,	D to H +3.73 m,
H to J − 6.59 m,	J to K −7.62 m,	K to A −11.88 m.

Determine the best values of the levels at all points B, C, D, etc., taking that of A to be zero and all values above to be equally reliable.

[London]

Answer level of D = 22.30 m

26 In the course of the precise levelling of a certain area the results in Table 8.35 were obtained. Determine the most probable level of B, C and D above A.

Table 8.35

Level line	Level difference	Length (km)
A to B	Rise 4.727 m	10
B to C	Rise 1.580 m	10
C to D	Rise 3.540 m	20 (read twice) .
D to A	Fall 9.846 m	15 (read three times)
B to D	Rise 5.125 m	16
A to C	Rise 6.315 m	20

[Salford]

Answer B 4.727 m; C 6.309 m; D 9.848 m

27 In a round of levels the results in Table 8.36 were obtained. Determine the most probable level of B, C and D above A.

Table 8.36

Level line	Rise (+m) Fall (−m)	Weight
A to B	+2.275	3
B to C	−1.216	2
C to D	−4.342	1
D to B	+5.508	1
D to A	+3.263	1

[Salford]

Answer B 2.273 m; C 1.066 m; D −3.258 m

28 An angle was measured 15 times and the following data recorded. What is the most probable value of the angle?

18° 59′ 59″, 19° 00′ 00″, 18° 59′ 58″
19° 00′ 00″, 18° 59′ 59″, 19° 00′ 01″
18° 59′ 58″, 19° 00′ 06″, 18° 59′ 59″
19° 00′ 00″, 18° 59′ 58″, 18° 59′ 59″
19° 00′ 00″, 18° 59′ 58″, 18° 59′ 57″   [Salford]

Answer 18° 59′ 59″ (reject 19° 00′ 06″)

29 Use the method of variation of co-ordinates to show how the co-ordinates of a single point may be obtained by intersection from two known points and by resection from three known points.

Explain how the strength of such fixations may be obtained from an analysis of the models used.   [RICS]

30 In a triangulation survey ABCD forms a quadrilateral with centre station O. The following angles were measured, all have equal weight.

OAB	=	57° 55′ 8″
OBA	=	38° 37′ 27″
OBC	=	62° 36′ 16″
OCB	=	34° 15′ 39″
OCD	=	36° 50′ 25″
ODC	=	51° 54′ 24″
ODA	=	27° 57′ 23″
OAD	=	49° 52′ 50″
AOB	=	83° 27′ 17″
BOC	=	83° 8′ 6″
COD	=	91° 15′ 9″
DOA	=	102° 9′ 32″

Correct the data by the method of equal shifts.   [Salford]

Answer OÂB = 57° 55′ 15″

31 Calculate, by least squares, the best estimate of the centre co-ordinates and radius of the circle to fit the following co-planar points

$$(12.0, 2.0), (36.8, 10.5), (30.5, 37.0), (2.5, 28.5).$$

Use the point (21, 19) as the provisional position of the centre and 20 for the provisional radius.   [London]

Answer 20.0, 20.2; radius = 19.6

The computer programs

Introduction

The computer programs have been written in standard BASIC which should be suitable for most home computers. The exception to this is the omission of LET from the assign statements, for example LET $X=X+1$ is listed as $X=X+1$. With reference to the manual for your computer it is possible to enhance the programs, especially in the area of data input and output.

Angles

All angles are input and output in degrees, minutes, seconds. The program logic uses the angle as one integer value in seconds, this minimizes the rounding error that occurs if degrees are used. BASIC's trigonometric functions process the angle in radians and the conversion factor

$$\frac{\pi}{(180 * 3600)} = \frac{1}{206\ 264.8}$$

is used in the programs.

Formulae

Assign statements in the program listing are laid out in the same format as the formulae in the text, this can lead to unnecessary brackets and hence inefficient programming. The experienced computer user can remove unnecessary brackets when entering the program into the computer.

Arc sine

Standard BASIC has no function to determine $\sin^{-1}$ of an angle, this is a problem when using the sine rule. It can be overcome using two trigonometric rules

$$\cos^2 \theta = 1 - \sin^2 \theta$$

and

$$\tan \theta = \frac{\sin \theta}{\cos \theta}$$

so

$$\theta = \tan^{-1}\left[\frac{\sin \theta}{\sqrt{(1 - \sin^2 \theta)}}\right],$$

where $\tan^{-1}$ can be represented by BASIC's function ATN.

Programmable calculators

Many readers will wish to use the programs on programmable calculators, and with this in mind the memory size needed to run the programs has been kept to a minimum and many of the program 'trimmings' have been left out.

It may be necessary to DIMension arrays with the lowest possible value and these are quoted for the examples in this book. The reader will have to reformat the output from some programs to suit the limited screen display. However, working with angles is simpler since these machines usually support the MODE statement that allows direct input in degrees, minutes and seconds in the format DD.MMSS (or in decimals of degrees) and the direct processing of $\sin^{-1}$ with ASN and $\cos^{-1}$ with ACS.

Rounding

Output data is rounded using the integer routine $X = INT ((X * 100) + 0.5)/100$ to give two decimal places. Most computers have inbuilt functions like ROUND or PRINT USING which will perform this job automatically.

IF/THEN statements

The format of IF/THEN GOTO statements varies widely from machine to machine. Some allow other commands to be nested on the same line often with an ELSE statement. In the listings a basic form of IF .... THEN GOTO x has been used. In some programs the GOTO statement has been omitted and this should be inserted if the computer will not recognise the abbreviated command IF ... THEN x.

Logic functions AND or OR

Some of the programs use the logic functions AND and OR in IF/THEN GOTO statements. Some computers do not support these functions and alternative program statements are given with the introductions to the appropriate programs.

Data checking

It is good programming practice to check the data when it is entered for non-sensible or illegal values. The programs can be enhanced by the inclusion of data checking subroutines such as the example below for checking angles.

```
210 INPUT "INPUT ANGLE DEG, MIN, SEC ",D,M,S
220 GOSUB 900
230 IF X=-1 THEN PRINT "RE INPUT THE ANGLE":GOTO 210
    .
    .
900 REM SUBROUTINE TO CHECK ANGLES
910 X=0
920 IF D<0   THEN PRINT "DEGREES TOO SMALL":X=-1:RETURN
930 IF D>360 THEN PRINT "DEGREES TOO LARGE":X=-1:RETURN
940 IF M<0   THEN PRINT "MINUTES TOO SMALL":X=-1:RETURN
950 IF M>59  THEN PRINT "MINUTES TOO LARGE":X=-1:RETURN
960 IF S<0   THEN PRINT "SECONDS TOO SMALL":X=-1:RETURN
970 IF S>59  THEN PRINT "SECONDS TOO LARGE":X=-1:RETURN
980 RETURN
```

Index